Fig. 1.18 The three main classes of emission nebulae: (a) the Great Nebula Orion; the ionization is due to several very hot stars deep within the nebula (© Association of Universities for Research in Astronomy, Inc., The Kitt Peak National Observatory).

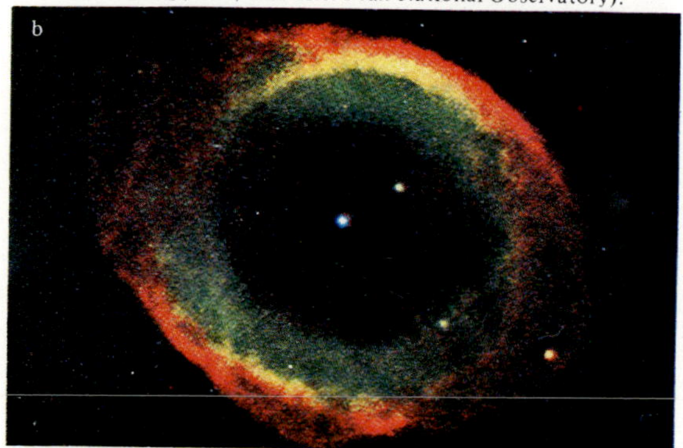

(b) Planetary Nebula in Lyra. The star in the center is responsible for the ionization (Lick Observatory photograph).

(c) Veil Nebula in Cygnus, central section. The remnant of a supernova that exploded 50,000 years ago (Lick Observatory photograph). See black and white reproductions on pages 32 and 33.

Physics of Stellar Evolution and Cosmology

Physics of Stellar Evolution and Cosmology

HOWARD S. GOLDBERG

University of Illinois at Chicago Circle

MICHAEL D. SCADRON

University of Arizona

GORDON AND BREACH SCIENCE PUBLISHERS
New York London Paris Montreux Tokyo Melbourne

Published 1981
Second Printing 1986
Third Printing with corrections 1987

Gordon and Breach Science Publishers

Post Office Box 197
London WC2E 9PX
England

58, rue Lhomond
75005 Paris
France

Post Office Box 161
1820 Montreux 2
Switzerland

14-9 Okubo 3-chome
Shinjuku-ku, Tokyo 160
Japan

Camberwell Business Center
Private Bag 30
Camberwell, Victoria 3124
Australia

Library of Congress Cataloging in Publication Data

Goldberg, Howard S., 1936–
Physics of stellar evolution and cosmology.

Bibliography: p.
Includes index.
1. Stars—Evolution. 2. Cosmology. I. Scadron, Michael D. II. Title.
QB806.G64 1982 523.01 81-7179
ISBN 0-677-05540-4 (hardcover) AACR2
ISBN 0-677-21740-4 (paperback)

To our wives,
Natalie and Arlene,
and our daughters,
Deborah, Stephanie, Kari, and Lisa

Contents

Preface

This intermediate-level one-semester text on astrophysics places primary emphasis on physical principles. We have tried to explore astrophysical phenomena at a very fundamental level by stressing the basic laws of physics as learned in an introductory university or college physics course. In order to focus on such physical principles, we have avoided the use of calculus as much as possible. Newton's laws, basic ideas of electricity and magnetism, and the fundamentals of modern physics constitute the building blocks we use to construct an organic and coherent edifice of astrophysics.

Indeed, the joy of integrating physical ideas around a single topic is something we have tried to transmit through the story of stellar evolution and cosmology. We have not, however, attempted to compile an encyclopedia of astronomical information. Our goal is for the general reader to develop intuition for the organization, logic, and beauty of a modern science. We have, therefore, written this intermediate text for a diverse audience: the undergraduate liberal arts or engineering student who has been bitten by the physics bug and who would like to explore the relationship between physics and astronomy; the graduate science student who wants to obtain a solid foundation for future astrophysical studies; the amateur astronomer who would like to understand more about what he sees; and finally, for all of us who love a good detective story with a great plot—the birth, evolution and death of stars and the universe.

This book is organized in the following manner. With no fanfare, chapter 1 plunges into a series of topics designed to give the reader a better grasp of the techniques and tools of astrophysics. The topics include the ideal gas law, the meaning of temperature, black body radiation, as well as a review of discrete spectra and the Doppler effect. We then use these tools to probe the interstellar medium in chapter 2. We discuss in detail the origins of 21-cm radiation and the method of using it to map the galaxy, and then turn our attention to the dust particles and molecules that populate interstellar space. We conclude the chapter on a disquieting note: taken superficially, the laws of physics seem to deny the possibility of stellar evolution! Undaunted, we resolve these problems in our best Sherlock Holmes fashion at the beginning of chapter 3. After analyzing those laws of physics needed to describe stellar evolution and the structure of stars, we follow the evolution of a star from the interstellar breeding ground to its birth on the main sequence. The evolution and nuclear fuel generation of such main sequence stars are discussed in chapter 4 along with their transition to the red giant phase and the stellar generation of heavy elements. Midway in the chapter we pause in telling this story long enough to develop those quantum-mechanical concepts necessary to understand phases of evolution from the red giant stage and beyond. Chapter 5 gives an updated account of the several possible end points of stellar evolution—white dwarf stars, neutron stars, and black holes. These mystical names have intrigued the imagination; therefore, we have taken pains to clarify the physical principles involved in defining and justifying these final stages of stellar evolution. The sleuthing required to uncover the existence of these final forms of matter is most exciting. Having reached the end, in a sense it is time to go back to the beginning. Thus in chapter 6 we develop the basic principles and kinematics of cosmology. We discuss Hubble's constant, the expanding universe, and conclude with a discussion of the steady state universe and with the discovery of the background microwave radiation. We then come full circle, from "the beginning" to "the end," in chapter 7, by deducing the dynamical consequences of the presently accepted "big bang" model of cosmology. We carefully progress up the primordial time scale through the eras: the early universe, the lepton era, the radiation era, and the matter era. Along the way, we explore primordial nucleosynthesis, the recombination of hydrogen, the formation of the galaxies, the cosmic fireball radiation, and finally speculate about the future: a cold end to the big bang, or a hot big crunch.

As elementary particle physicists, we share the sense of wonder and discovery that is taking place in astronomy and astrophysics today. We have tried to impart some of this excitement and adventure, and through the ideas presented in this book, we hope you will be able to follow and appreciate the great drama as it unfolds in the future.

The authors gratefully acknowledge the comments and criticisms of Professors D. Barker, J. Cocke, and R. Weymann. Also, we appreciate the feedback from the students at the University of Arizona and the University of Illinois at Chicago Circle who suffered through the initial notes and lectures on this material and whose stimulating responses motivated the writing of this book.

CHAPTER 1

The Language of Astrophysics

Have ever you heard of the Land of Beyond,
That dreams at the gates of the day?
Alluring it lies at the skirts of the skies,
And ever so far away;
Alluring it calls: O ye the yoke galls,
And ye of the trail overfond,
With saddle and pack, by paddle and track,
Let's go to the Land of Beyond!

Robert Service, *The Land of Beyond*

We begin by explaining a few of the fundamental ideas and tools of physics as they pertain to a study of the stars and galaxies.

1.1. A Window on the Universe

Most of our information from outer space comes to us in the form of electromagnetic (EM) radiation or, in the sense of the dual nature of particles and waves, as photons. The relation between the energy of the photon, E_γ, and the frequency, ν, of the equivalent EM radiation is ($\nu = c/\lambda$)

$$E_\gamma = h\nu$$
$$= h\frac{c}{\lambda},$$

where $h \approx 4 \times 10^{-15}$ eV-sec; $c \approx 3 \times 10^8$ m/sec; $hc \approx 12 \times 10^{-7}$ eV-m = 1.2×10^4 eV-Å; and where 1 Å = 10^{-10} m. [We shall use the MKS

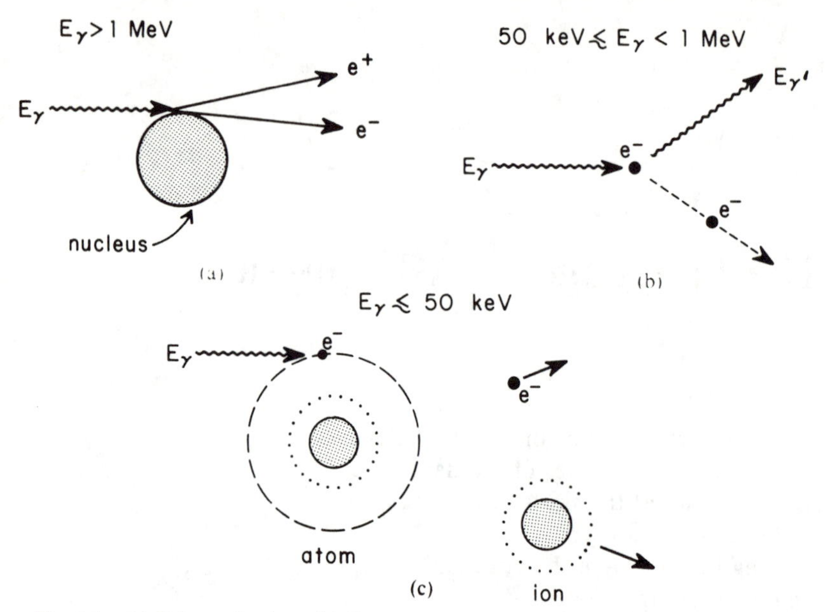

Fig. 1.1. (a) Pair production; (b) Compton scattering; (c) Photoelectric ionization.

(meter-kilogram-second) system combined with a measure of energy in electron volt units, where 1 eV = 1.6×10^{-19} joules (J).] Fortunately we are buried under 500 km of atmosphere which protects us from ultraviolet radiation and is made up of atoms and molecules including mainly oxygen and nitrogen along with other gases in much smaller amounts (argon, carbon dioxide, water vapor, neon, helium, krypton, and xenon). Because of these gases, photons have a difficult time penetrating the atmosphere due to three basic ''particle''-like interactions described in Fig. 1.1.

In the case of a photon whose energy is greater than 1 MeV, the dominant process according to the laws of quantum mechanics, will be the annihilation of the photon into an electron–positron pair. This will happen when the photon passes close to a nucleus as shown in Fig. 1.1a. On the other hand, the dominant process for a photon whose energy is in the range 50 keV to 1 MeV will be Compton scattering off an electron in an atom, as shown in Fig. 1.1b. The electron may Compton scatter several times, losing energy each time it scatters until it drops into an energy region ($\lesssim$ 50 keV) where the probability of photoelectric ionization (Fig. 1.1c) is very high. The photon then annihilates, leaving behind an electron and an ion, and hence an ionosphere.

But for photons between the energies of about 1.7 and 3 eV, a small miracle takes place. The gap between the ground-state energy levels of the *atoms* and the point where the electrons will be free is more than 3 eV. That is, the magnitude of the binding energy of the outermost electrons is greater than 3 eV—compared to the binding energy of hydrogen at 13.6 eV (see Sec. 1.5). Photons with energies of less than 3 eV therefore cannot ionize atoms, and so pass through the atmosphere undisturbed. Because these photons pass through freely, this region is called an energy window. Above 3 eV the window is closed. Between approximately 1.7 and 3 eV it is open. Below 1.7 eV down to 10^{-4} eV, the window is only partially open because the small *molecular* binding energies "take over" annihilation. Between 10^{-4} eV and 10^{-8} eV, corresponding to wavelengths between 1 cm and 100 m, photons can also pass through undisturbed—but that is it, for below 10^{-8} eV the window is permanently closed. The reason is that the ionosphere, which is a layer of electrons, reflects the longer wavelength radiation as if it, the ionosphere, were a metal. (You may recall that for a metal, the incoming electromagnetic radiation sets the free electrons oscillating. This acceleration results in a "scattered" electric field that is opposite to the "incident" one, resulting in a cancellation of the two in the forward direction and a single scattered or reflected wave in the backward direction. We shall discuss these ideas in more detail in chapter 2 and chapter 5.) In Fig. 1.2 we note that the two clear windows correspond to visible light between $\lambda \approx 4000$ Å (3 eV, far violet) and $\lambda \approx 7200$ Å (1.7 eV, deep red)—hence the name, the visible window, and to radio waves with $\lambda \sim 1$ cm $\sim 10^2$ m and hence the name, the radio window. Because we will use it often, we note the special region between 1 mm and 1 m is called the microwave region. The lower part of the microwave region is partially transparent, and as we shall see in chapter 2, this is of great significance for astrophysics.

In "light" of this, it should not come as a surprise that our eyes have developed to detect the radiation that passes through the atmosphere and illuminates our surroundings. In that sense the word "visible" radiation is

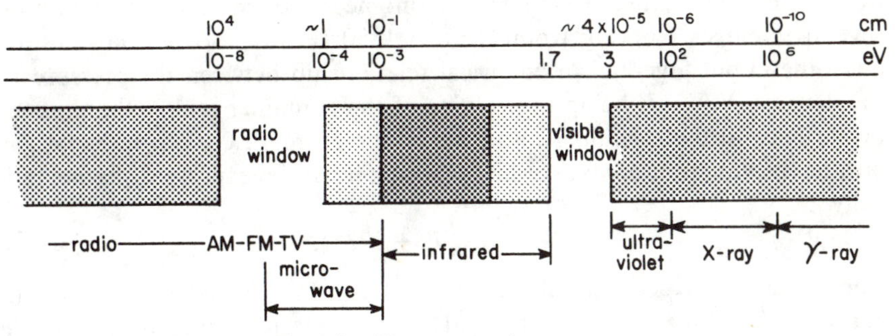

Fig. 1.2. Electromagnetic spectrum.

really putting the cart before the horse. Assuming the laws of physics are the same everywhere, so that in different parts of the universe the atoms have the same energy levels, then "humanoids" living under atmospheric conditions like our own should have "detectors" like ours as well. In case they are looking at us now they would detect the dominant radiation that can escape the Earth—our television shows!! Heaven help them . . . .

Within the last 10 years we have gotten above the atmosphere with ultraviolet ($\sim$10 eV), X-ray ($\sim$10 keV), and γ-ray ($\sim$MeV) satellite detectors. But we will travel an historical path so that for most of this book we will look through the eyes of our predecessors—using their tools— and that means through those narrow windows between 1.7 and 3 eV and 10^{-4} and 10^{-8} eV.

1.2. An Ideal Gas and the Ideal Gas Law

An ideal gas will have the following properties:

i. Constituents are in *random* motion.
ii. The total number of constituents is large.
iii. The volume of each constituent (δV) is very small compared to the volume of the container (V)

$$(\delta V \ll V).$$

iv. No appreciable force acts on the constituents except during collisions with the walls—there is no potential energy (PE) between the particles.
v. All collisions are elastic and therefore conserve not only momentum but kinetic energy (KE) as well.

If a large number of such particles are placed in a container, they will exert a pressure P on the walls and that pressure will be a function of something called the temperature. Using Hg thermometers to measure temperature, several scientists working from 1660 to 1850 (Power, Boyle, Gay-Lussac, and others) put together an empirical relationship between the pressure P, the density, N/V (V being the volume of the container and N the number of particles in the container), and the temperature T. Called the ideal gas law, or the equation of state of an ideal gas, it can be expressed mathematically as

$$PV = Nk(T_c + 273°),$$

where k is the experimentally determined constant called the Boltzmann constant, $k \approx 1.4 \times 10^{-23}$ J/°K, and where T_c is measured in centigrade degrees (100 units are defined between melting ice at 0 °C and boiling water at 100 °C under atmospheric pressure). For a more accurate tabulation of k and other important physical parameters see Appendix A.

It is clear from this equation that the pressure goes to zero at $T_c = -273$ °C, so that it makes sense to define a new temperature scale where doubling the temperature will correspond to a doubling of the pressure. This scale, called the scientific or Kelvin scale, is related to the centigrade scale via $T_K = T_c + 273$ so that the equation of state becomes

$$PV = NkT_K \tag{1.1}$$

and it is this form, that we shall use for the ideal gas law (we shall henceforth refer to T_K simply as T, it being understood that temperature is always in degrees Kelvin). For the first part of this book we will be able to consider particles in outer space and in all but the central-most portion of stars as components of an ideal gas. Later, however, when we study neutron stars and white dwarfs, we will enter a pressure region where $PV = NkT$ no longer holds and a new quantum mechanical equation of state will be necessary.

1.3. The Meaning of Temperature

The equation of state is a macroscopic equation since it relates the gross properties of the system—pressure, volume, and temperature. But it does not explain what causes the pressure. The latter is related to the microscopic properties of the ideal gas: the particles colliding with the walls, transferring momentum, and thereby creating a force. One can attack the problem in this way, using the laws of mechanics, $F = ma$ etc., and theoretically derive a relation between the pressure on the walls and the change in momentum of the internal particles. Such a fundamental kinetic theory equation was first derived by Bernoulli around 1750 but the detailed calculations were not carried out until 1850 to 1890 by Maxwell, Boltzmann, and Gibbs. Their final result, which we derive in detail in Appendix B, relates the pressure on the walls P, to the speed of each of the N particles inside the container according to

$$PV = \frac{m}{3} \sum_{i=1}^{N} v_i^2 \tag{1.2}$$

where V is the volume of the container, v_i the speed of the ith particle and m the mass of the individual particles. Notice that the particles do not have to have the same speed, and it is the square of the speed that is summed. As we show in Appendix B, the number $\frac{1}{3}$ arises because of the random motion of the particles in three dimensions. If we multiply top and bottom of (1.2) by N we obtain

$$PV = \frac{mN}{3}\left[\frac{\sum v_i^2}{N}\right].$$

The term in the bracket is a very special kind of average because we square the velocity before summing and dividing by N. It is called the mean square average (or sometimes the root-mean-square-average-squared) and is denoted by v_{rms}^2, so that

$$PV = \frac{mNv_{\mathrm{rms}}^2}{3} \tag{1.3}$$

This is the fundamental kinetic theory equation. It relates the macroscopic quantities of pressure and volume to the microscopic properties of the ideal gas particles—their mass and their velocities. We further note that the average nonrelativistic translational kinetic energy of the particle in the box is

$$\langle \mathrm{KE} \rangle = \frac{\sum \frac{1}{2} m_i v_i^2}{N} = \tfrac{1}{2} m \frac{\sum v_i^2}{N} = \tfrac{1}{2} m v_{\mathrm{rms}}^2 .$$

With this equation in mind, we modify (1.3) to read $PV = \frac{2}{3} N(mv_{\mathrm{rms}}^2/2)$, which then can be written as the final form of the nonrelativistic fundamental kinetic theory equation,

$$PV = \tfrac{2}{3} N \langle \mathrm{KE} \rangle . \tag{1.4}$$

It is this equation we compare to Boyle's law, $PV = NkT$. We now see that temperature on a microscopic level is quite definitely defined, i.e.,

$$\tfrac{2}{3} N \langle \mathrm{KE} \rangle = NkT,$$

so that the average translational kinetic energy per particle is

$$\langle \mathrm{KE} \rangle = \tfrac{3}{2} kT. \tag{1.5}$$

Thus temperature is merely a representation of our ignorance about the internal workings of an ideal gas. It is nothing but a measure of the average random KE of these particles. To be a true measure of temperature, the fluctuations in the average kinetic energy must be small. This means we need a large number of particles. We cannot define the temperature for one particle or even a hundred, for that matter. For with too few particles, the reading of the thermometer would fluctuate rapidly. But even a large number of particles must be moving *randomly*. A block of 10^{23} particles translating with a *total KE*, $\mathrm{KE_{tot}}$, of 1 J does *not* have a temperature of $T = \frac{2}{3}(\mathrm{KE_{tot}}/Nk) \sim \frac{1}{2}$ °K; no temperature can be defined from just translational motion. A freely falling body's KE increases as it falls, but this translational motion cannot affect its temperature, much less make T increase. Recall that for translational motion we can always get into a reference frame that rides alongside the object so that $KE_{trans} = 0$, in which case the object would have no temperature at all! In reality, however, the translating object also has *internal* random motion so that it does have a unique temperature, regardless of reference frame.

It should be clear that with the equation $\langle KE \rangle = \frac{3}{2}kT$, we do not need temperature at all! We could just as well talk about joules rather than degrees. Convention, however, dies hard and T is used a great deal—but to solve problems we need $\langle KE \rangle$. Note that Eq. (1.5) implies that as $\langle KE \rangle \to 0$, $T \to 0$. This is not strictly true, but we require quantum mechanics to understand why. We will examine the conditions at $T = 0$ when we take up white dwarfs and neutron stars in chapter 5.

We now have a set of related quantities v_{rms}, $\langle KE \rangle$, and T. It is useful to get a feel for how one quantity compares to the other. As we shall see, for example, clouds of gas in interstellar space have $v_{rms} \approx 1.5$ km/sec. If the gas is made up of hydrogen atoms with $m \approx 1.7 \times 10^{-27}$ kg, this corresponds to a T of

$$T = \frac{2}{3}\frac{1}{k}\frac{m}{2} \cdot v_{rms}^2$$

$$\approx \frac{2}{3}\frac{1}{1.4 \times 10^{-23}}\frac{1.7 \times 10^{-27}}{2}\left(\frac{3}{2} \times 10^3\right)^2 \sim 90\ ^\circ\mathrm{K}.$$

For a gas cloud surrounding a star, we might find $T = 10{,}000\ ^\circ$K, so that from (1.3)–(1.5)

$$\tfrac{3}{2}kT = \langle KE \rangle = \tfrac{1}{2}mv_{rms}^2$$

$$v_{rms} = \sqrt{\frac{3kT}{m}} \sim 16 \text{ km/sec.}$$

Notice that the velocity increases as the square root of the temperature, so that large increases in T correspond to much smaller increases in v_{rms}. As a bonus, we can also relate the total random KE to the temperature provided we know the number of particles:

$$KE_{tot} = N\langle KE \rangle = \tfrac{3}{2}NkT.$$

So far we have no means of determining either T or $\langle KE \rangle$ for stars and clouds. That will be one of our next tasks.

A word of caution. In all of these relations, you are safe if you use MKS units. That means $k \approx 1.4 \times 10^{-23}$ J/°K. When calculating KE_{tot} or $\langle KE \rangle$, however, it is useful to express these numbers in eV. To do this recall that 1 eV $= 1.6 \times 10^{-19}$ J. Converting Boltzmann's constant to eV units, we have $k = 0.86 \times 10^{-4}$ eV/°K. Thus at $T = 10{,}000$ °K,

$$\langle \text{KE} \rangle = \tfrac{3}{2}kT \approx \tfrac{3}{2}(0.86 \times 10^{-4}) \times 10^4 \sim 1.3 \text{ eV}.$$

So if we have a container of "ideal gas" particles at temperature T in a volume V, then we know

i. the pressure on the walls

$$P = \frac{NkT}{V};$$

ii. the average kinetic energy of a particle

$$\langle \text{KE} \rangle = \tfrac{3}{2}kT,$$

and the total KE

$$\text{KE}_{\text{tot}} = \tfrac{3}{2}NkT.$$

But knowing this does not tell us how these N particles vary with KE. Knowing the average KE is 1.3 eV at 10^4 °K does not tell us if all the particles are at 1.3 eV—or if they are spread out around 1.3 eV—or if there are lots of particles with small KEs and few with large KEs; the possibilities are shown in Fig. 1.3. Indeed all instruments have a finite energy "window" so that all we can determine experimentally is the number of particles in a given energy range. Thus, experimentally we have

$$\text{KE}_{\text{tot}} = \sum N_i \times \text{KE}_i$$

$$\langle \text{KE} \rangle = \frac{\sum N_i \times \text{KE}_i}{\sum N_i},$$

where KE_i is the kinetic energy at the *center* of the ith bin which contains N_i particles.

If all the particles occupy one bin, as in Fig. 1.3a, then $N_i \times \text{KE}_i$ will also occupy one bin as in Fig. 1.4a. If N_i is spread out around 1.3 eV, as in Fig. 1.3b, then $N_i \times \text{KE}_i$ will also be spread out around 1.3 eV, as in Fig. 1.4b.

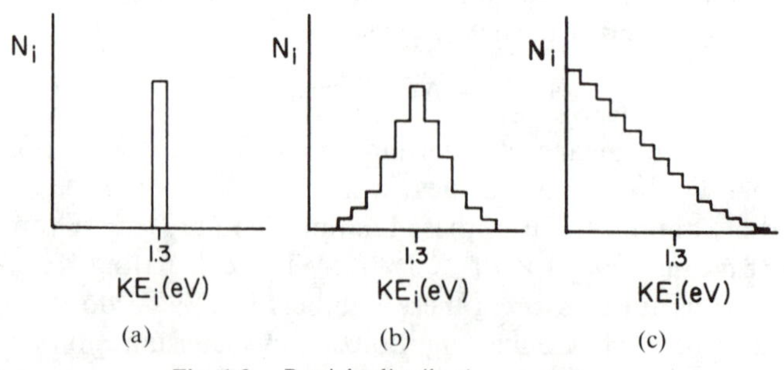

Fig. 1.3. Particle distribution vs. KE_i.

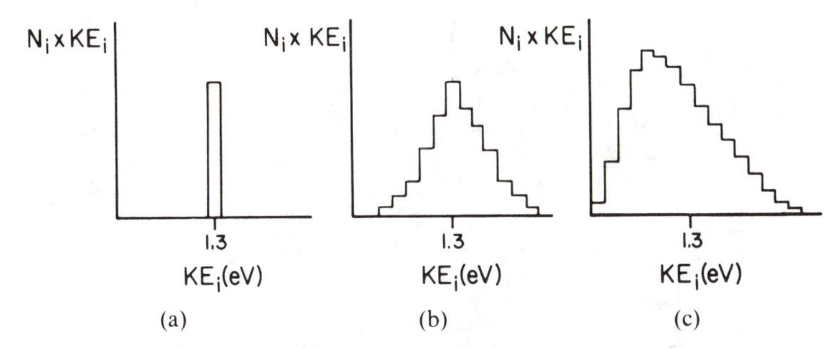

Fig. 1.4. Energy distribution vs. KE_i.

Figure 1.3c, however, is the most interesting case. We have lots of low-energy particles but since their KE is small, the product $N_i \times \mathrm{KE}_i$ is small too. This is reflected in the lower portion of Fig. 1.4c, which falls to zero. Even though there are only a few high-energy particles, they contribute significantly to the product $N_i \times \mathrm{KE}_i$, thus holding up the high end of Fig. 1.4c. But eventually $N_i \rightarrow 0$, so that the high end of 1.4c goes to zero too. The result is another bell-shaped curve—peaking near 1.3 eV. Thus all three curves of $N \times \mathrm{KE}$ give us the same average kinetic energy, 1.3 eV, even though the N_i distributions are different. Which curve then is the correct one? How are the particles distributed in an ideal gas?

It turns out that the theory of probability and statistics is sufficient to determine the correct particle distribution, called the Maxwell–Boltzmann distribution in a gas in thermal equilibrium. One can also, however, do so experimentally by cutting a small hole in the box containing an ideal gas at a temperature T, and counting the number of particles that come out in a given energy range. What we find is a curve similar to Fig. 1.3c. The curve is an exponential with the number of particles vs. temperature falling off as $N_i = N_0 e^{-\mathrm{KE}_i/kT}$. This means that every time we increase KE by a fixed amount (say 1 eV), the curve drops by a fixed percent (say 50%); thus if the number of particles drops by 50% in going from 0 to 1 eV, it will drop by another 50% in going from 1 to 2 eV and another 50% from 2 to 3 eV. If we had 20,000 particles in the very lowest energy bin, then in this arbitrary example we would have only 2500 particles in the energy bin around 3 eV. How fast the N_i vs. KE_i curve falls off is given by kT. *This is very important.* For example, at $T_1 = 17{,}500$ °K, $kT \approx 1.5$ eV so that in the first 1.5 eV the N_i vs. KE_i curve drops by $1 - e^{-1} = 0.632$ or 63% (recall that $e \cong 2.72$). For $T_2 = 35{,}000$ °K, or $kT = 3$ eV, it takes 3 eV for it to drop by 63%, but if the *total number* of particles (measured by the area under the N_i curve) is the same, the higher temperature, T_2, implies fewer lower-energy particles and more high-energy particles. These two N_i vs. KE_i curves have the form shown in Fig. 1.5a and they produce the energy distributions (N_i KE_i vs. KE_i) like the curves in Fig. 1.5b. With fewer low-energy particles and more

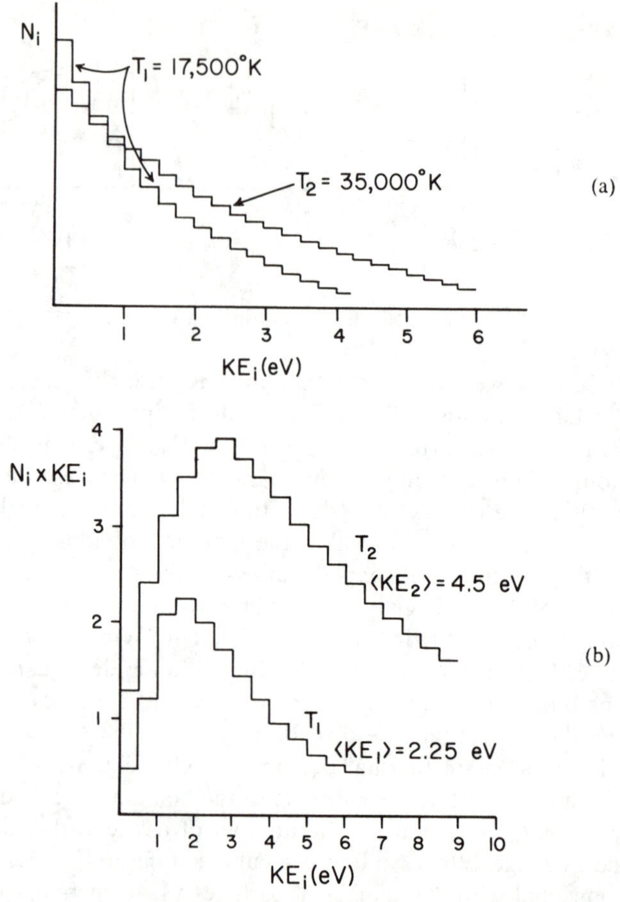

Fig. 1.5. (a) Particle and (b) energy distributions vs. KE_i at two different temperatures.

high-energy ones, the container at temperature T_2 will have a higher average kinetic energy than the one at T_1. This is, of course, also reflected in the kinetic theory equation, $\langle KE \rangle \simeq \frac{3}{2}kT$. You must understand that any gas at a given temperature will have some particles with very high KEs and can initiate reactions you might not expect from $\langle KE \rangle$ alone. Thus temperature is not only a measure of the average kinetic energy of the gas, but of the distribution of particles as well.

1.4. Black Body Radiation

The Maxwell–Boltzmann distribution, $N_i \propto e^{-KE_i/kT}$, is valid for nonrelativistic particles at high temperatures. For relativistic particles moving with velocities near the speed of light, the distribution of particles depends upon the angular momentum quantum nature of the particles. Quantum statistics, however, is beyond the level of this text and so we again follow the path of history and investigate the distribution of (relativistic) photons, called black body radiation, from the viewpoint of a "black body" cavity.

1.4.1. Formal Definition

We evacuate a cavity so that there are no particles inside. Then we heat the walls to a temperature T and wait until equilibrium is reached so that all parts of the wall are at the same temperature. Next we cut a small hole of area δA in the wall. Lo and behold, EM radiation comes out as shown in Fig. 1.6. With a suitable radiation detector (a heat sensing thermocouple wrapped in a black metal foil plus an appropriate "prism" to select frequencies), we find two remarkable facts about this radiation:

i. The energy coming out per unit area per unit time (called the energy flux $\mathcal{L} \equiv \mathrm{J/m^2\text{-}sec}$) is related to the temperature of the walls via

$$\mathcal{L} = \sigma T^4, \tag{1.6}$$

where $\sigma = 5.7 \times 10^{-8}\ \mathrm{J/m^2\text{-}sec\text{-}\,^\circ K^4}$.

ii. If we divide up the radiation energy being emitted by the black body into frequency bands, we find that the shape of $\mathcal{L}_\nu \equiv \Delta\mathcal{L}/\Delta\nu$ (the flux per unit frequency range) plotted vs. ν is a universal curve that depends only on the temperature as shown in Fig. 1.7. Then $\mathcal{L}_\nu$ peaks at a frequency (measured in cycles per second)

$$\nu_{\text{peak}} \approx 6 \times 10^{10}\ T,$$

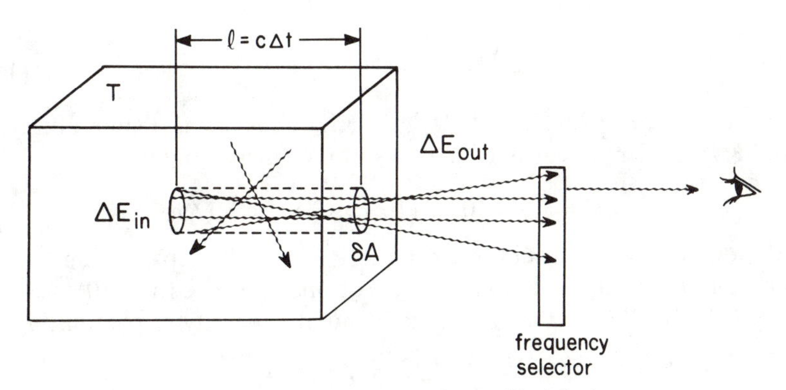

Fig. 1.6. Photons escaping from a black body.

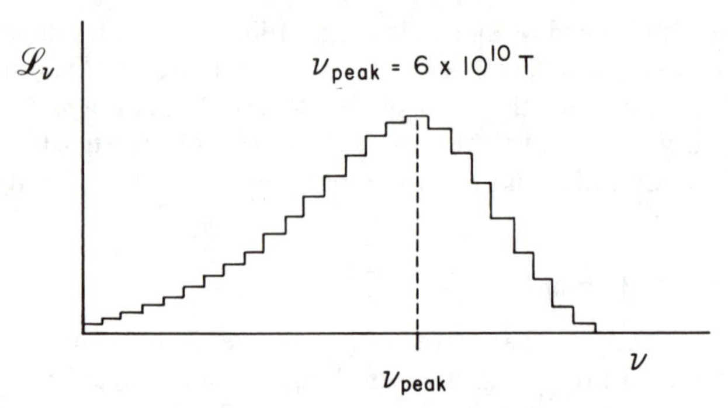

Fig. 1.7. Photon energy flux vs. frequency.

or alternatively,

$$\lambda_{\text{peak}} \approx 0.51 \text{ cm}/T, \qquad h\nu_{\text{peak}} \approx 3kT.$$

The area under the $\mathfrak{L}_\nu$ curve in Fig. 1.7 is the total energy flux $\mathfrak{L}$.

An important and useful quantity to know is the energy per unit volume inside the container. The total energy coming out of the hole in time Δt is

$$\Delta E_{\text{out}} = \mathfrak{L}\,\delta A\,\Delta t.$$

The energy is coming from a cylindrical volume of area δA and length $l = c\Delta t$, as shown in Fig. 1.6. Because the photons are heading in all directions, it turns out that only $\frac{1}{4}$ of the photons in the cylinder make it out of the hole. Thus,

$$\tfrac{1}{4}\Delta E_{\text{in cyl}} = \Delta E_{\text{out}} = \mathfrak{L}\,\delta A\,\Delta t.$$

The energy per unit volume in the container therefore is

$$\frac{E}{V} = \frac{\Delta E_{\text{in cyl}}}{\delta V_{\text{cyl}}} = \frac{4\mathfrak{L}\,\delta A\,\Delta t}{\delta A c \Delta t} = \frac{4\mathfrak{L}}{c}$$

$$= \frac{4\sigma T^4}{c} \equiv aT^4, \tag{1.7}$$

where $a = 4\sigma/c = 7.6 \times 10^{-16}$ J/m^3-°K^4. Thus at a temperature of 1000 °K, the energy density of photons in the cavity would be

$$E/V = 7.6 \times 10^{-16}\,(10^3)^4 \approx 8 \times 10^{-4}\ \text{J/m}^3.$$

The radiation curve would peak at $\nu_{\text{peak}} = 6 \times 10^{13}$ cps, giving us an approximate average photon energy of $h\nu_{\text{peak}} \approx 3 \times 1.4 \times 10^{-23} \times 10^3 = 4 \times 10^{-20}$ J. We can then estimate the number density of photons, n_γ, as

$$n_\gamma \approx \frac{E/V}{h\nu_{\text{peak}}} \approx \frac{8 \times 10^{-4}}{4 \times 10^{-20}}\ \frac{1}{\text{m}^3} \approx 2 \times 10^{16}/\text{m}^3 = 2 \times 10^{10}/\text{cm}^3.$$

Our cavity is clearly loaded with photons. The pressure exerted by a black body photon gas then follows from the general fundamental kinetic theory equation $PV = mNv_{\mathrm{rms}}^2/3$, with v_{rms}^2 replaced by c^2, the velocity of light squared, and the photon "mass" m replaced by the Einstein relation E_γ/c^2. Then NE_γ/V is the energy density of the photon gas, E/V, giving us

$$P = \frac{1}{3}\left(\frac{E_\gamma}{c^2}\right)\frac{Nc^2}{V} = \frac{1}{3}\frac{E}{V} = \frac{1}{3}aT^4. \tag{1.8}$$

At 1000 °K, the photon pressure on the walls of the cavity would be $P = \frac{8}{3} \times 10^{-4}\ \mathrm{N/m^2} = 2.7 \times 10^{-4}\ \mathrm{N/m^2}$. Thus if we raise the temperature of a black body, more and more energy per unit area per unit time comes out of the hole, the energy density and pressure inside increase, and the frequency at which the energy peaks moves toward higher and higher frequencies (i.e., the EM radiation will have a higher average energy).

The curve we have been describing was obtained by measuring the *frequency* of the black body radiation. If instead we measure the *wavelength* of the radiation, as shown in Fig. 1.8, the curve has a slightly different shape and peaks at a wavelength given by

$$\lambda_{\mathrm{peak}} \approx \frac{2.9 \times 10^7}{T}\ \text{Å} = \frac{0.29\ \mathrm{cm}}{T},$$

corresponding to a frequency and photon energy of

$$\nu_{\mathrm{peak}} \approx 10^{11}\ T; \qquad h\nu_{\mathrm{peak}} \approx 5\ kT.$$

Notice that these numbers differ from the ones for the $\mathcal{L}_\nu$ curves of Fig. 1.7. The area under both curves, however, is the total energy flux $\mathcal{L}$. If we know the temperature of the black body source, we can predict the peak wavelength and frequency for the particular type of measurement we make. But we must specify whether we are measuring wavelength or frequency. Likewise if we are to measure the temperature of a black body emitter, we must specify how we are to make the measurement. If the instrument

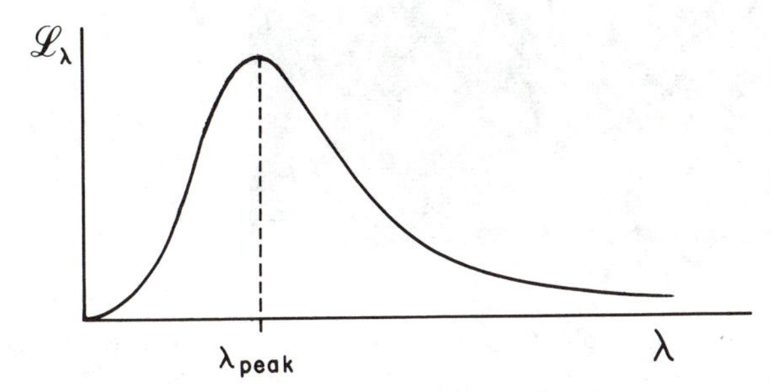

Fig. 1.8. Photon energy flux vs. wavelength.

measures frequency, then

$$T = \frac{\nu_{\text{peak}}(\text{cps})}{0.6 \times 10^{11}} \, (^\circ\text{K}). \tag{1.9}$$

If the instrument measures wavelength, then

$$T = \frac{0.29}{\lambda_{\text{peak}}(\text{cm})} \, (^\circ\text{K}). \tag{1.10}$$

As we have already implied, any object that emits EM radiation in the form of such a curve is called a black body. The reason we call it a black body is because any incident radiation that *enters* the hole has almost no chance of getting out since it rattles around inside the cavity. Since any radiation that falls on a truly black object is completely absorbed, i.e., does not "come back out," we call the cavity an equivalent black body. (An example of the cavity effect is having your window open a crack when driving through a polluted or smokey area. The particles or odor enter quickly but, as one can attest from the smell, they leave very very slowly. Your car is behaving like a black body.) Actually the object need not appear black since clearly the hole would "shine," dominated by the peak wavelength. *We only demand that most of the energy remain inside the "cavity."* The hot iron in Fig. 1.9 is a beautiful example of an infrared black

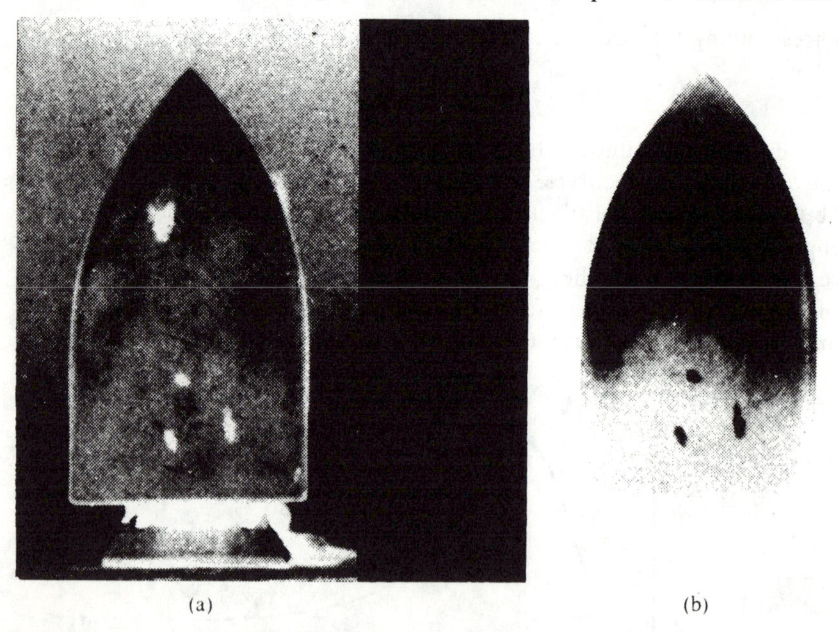

(a) (b)

Fig. 1.9. Infrared photograph of (a) hot iron; (b) Iron at room temperature (Mt. Wilson Observatory).

body emitter: its spectrum has the black body shape and peaks in the infrared. When the iron is at room temperature, however, as shown in Fig. 1.9b, the spots that previously glowed now absorb light; they are quite literally black body cavities. A tungsten filament like the kind found in light bulbs also produces a black body spectrum. When heated to 10^4 °K, its spectrum peaks in the visible. Your skin is a pretty decent infrared detector, as you can verify by bringing your hand near the iron, while your eye is the visible detector. One of the important characteristics of a black body emitter is that the spectrum is independent of the kind of object doing the emitting; a piece of steel and a piece of tungsten both at the same temperature emit radiation having the identical black body spectrum.

1.4.2. A Classical Quandary

You would think that a curve that is so universal, so simple, and independent of the kind of material being heated would also have a simple theoretical explanation. But alas, after black body radiation was discovered in the 19th century its justification became one of the great theoretical problems. In a sense, it seemed straightforward enough: The walls are at temperature T, the molecules inside the walls are moving randomly with $\langle \mathrm{KE} \rangle \sim kT$ (since the walls are not a perfect gas, we do not expect $\langle \mathrm{KE} \rangle = \frac{3}{2}kT$). These randomly moving particles—vibrating about fixed lattice sites—should accelerate, and classically we know accelerating charges emit EM radiation and, voilà, radiation of all frequencies should fill the cavity. We should even be able to understand why the spectrum is independent of the material of the walls of the cavity. An oscillator that emits EM radiation must lose energy and ultimately stop oscillating. But the emitted radiation can interact with other oscillators by being absorbed and thus pump energy back into the system, causing further oscillations and more emission. The molecules can now absorb energy from the collisions set up by the thermal source and from the radiation in the cavity. Once equilibrium is established, equal amounts of radiation are emitted and absorbed at all frequencies and T remains constant. Once equilibrium has been reached, the kind of oscillator does not enter the picture; all that is relevant is that only equal amounts of energy, at any frequency, are emitted and absorbed. Thus black body radiation cannot depend on the kind of material—and it does not.

By 1900, however, only one physicist, Lord Rayleigh, using classical techniques, had come close to deriving a curve that fits the black body spectrum and his was a disaster as shown in Fig. 1.10; all he could fit was the low frequency part—at high frequencies his curve went to infinity! Now it is not hard to understand classically why the low-frequency spectrum

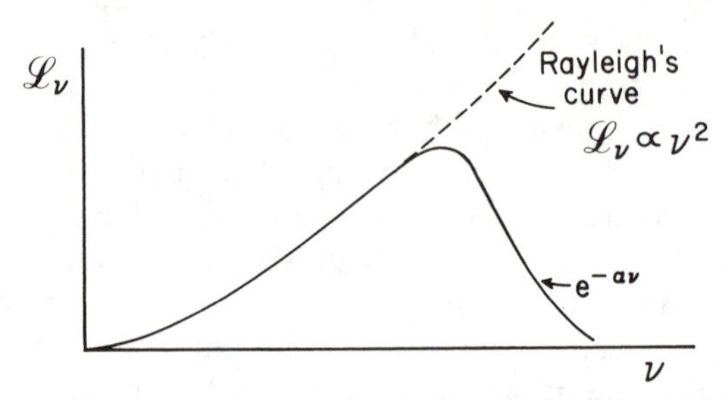

Fig. 1.10. Deviation of energy flux distribution from Rayleigh's law.

goes to zero; only wavelengths that are smaller than the size of the box can fit inside and be in equilibrium with the walls. Low-frequency radiation corresponding to wavelengths larger than the cavity cannot be fitted into the box, with the result that the low frequency spectrum is suppressed. Rayleigh found that $\mathfrak{L}_\nu \propto \nu^2$, in perfect agreement with the low-frequency part of the curve. Since increasing frequency corresponds to smaller and smaller wavelengths, there should be plenty of high-frequency radiation. Yet instead of continuing to increase as ν^2, the black body curve turns over and heads down exponentially according to $\mathfrak{L}_\nu \propto \nu^3 e^{-\alpha\nu}$. In Rayleigh's day this lack of high-frequency radiation was called the ultraviolet catastrophe; a catastrophe indeed if classical physics could not explain the shape of the spectrum.

Enter the German physicist Max Planck, with a very clever strategy. Knowing the shape of the black body spectrum as a function of frequency and temperature, he adjusted parameters until he arrived at an equation ($\mathfrak{L}_\nu = F(\nu,T)$) that fits the empirical curve. Then he tried to figure out how to derive this equation. (His technique should be dear to the heart of all students—look up the answer and then try to figure out how to do the problem). In a letter to Professor R. W. Wood, Planck wrote

> . . . for six years (from 1894 on) I had been doing battle with the problem of the equilibrium between radiation and matter without success. I knew that the problem is of fundamental importance for physics. I knew the formula that reproduced the energy distribution is the normal spectrum; a theoretical interpretation *would have to be* found at any cost, no matter how high. Classical physics was not adequate, that was clear to me . . .

On December 14, 1900 Planck announced that he had found the solution to the problem and indeed that his whole theory was based on one radical and far-fetched assumption: that radiation and matter did not interact continuously but in discrete packets of energy he called "quanta" (from the Latin "how much") and that in each interaction, the energy exchanged is $E_\gamma = h\nu$. Planck insisted, however, that the radiation in the cavity was still continuous and that only in its interaction with matter—for some mysterious reason—was it quantized. It was Einstein who made the bold hypothesis that the radiation itself was quantized. To the day he died Planck never accepted Einstein's extension of his work.

With Einstein's interpretation of Planck's results, one can explain the reason for the high-frequency cutoff. Since a photon quantum contains a fixed amount of energy, a charged particle cannot emit a quantum whose energy is greater than its own KE. The particle may emit many low-energy quanta or a few or even one high-energy quantum, but in all cases the energy of the emitted photon quanta must be equal to or less than the particle's KE. Since we showed in Sec. 1.2 that a wall at temperature T has few high-energy particles, the resultant spectrum will contain few high-energy photon quanta. The exponential cutoff at high black body frequencies reflects this fact of life. The ultraviolet catastrophe then was not a catastrophe after all, once one understood the nonclassical nature of the radiation.

1.4.3. Discrete Energy Levels in Atoms: Bohr Model

We have already alluded to the discrete energy levels of electrons in atoms and the magic number of the binding energy of hydrogen, 13.6 eV. These were unknown facts when Planck put forth his notion of quantized (electron) oscillator modes in a black body cavity in order to circumvent the ultraviolet catastrophe. After the turn of the century, however, events moved more quickly, as the experiments of Rutherford and Franck-Hertz suggested a solar system-type model for atoms with discrete electron energy levels. In 1916, Bohr linked this picture with Einstein's explanation of the photoelectric effect and the observations of the spectroscopists Lyman, Balmer, et al. Bohr proposed that the electrostatic force between the negatively charged (planetary) electrons and the positively charged (solar) nucleus acts in a classical manner except that the angular momentum of each electron can assume only integral multiples of a small but fundamental value. To fit the binding energy of hydrogen, Bohr discovered that this fundamental angular momentum L was none other than Planck's constant h (which has the dimensions of angular momentum) divided by 2π, i.e., $L = n\hbar = \hbar, 2\hbar, 3\hbar, \ldots$ where $\hbar = h/2\pi$.

More specifically for the hydrogen atom, Bohr assumed the electron circles the proton with the electrostatic force, $F_E = k_0 e^2/r^2$, acting as the net force $F_{\text{net}} = m_e v^2/r$. Setting $F_E = F_{\text{net}}$ then leads to the classical relation between the electron's v and r, in analogy with planetary motion ($k_0 \approx 9 \times 10^9$ in MKS units),

$$m_e v^2 = k_0 e^2 / r.$$

If this relation is multiplied by $\frac{1}{2}$, it says that the electron's KE is $\frac{1}{2}$ the magnitude of its (electrostatic) PE, just as in the case of (gravitationally bound) planets. Both situations are a special case of the virial theorem, which will be discussed in greater detail in chapter 3. The total energy of the orbiting electron is then

$$E = \text{KE} + \text{PE} = -\tfrac{1}{2} m_e v^2 = -\tfrac{1}{2} \frac{k_0 e^2}{r}. \tag{1.11}$$

When this classical result is combined with Bohr's quantum condition $L = mvr = \hbar, 2\hbar \cdots \to n\hbar$ we learn that

$$v \to v_n = \frac{k_0 e^2}{n\hbar} = \frac{\alpha c}{n} \approx \frac{1}{n} \frac{c}{137}$$

where $\alpha \equiv k_0 e^2 / \hbar c \approx 1/137$ is the dimensionless "fine structure constant," and also

$$r \to r_n = \frac{n^2 \hbar^2}{m_e k_0 e^2} = n^2 a_0,$$

where

$$a_0 = \frac{\hbar^2}{m_e k_0 e^2} = \frac{\hbar}{m_e c} \frac{1}{\alpha} \approx \frac{1}{2} \times 10^{-10} \text{ m}$$

is called the Bohr radius which is indeed the size of an atom. By way of comparison, in the ground state $n = 1$, the electron's velocity is $c/137$, very fast indeed, but still nonrelativistic. Since the proton's radius R_p is $\sim 10^{-15}$ m, the electron Bohr radius is $\sim 10^5\ R_p$. Contrast this atomic picture with the solar system itself: the Earth travels around the Sun with speed $v \sim 10^{-4} c$ at a distance of about 200 $R_\odot$ (where $R_\odot$ is the solar radius); Pluto has $v \sim 10^{-5} c$ at $10^4\ R_\odot$.

Substituting this quantized v or r into the classical virial theorem result, we see that the total energy of the electron is also quantized, with

$$E \to E_n = -\frac{\frac{1}{2} m_e (\alpha c)^2}{n^2} = -\frac{k_0 (e^2/2) a_0}{n^2}$$

Since the electron rest energy is $m_e c^2 \approx \frac{1}{2}$ MeV, we obtain

$$E_n = -\frac{\frac{1}{2} \times \frac{1}{2} \times 10^6 (1/137)^2}{n^2} \approx -13.6 \text{ eV}/n^2. \tag{1.12}$$

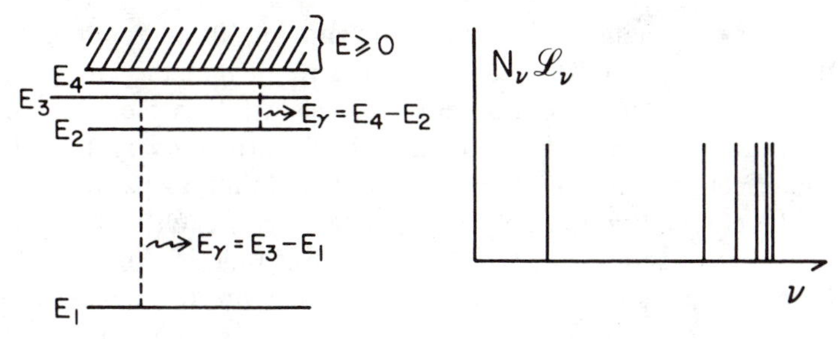

Fig. 1.11. Atomic electron transitions and photon spectral lines.

The ionization energy E_I corresponds to the energy needed to remove the electron from the ground state $n = 1$; i.e., $E_I = 13.6$ eV. The observed discreteness of spectral (frequency) lines emanating from hydrogen is then a statement of energy conservation $E_i = E_f + E_\gamma$ or

$$E_\gamma = h\nu = E_i - E_f \approx 13.6 \text{ eV}\left(\frac{1}{n_f^2} - \frac{1}{n_i^2}\right) \tag{1.13}$$

The transitions and spectral lines are depicted in Fig. 1.11. So we see that Planck was correct to assume that photons are quantized in their interaction with bound electrons. In present day quantum theory, all of the above formulae are valid, but $n = 1, 2, \ldots,$ must be reinterpreted as the "principal quantum number" with $n > l$, l being the orbital angular momentum quantum number in $L = l\hbar$ such that $l = 0, 1, \ldots, n - 1$.

1.4.4. Why a Continuous Black Body Spectrum?

One of the questions about black body radiation that we should try to answer is: How can atoms whose energy levels are discrete produce a continuous spectrum of photons? As we have just shown (see Sec. 1.4.3), since only certain E_ns are allowed, the photon must have a fixed set of frequencies which in turn produce a discrete line spectrum, as shown in Fig. 1.11, not a continuous one. The dynamical expression for the energy levels of hydrogen is $E_n = \text{KE}_n + \text{PE}_n = -13.6/n^2$ eV. Let us focus on the potential-energy term which for hydrogen is $\text{PE}_n = -k_0 e^2/r_n$, where r_n is the electron's quantized distance from the proton. If that atom should get close enough to another so that it interacts, i.e., "sees" the electric and magnetic fields that the other atom produces, then we have to modify E_n by adding other potential energy terms to account for the new conservative forces,

$$E_n = \text{KE}_n + \text{PE}_n + \sum \text{PE}'.$$

If you now imagine that the two atoms are moving about randomly, you will see that the $\sum PE'$ will fluctuate, with the result that the energy levels will fluctuate as well. The outermost energy levels will be affected the most. If we now add N atoms all interacting (those further away to a lesser degree), the effect becomes one of producing a continuous band of energy levels and therefore a continuum of photon energies each still due, however, to a discrete transition. The effect will be a function of the average spacing between atoms since the further apart they are the weaker are the interacting forces, but for solids where the average spacing are about the same, this will be a minor distinction and so all solids will begin to look alike in terms of the continuous photon spectrum. In fact, the only variable will be temperature, for as T increases, the collisions between atoms can transfer more energy (remember $\langle KE \rangle \sim kT$), causing excitations to higher energy levels. This produces in turn more photons of a higher average E_γ, which is exactly what we see. When equilibrium is reached, the number of absorbed photons equals the number of emitted photons in each energy band and we will have continuous black body radiation.

1.4.5. But Black Body Radiation from a Star?

Clearly most stars are not solid. The average spacing between stellar atoms will be large compared to the spacing between atoms in a solid. Stars, therefore, will not have continuous bands of energy levels but will maintain their discrete spectra. But in a star there is a different way of producing the black body radiation; the interaction between ions (say protons) and free electrons. It is called "bremsstrahlung"—the German word for braking radiation. Again it is accelerating electrons that produce the radiation, but now that acceleration is caused by the electrostatic deflection of electrons by ions. Each deflection causes a photon to be emitted with

$$E_\gamma = \Delta KE = KE_{\text{initial}} - KE_{\text{final}} = h\nu,$$

where ΔKE is the kinetic energy lost each time an electron is deflected by an ion. Since this is a random process, only the laws of probability determine how many encounters there will be and the energy lost in each encounter. Thus there will be a random distribution of KEs and therefore a continuous photon spectrum even though the photons have discrete energies. The inverse bremsstrahlung process, in which a photon is absorbed by a free electron, increasing the electron's KE, also exists. Thus equilibrium can be established with an equal number of photons being emitted as absorbed in a continuous fashion, each being a discrete quantum. Equilibrium is preserved because these bremsstrahlung photons are effectively trapped in the stellar interior, taking about 10^6 years to escape (we shall verify this later). A star is therefore almost an ideal black body cavity, the stellar surface temperature being the temperature of the black body

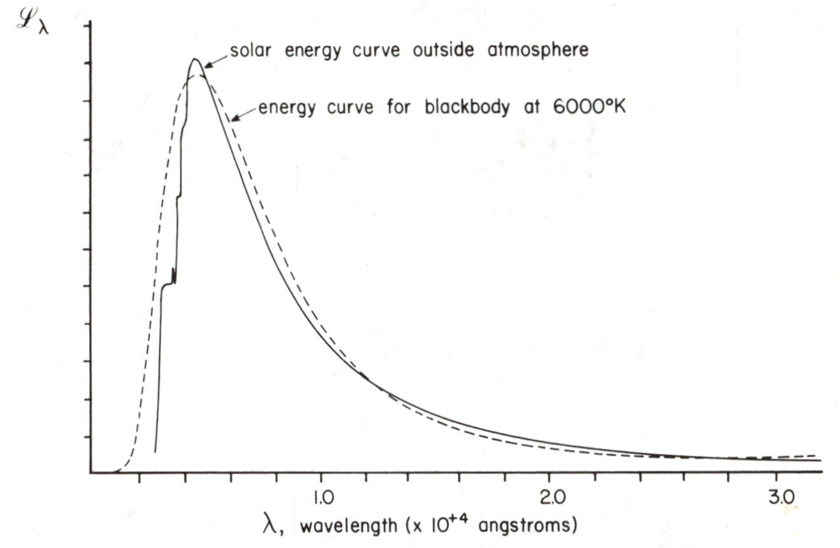

Fig. 1.12. Comparison of solar spectrum with black body spectrum.

"walls." Although the mechanism for producing the radiation is different, the end result is the same as that for a solid. Even the same temperature dependence is there too. The higher the temperature, the higher the average KE of the randomly moving electrons. This means they can undergo more interactions before stopping ($KE_{\text{initial}} - 0 = \sum KE_i$) or lose more KE per interaction. This in turn means that

i. more photons on the average can be emitted (assuming one per interaction), or
ii. those that are emitted will, on the average, have more energy.

Either way, as T increases the number of photons will go up and the average photon energy will move toward higher frequencies.

Thus Fig. 1.12, the spectrum of the Sun with $\mathfrak{L}_\lambda$ taken above the atmosphere and compared with an ideal black body radiator should come as no surprise. The Sun is a black body emitter, as are all the stars. But what good is it to know the star is a black body? That is our next topic.

1.4.6. The Surface Temperature, Luminosity, and Radius of a Star

The "beauty" of black body radiation is this: If we can find the peak of the $\mathfrak{L}_\nu$ (or $\mathfrak{L}_\lambda$) curve, then we can find the surface temperature of the star from

$$h\nu_{\text{peak}} \approx 3\,kT \left(\text{or for } \mathfrak{L}_\lambda,\ h\frac{c}{\lambda_{\text{peak}}} \approx 5\,kT\right).$$

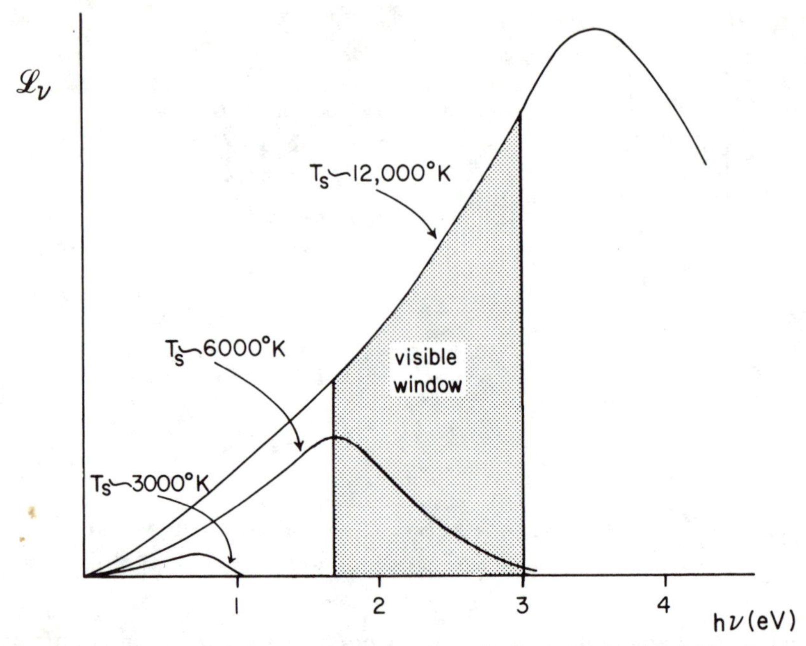

Fig. 1.13. Comparison of characteristic energy flux distributions with the visible window.

We stress the surface temperature because in an ideal black body, only a small fraction of the energy can escape (remember it must be in equilibrium and thus an almost perfect absorber). What does escape will come mainly from the surface. If the star's density is too low, we may see some lower layers and thus not obtain an unique black body curve but some kind of folded composite. If we could stick a detector inside and measure the black body curve at various layers, we could find the temperature of the star at each radius. Clearly we cannot do this. Of course, all that ground-based telescopes can detect is the visible part of the spectrum, but that is enough, for if we know the shape over a portion of the curve and we know what shape (or function) the curve should be, then we can find the whole curve, the peak, and thus T_s. The surface temperature of stars range between 3000 °K and 80,000 °K (our sun has $T_s \approx 5800$ °K, i.e. $\lambda_{\text{peak}} \approx 5000$ Å). We have plotted $\mathcal{L}_\nu$ for three different temperatures in Fig. 1.13. From these curves we can see that the energy emitted by hot stars will be outside the visible region.

A word about our visible window is now in order. For $T_s \sim 5800$ °K, the peak of the $\mathcal{L}_\nu$ curve is at

$$h\nu_{\text{peak}} \approx 3\,kT = 1.5 \text{ eV}.$$

These solar photons are in the infrared region, slightly below the visible window (1.7–3 eV). Together, the three photoreceptor cone cells in our retina are most sensitive to yellow–green light of 5550 Å ($h\nu \sim 2.2$ eV). Since $\mathcal{L}_\nu$ is skewed toward the infrared, we see the sun as yellow–orange. The light that plants absorb for photosynthesis, through the pigment chlorophyll, is at each end of the visible spectrum; in the blue–violet between 4000 and 4800 Å ($h\nu \sim 3$ eV), and in the yellow–red between 5700 and 7000 Å ($h\nu \sim 2$ eV). (Plants are green because green is the dominant color that remains to be scattered back after the others have been absorbed.) Is it natural selection that has placed the needs of plants just within the bounds of the visible window? What is more amazing is that the Earth's visible window, due purely to atomic binding energies, almost perfectly matches the maximum output of the "black body" Sun, due entirely to its surface temperature. Or then again, maybe we would not be here if it were not so well matched.

Once we know the surface temperature of a star, we also know the energy per unit area being emitted by the star each second, $\mathcal{L} = \sigma T^4$. If we further assume that the star is a sphere, then the *luminosity* (power) of the star, the energy coming out each second, L, is just

$$L = \mathcal{L} 4\pi R^2 = 4\pi R^2 \sigma T^4 (\mathrm{J/sec}), \tag{1.14}$$

where R is the radius of the star in meters, with $\sigma = 5.7 \times 10^{-8}$ J/sec-m^2-°K^4. If, and it is a big if, one can measure L, then one can determine the radius of the star—an object perhaps so distant that it appears as no more than a speck on the most powerful telescope!

Unfortunately we cannot measure L directly. While L in the above expression is the total energy being radiated away each second, only a tiny fraction of it reaches us. If ΔA is the area of the Earth-bound detector and d is the distance to the star, then the fraction of L reaching the detector is $\Delta A/4\pi d^2$. This is further reduced since we can detect only the visible portion of the radiation that reaches us. Since it is black body radiation, however, knowing the visible portion is sufficient to give us the total energy per unit time, L_E, hitting the Earth detector. Then the luminosity of the star, a distance d away is

$$L = L_E \frac{4\pi d^2}{\Delta A} \equiv \mathcal{L}_E 4\pi d^2, \tag{1.15}$$

where $\mathcal{L}_E = L_E/\Delta A$ is the energy flux received by the Earth from the star. There is usually no problem in measuring $\mathcal{L}_E$. The difficulty is in measuring d. Out to ~500 LY (1 LY is the distance that a beam of light moving with a speed of 186,300 mile/sec would travel in one year), we can use the method of stellar parallax which we shall describe in the next section. But the Galaxy is 100,000 LY across, so we can only find the luminosity of a few thousand of its 10^{11} stars this way. In subsequent chapters we shall discuss other techniques for extending the distance measurement, but none are as direct and accurate as stellar parallax. We should also emphasize that we are measuring the luminosity of the star, not as it is now, but as it was many years ago when the light first left the star; 5,000 years ago if the star is 5,000 LY away. This time lag means that we do not really know the present luminosity of any star. Also we emphasize that determining L alone does not tell us either the size or temperature of the star. A low value of L may be due to a high T_s coupled to a small star, or a low T_s and a large star. Only with an independent measurement of T_s can we determine the size of the star.

From the Sun, we measure on Earth an energy flux of about $\mathcal{L}_E \sim 1300$ J/sec-m^2, called the solar constant, which corresponds to heating up 1 gm water 2 °K/min/cm^2. Since the distance between the Earth and Sun is known to be $d = R_{\mathrm{ES}} = 1.5 \times 10^{11}$ m, the above $\mathcal{L}_E$ implies that the luminosity or total power output of the Sun is

$$L_\odot = \mathcal{L}_E 4\pi d^2 \approx 1.3 \times 10^3 \times 4\pi (1.5 \times 10^{11})^2 \approx 4 \times 10^{26} \text{ J/sec.}$$

(The symbol $\odot$ is standard astronomical notion for our Sun). On the surface of the Sun, this stellar luminosity can be converted to an energy flux

$$\mathcal{L}_\odot = \frac{L_\odot}{4\pi R_\odot^2} \approx \frac{4 \times 10^{26}}{4\pi (7 \times 10^8)^2} \sim \frac{2}{3} \times 10^8 \text{ J/sec-m}^2,$$

and since $\mathcal{L}_\odot = \sigma T^4$, this latter number leads directly to the surface temperature of the Sun $T_\odot \sim 5800$ °K, consistent with the λ_{peak} determination.

1.5. Measuring Stellar Distances by Trigonometric Parallax

The technique of stellar parallax is described as follows:

i. We observe a star on a given date (say January 1) and point the telescope so that the image of the object falls exactly at the center of the film, as depicted in Figs. 1.14 and 1.15a.
ii. Then we wait 6 months (until July 1) and again record the object on film, but we keep the telescope at the same angle relative to distant stars as it was on January 1. We do this by aiming the telescope at the same cluster of very distant stars ($d \approx \infty$). Having done this, our film will show that the image of the object has moved a distance a, as depicted in Fig. 1.15b.

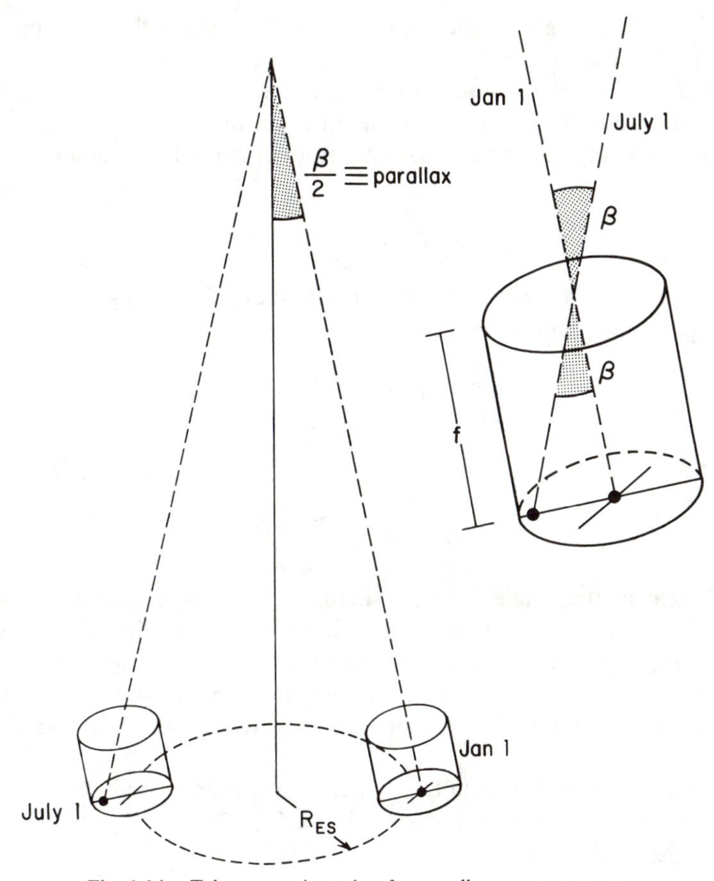

Fig. 1.14. Telescope orientation for parallax measurements.

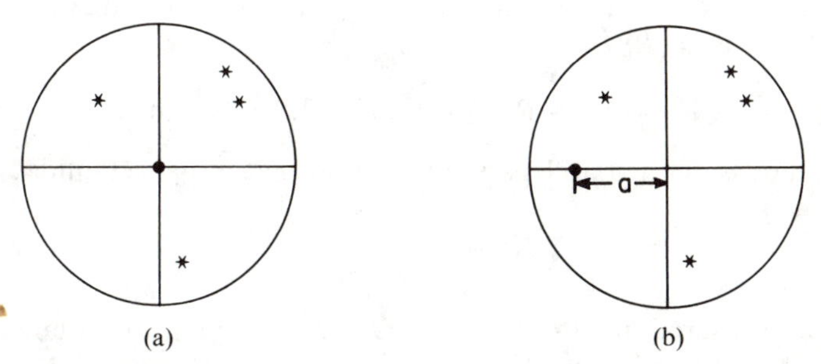

Fig. 1.15. Illustration of parallax motion with respect to distant background stars: (a) film on January 1; (b) film on July 1.

Now when one designs a telescope, one uses a lens with a certain focal length, f. What focal length means is that *parallel* rays of light will be brought to a focus at a distance f behind the lens. This, of course, is where we place the film. Therefore, with this in mind and referring to Fig. 1.14, we see that the angular spread, β, between the light entering January 1 and July 1 is

$$\tan \beta = \frac{a}{f}.$$

But, because the telescope moved during these 6 months we have, by triangulation and the fact that $d \gg R_{ES}$,

$$\tan\left(\frac{\beta}{2}\right) = \frac{R_{ES}}{d}.$$

This gives us

$$d = \frac{R_{ES}}{\tan(\beta/2)} \approx \frac{R_{ES}}{\beta/2},$$

where, because the angles are so small, we can write $\tan(\beta/2) \approx \beta/2 \approx a/(2f)$. The only limitation to this method is our ability to resolve the spot (i.e., separate the images of the object) between January 1 and July 1. To this end it should be obvious that a long focal length helps by producing larger image separations ($a = f \tan \beta$), hence telescopes with folded light paths.

It turns out that $\beta/2$ is usually so small that it is expressed not in radians but in seconds of arc: 1 radian $\approx 57° \approx 57 \times 3600'' = 206265''$ (where $'' \equiv$ seconds of arc). Thus,

$$d = \frac{R_{ES} \times 206265}{(\beta/2)''},$$

where $\beta/2$, often called the parallax angle or just parallax, is now measured in seconds of arc. For a parallax angle of $\beta/2 = 1''$, the radial distance of the star from us is ($R_{ES} = 1.5 \times 10^{11}$ m)

$$d \equiv 1 \text{ pc} = R_{ES} \times 206265 = 3.1 \times 10^{16} \text{ m}.$$

This distance is defined as 1 parsec (1 pc), so that in parsec units

$$d(\text{pc}) = \frac{1}{(\beta/2)''},$$

with $\beta/2$ measured in arc seconds. Other astronomical distance units include astronomical units, 1 AU $= R_{ES} = 1.5 \times 10^{11}$ m, and light years

(LY), the distance light travels in 1 year $\approx \pi \times 10^7$ sec or

$$1 \text{ LY} = c \times 1 \text{ year} \approx 3 \times 10^8 \times 3 \times 10^7$$
$$\approx 0.9 \times 10^{16} \text{ m} \approx 0.3 \text{ pc},$$

or 1 pc $\approx$ 3.26 LY. As an example, we note that the nearest star to us, α-Centauri, has a parallax angle of $\beta/2 = 0.754''$, corresponding to $d = 1/(0.754)'' = 1.32$ pc $= 4 \times 10^{16}$ m $= 2.6 \times 10^5$ AU $= 4.2$ LY. It is interesting to note that, within a distance of 10 LY from our Sun, there are only 20 stars. Present telescopes allow us to resolve angles no smaller than $\beta/2 = 0.005''$ or stars 200 pc away. Thus ideally, the parallax method can be used out to a maximum distance of $d_{\text{max}} = 200$ pc $= 6 \times 10^{18}$ m $= 4 \times 10^7$ AU $= 650$ LY. This minimum resolution, however, also means that we have a built-in error of $\pm 0.005''$, so that a measured value of $\beta/2 = 0.2''$ must be written as $0.2'' \pm 0.005''$, giving us an error of $0.005/0.2 = 2\frac{1}{2}\%$. If we decide to accept only those measurements whose accuracy is better than 10%, we are limited to $\beta/2 > .05''$ or $d < 20$ pc ≈ 65 LY.

Although we find the distance to only a relatively few stars by the trigonometric parallax method, the technique is important, because it acts as a stepping stone for other distance-measuring procedures. For example, when we determine the *absolute* luminosity of nearby stars using stellar parallax ($L = \mathfrak{L}_E 4\pi d^2$), we find, as we shall see in the next chapter, that their luminosities plotted against surface temperature fall in a peculiar pattern. If we observe the *apparent* energy flux of other groups of stars, $\mathfrak{L}'_E$, and find that certain groups have the same pattern, then we can assume their *absolute* luminosities are the same as that of the local stars, $L = L'$, and therefore,

$$d' = \sqrt{\frac{L}{4\pi \mathfrak{L}'_E}} \,.$$

Finally, we note that if the object we are looking at gives us a measurable spot on the film—say a width w—then the angular size α of the object (the angle it subtends at the telescope) can be obtained from the expression

$$\tan \alpha = w/f,$$

as shown in Fig. 1.16.

Of course we would like to know the linear size of the object. This can be obtained if we know how far away the object is. Then, the width D is

$$D = d \tan \alpha = d(w/f).$$

Of course we cannot find the diameters of stars this way—D/d is far too small. But we can find the size of the huge and often spectacular interstellar

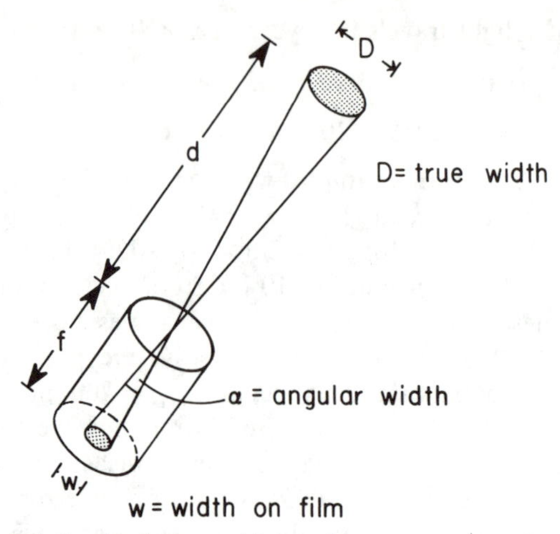

Fig. 1.16. Relationship between object and image parameters for a telescope.

clouds of hydrogen gas and the diameter of the equally spectacular ring nebulae. Indeed, some of these clouds or nebulae (Latin for cloud) are over 10 LY across.

1.6. Discrete Spectra

You should be very familiar with the concept of excitation—the raising of an atom from its ground state to one of the higher allowed energy levels. The lowest hydrogen levels are enumerated in Fig. 1.17. The modes of excitation are (1) collisions and (2) photon absorption, with conservation of energy determining to what level the atom will be excited. Then in $\sim 10^{-8}$ sec the electrons return to the ground state, either directly or via intermediate excited states, conserving energy and angular momentum (remember, the photon carries one unit of angular momentum, $1(h/2\pi) = 1\hbar$) at each step. The result is not a continuous spectrum of photon energies, but a unique spectral pattern that depends upon the degree of excitation (i.e., how much we excite the atom) and the type of atom, as discussed in Sec. 1.4. In Fig. 1.17 we have indicated the energy levels which correspond to various values of l, where, as mentioned in Sec. 1.4 the total angular momentum L is equal to $l\hbar$. Since the photon carried away 1 $\hbar$, l must change by 1 in each transition; hence the great variety of transitions shown in Fig. 1.17.

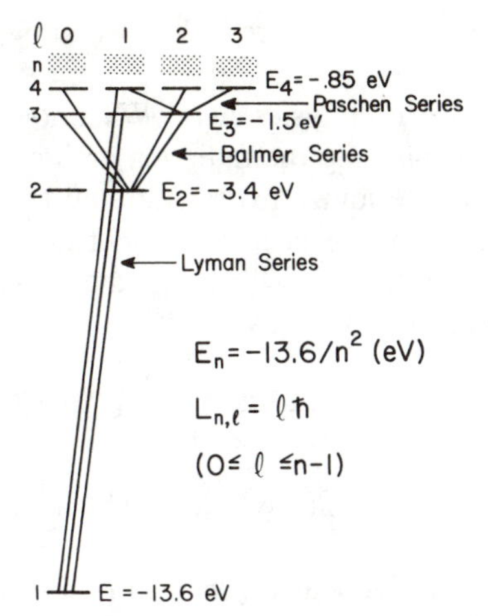

Lyman Series		Balmer Series		Paschen Series	
$n_i - n_f$	λ(Å)	$n_i - n_f$	λ(Å)	$n_i - n_f$	λ(Å)
2-1	1216	3-2	6563	4-3	18,750
3-1	1026	4-2	4861	5-3	12,821
4-1	931	5-2	4341	.	.
⋮	⋮	6-2	4101	.	.
ultraviolet		visible		infrared	

Fig. 1.17. Energy level diagram of hydrogen and wavelengths of selected transitions.

1.6.1. Emission Spectra

An emission nebula is one of the most spectacular producers of an emission spectrum. It is caused by gas clouds that are close to, or even surrounding a star. The large photon flux from the star's continuous spectrum, if energetic enough, is capable of causing appreciable ionization of the atoms in the cloud. After ionization, a freed electron eventually will be recaptured (remember, + and − charges attract), but in an excited state of the atom. A typical chain of events for hydrogen is then

$$\gamma_{\text{star}} + \underset{\text{ground state}}{\text{H}} \underset{\text{ionization}}{\rightarrow \text{H}^+ + e^-} ; \qquad \underset{\text{capture}}{e^- + \text{H}^+} \rightarrow \underset{\text{excited state}}{\text{H}^*} \rightarrow \underset{\text{ground state}}{\text{H}} + \sum \gamma$$

where after capture the atom returns to the ground state, emitting a variety of γs along the way. For hydrogen, the dominant visible line will come from the $n_i = 3$ to $n_f = 2$ transition. With $E_\gamma = E_3 - E_2 = 1.9$ eV, the wavelength of this so called H_α photon will be 6563 Å—nice and red. The other H lines are either in the ultraviolet or infrared region, or cannot compete with the H_α line. If we see red, then we know there is hydrogen in the cloud.

We can even do a bit of physics here. We know that the continuous black body curve $\mathcal{L}_\lambda$, measuring the radiation emitted by the star, peaks at $5\ kT_s$. We also know that to ionize hydrogen requires 13.6 eV. Thus, if we want appreciable ionization, we need a star hot enough to supply a large number of photons whose energy will be greater than 13.6 eV. This will certainly happen when $5kT_s \geqslant 13.6$ eV or

$$T_s \geqslant \frac{13.6}{5k} \approx \frac{13.6}{5 \times 0.9 \times 10^{-4}} \approx 30{,}000\ ^\circ\mathrm{K}.$$

So not all stars will produce an emission nebula—they will certainly if $T_s \geqslant 3 \times 10^4\ ^\circ\mathrm{K}$.

Early on, spectroscopists were astonished to find a green emanation from nebulae. It corresponded to a unique wavelength of 5000 Å, but could not be associated with any known element—hence the proud discovery of "Nebulium." Later, however, much to the embarrassment of the Nebuliumists, it was realized that Nebulium was nothing more than the transition from the 1st excited state to the ground state of doubly ionized oxygen, O^{++}. The reason, at that time, that Nebulium had never been seen in the laboratory was because the 1st excited state of O^{++} is *metastable*. An example of a metastable excited state is the $n = 2$, $l = 0$ state of hydrogen. The ground state is $n = 1$, $l = 0$, and because the photon always carries away 1 unit of angular momentum ($l = 1$), the transition between the $n = 2$, $l = 0$ and $n = 1$, $l = 0$, due to conservation of angular momentum, is forbidden, i.e.,

$$\bar{L}_u = \bar{L}_l + \bar{L}_{\mathrm{photon}}$$
$$0\hbar \neq 0\hbar + 1\hbar.$$

While such "forbidden" transitions can occur, they do so with a very small probability, that is, the lifetimes of these states are much longer than 10^{-8} sec. Why then did they not see Nebulium in their Earth laboratories? The reason is that there is another process by which the atom can end up in the ground state without emitting radiation! If an atom in an excited state should collide with another atom (or electron), energy can be exchanged along with angular momentum. Both energy and angular momentum can thus be conserved and the atom can end up in the ground state without the emission of a photon. At laboratory densities, the long lifetime of the metastable state allows the collisional deexcitation process to dominate over "forbidden" emission, so that the few O^{++} atoms that are produced will end up in the ground state without producing radiation. But in the rarefied atmosphere of an emission nebula, the forbidden transition competes very favorably with collision deexcitation and we see the green line.

There are three main classes of emission nebulae:

i. A nebula, where ionized gas surrounds a hot star (or stars). (Fig. 1.18a) Such a juxtaposition of ionized gas and stars is called an HII region by astronomers. (An HI region is composed of neutral hydrogen.)
ii. A planetary nebula, where the gas has been expelled by a red giant star, leaving behind a small, hot star that is on its way to becoming a white dwarf. (Fig. 1.18b)
iii. A supernova remnant, where the gas has been explosively ejected by a star. (Fig. 1.18c)

We shall study all three processes in more detail later but for now it is enough to realize that the light from the nebula can tell us something about the composition of the cloud. For example, the inside edge of the beautiful planetary nebula in Fig. 1.18b is green because of the presence of O^{++}. Green shows up so well in this photograph because a dye-transfer process has been used to match the response of the film to that of the human eye. Since we see green much better than red (recall, our photoreceptors peak at $\lambda \sim 5500$ Å), the green O^{++} lines will dominate Fig. 1.18b even though hydrogen is the major component of the gas. But as the photons that ionize O^{+} (~ 34 eV) are used up, O^{++} can no longer be formed, the green disappears and the red from the H_α line of hydrogen can be seen. Besides H and O, lines of He, N, Ar, and even sulfur are seen in these emission spectra.

If gas clouds contain these elements, then stars should too, but we already have seen that the discrete spectra of stars will be washed out into the continuous black body radiation which is independent of the type of atom involved. Surrounding a star, however, is a less dense region called a chromosphere which will contain atoms that can emit a discrete spectrum. Unfortunately the discrete spectrum will be overwhelmed by the black body radiation from the underlying star. Here the emission spectrum is its own worst enemy since the chromosphere radiates in all directions, not just toward us. During a solar eclipse, however, when the solar chromosphere is made visible, one can actually see its discrete spectrum—a nice pink glow at 6500 Å.

1.6.2. Absorption Spectra

Even though most stars do not produce visible discrete emission spectra, they do, amazingly enough, produce visible discrete absorption spectra.

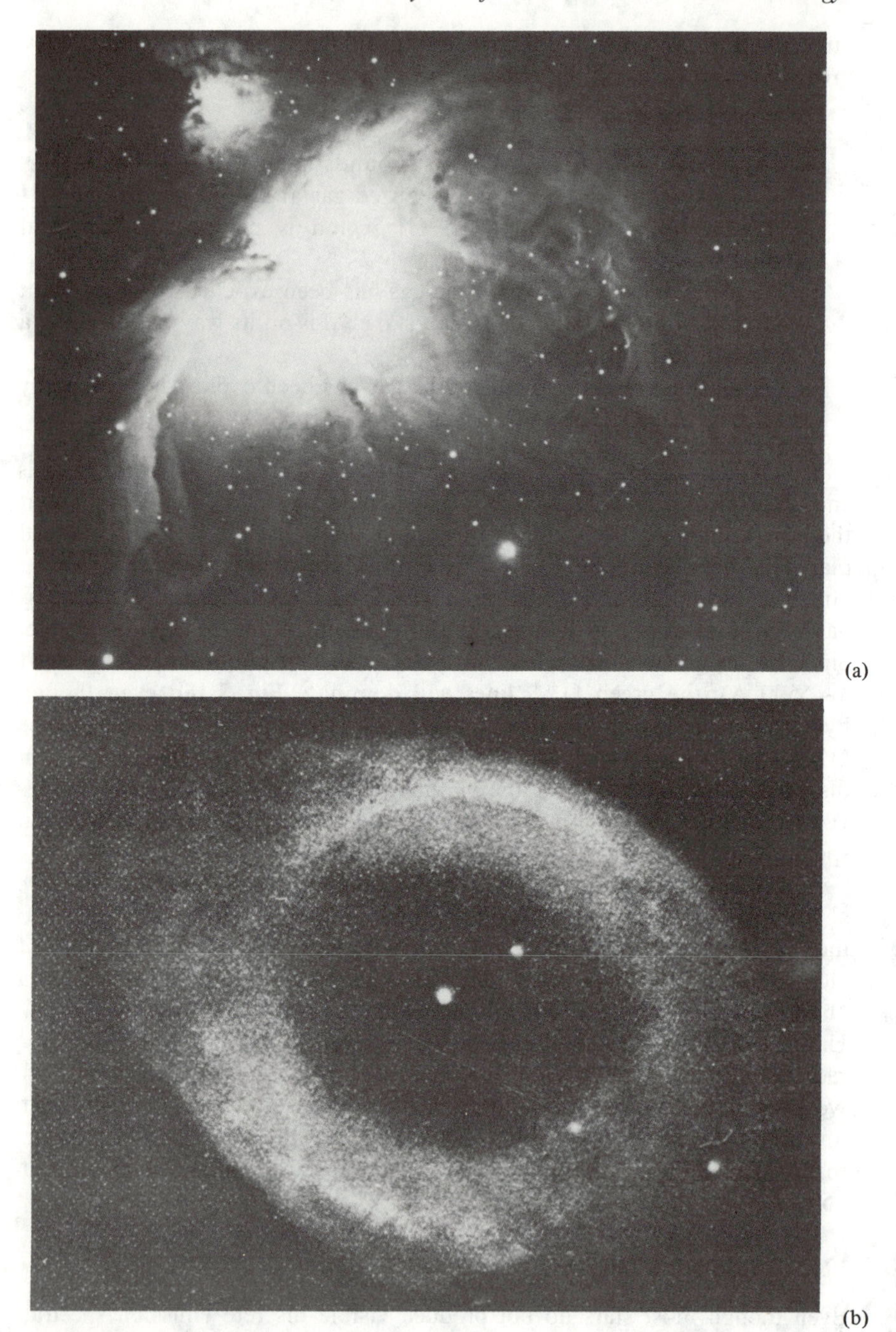

(a)

(b)

Fig. 1.18 (a), (b)

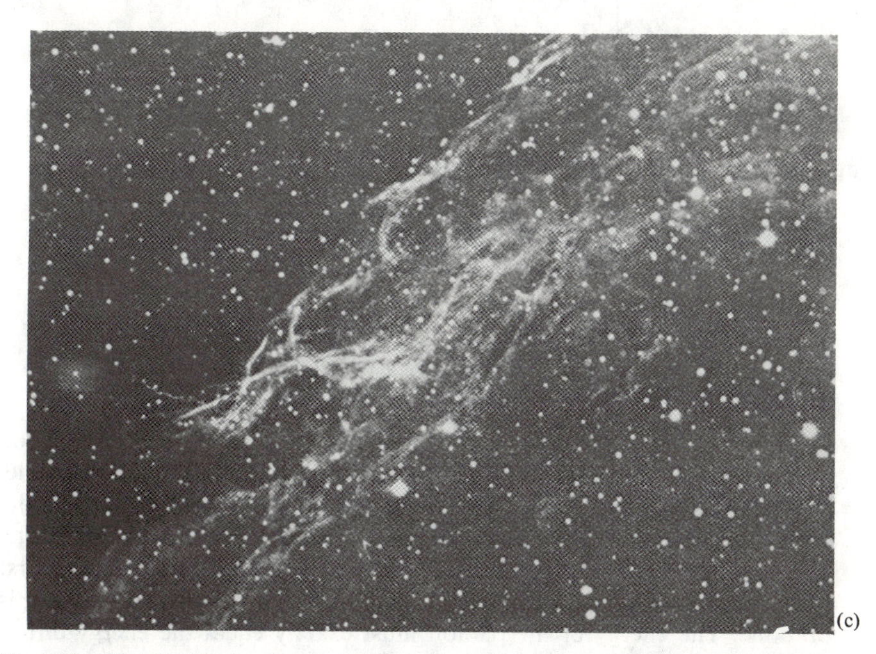
(c)

Fig. 1.18. The three main classes of emission nebulae: (a) the Great Nebula Orion; the ionization is due to several very hot stars deep within the nebula (© Association of Universities for Research in Astronomy, Inc., The Kitt Peak National Observatory). (b) Planetary Nebula in Lyra. The star in the center is responsible for the ionization (Lick Observatory photograph). (c) Veil Nebula in Cygnus, central section. The remnant of a supernova that exploded 50,000 years ago (Lick Observatory photograph). See figures in full color on frontispiece.

The mechanism is as follows: Say our gaseous envelope contains helium. Then in the continuous black body spectrum, some of the γs at the frequency ν_0, between the ground state and the 1st excited state of helium, will be photoelectrically absorbed (i.e., annihilated) and disappear. Let us assume that 1000 of such γs are heading *toward us*, from the Sun, and that all are absorbed. We now have 1000 He atoms in the 1st excited state. They will reemit photons of frequency ν_0, *but* these will be emitted in *all directions*, so that only a few of the original 1000 will reach us. Indeed, the measured spectrum will look like that in Fig. 1.19, with γs missing at the absorption frequencies, while those not precisely at these frequencies pass through undisturbed, giving us the usual black body spectrum. Since the radiation from the 1st excited state is usually in the ultraviolet, we would never "see" absorption spectra if it were not for one remarkable fact—photons can be absorbed while the atom is in an excited state. Normally the atom is in its ground state and only remains in an excited state $\sim 10^{-8}$

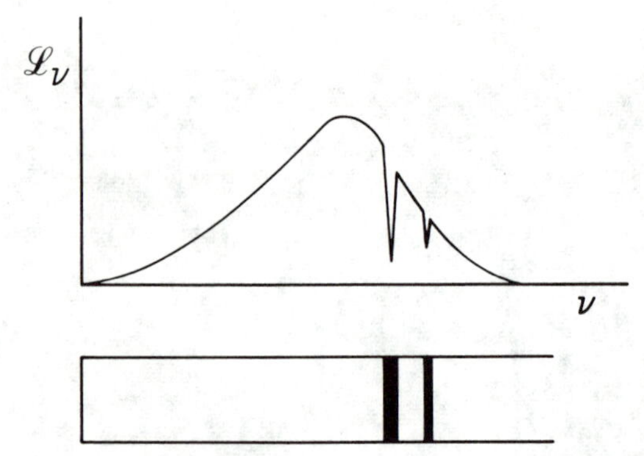

Fig. 1.19. Photon frequencies absorbed out of black body spectrum.

sec. But because the black body photon flux is high, there is a good probability that the atom will absorb a photon while still in an $n = 2$ state. In this way we can actually "see" the absorption of visible photons as the atom is raised from the $n = 2$ to $n = 3$ state or the $n = 2$ to $n = 4$ state, etc. Collisions are the primary means by which atoms are raised to the first excited state because particles are not as finicky as photons in getting this job done. The energy of the photon must exactly equal the energy difference between the ground state and the 1st excited state ($E_\gamma = E_2 - E_1$) if excitation is to occur, while a particle need only have its kinetic energy greater than this energy difference [$\mathrm{KE}_p > (E_2 - E_1)$]. As we shall see below, the temperature sensitivity of the $\langle \mathrm{KE} \rangle$ of the colliding particles ($\langle \mathrm{KE} \rangle \sim kT$) gives us a clue as to the temperature of the star.

The absorption lines that we do see, as shown in Fig. 1.20, will not only depend upon what elements are present, but also upon the surface temperature of the star. For example, if we see the absorption lines of *ionized* He (He^+), then the temperature must be high enough so that an appreciable number of particles have kinetic energies in excess of the He ionization energy. When these particles collide with the He atoms, they will ionize them. (Remember, at any temperature there are always some particles obeying the Maxwell–Boltzmann distribution with sufficient kinetic energy to ionize almost anything.) The rule of thumb here is that if the energy needed to ionize or excite is ΔE, then the temperature required to produce enough ions or excitations for the effect to be noticeable is $T > \Delta E/10\,k$. (We will explain this rule in Chapter 7.) Since we need 24 eV to ionize He, the presence of He^+ indicates that $T_s > 24/10\,k \approx 28{,}000$ °K. Below 12,000 °K, the lines of *neutral* helium disappear because there are not enough collisions with sufficient energy to raise the atoms to the $n = 2$ state, and prime them for photon absorption. The intensity of neutral hy-

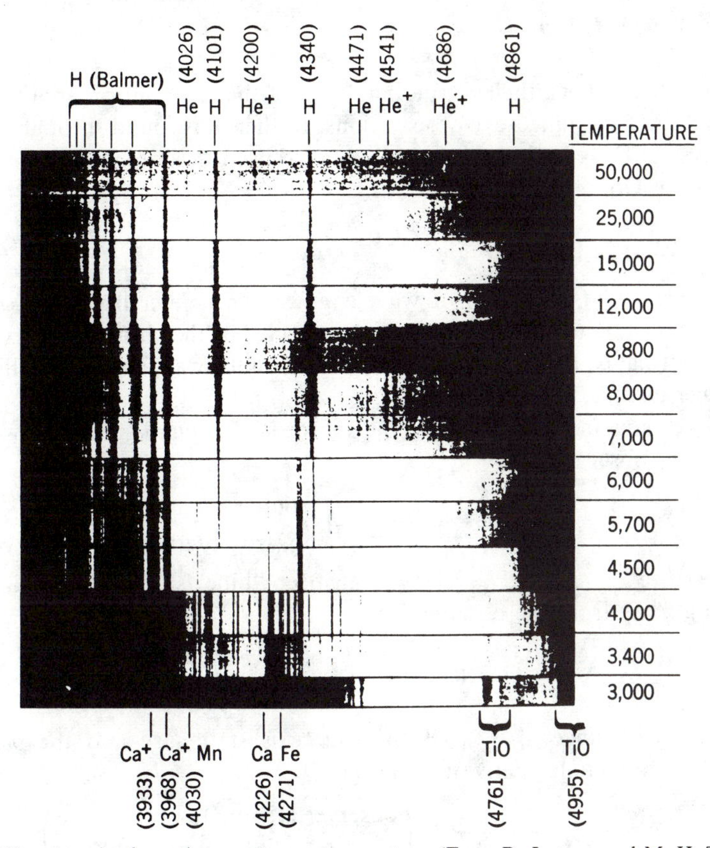

Fig. 1.20. Atomic absorption spectra vs. temperature. (From R. Jastrow and M. H. Thompson, *Astronomy: Fundamentals and Frontiers*, John Wiley and Sons, New York, 1974).

drogen lines, on the other hand, peaks around 8000 °K; above 16,000 °K. an increasing fraction of the atoms are ionized ($T_s > 13.6/10k$) and hence incapable of producing a spectrum of neutral hydrogen, while below 7000 °K, fewer and fewer collisions have the energy to lift the atoms to the 1st excited state. At these lower temperatures, however, the spectra of neutral metals, which have excitation energies on the order of a few eV, take over and dominate at $T_s \sim 4000$ °K. Taken together, these various clues give us a rough estimate of the approximate surface temperature of a star. An added bonus is that using this technique, the luminosity does not enter into the determination of T_s. As we shall now show you, coupled with the Doppler effect, emission lines also contain information about the motions of the star—translation and rotation.

Excitation, ionization → 2 different processes, situations

1.7. Doppler Effect

It is ironic but nevertheless true that most of the information about stars is not obtained by naked-eye observations. Rather, it is found through the use of physical tools such as the ones we have been describing. Another such tool is the Doppler effect.

1.7.1. What Is It?

If a source emits radiation of wavelength λ_0 and either the source or the observer is moving then the wavelength detected by the observer, λ', will not equal λ_0 (that is, $\lambda' \neq \lambda_0$). This shift in wavelength away from λ_0 is called the Doppler effect.

As we show in Appendix C, if the source is moving at speed u towards a stationary observer

$$\lambda' = \lambda_0(1 - u/c), \qquad \begin{pmatrix}\text{source moving}\\ \text{toward observer}\end{pmatrix}$$

where since $\lambda' < \lambda_0$, the wavelength has been "blue" shifted. If the source is moving away from the observer, then

$$\lambda' = \lambda_0(1 + u/c), \qquad \begin{pmatrix}\text{source moving}\\ \text{away from observer}\end{pmatrix}$$

a "red" shift since $\lambda' > \lambda_0$. If the source is stationary and the *observer* moves then a similar derivation shows

$$\lambda' = \lambda_0 \frac{1}{(1 + u/c)} \qquad \begin{pmatrix}\text{observer moving toward source}\\ \lambda' < \lambda_0\text{, still a "blue" shift}\end{pmatrix}$$

$$\lambda' = \lambda_0 \frac{1}{(1 - u/c)} \qquad \begin{pmatrix}\text{observer moving away from source}\\ \lambda' > \lambda_0\text{, still a "red" shift}\end{pmatrix}$$

Again we get the same kind of red and blue shift we got above when the source was moving. There are, however, some subtle differences between the two situations:

Source Moving	Observer Moving
$\lambda' = \lambda_0(1 \mp u/c)$	$\lambda' = \lambda_0 \dfrac{1}{(1 \pm u/c)}$

(1.16)

But for $u/c \ll 1$, the two results are the same: $\lambda' \cong \lambda_0(1 \mp u/c)$ where $(-)$ means they are moving toward each other and $(+)$ away from each other. That is, it is not important whether source or observer is moving—only the relative motion is physically significant.

However, for $u/c \cong 1$ (i.e., for velocities close to the speed of light), the wavelength for the source moving toward the observer ($\lambda' \to 0$) is very different from the wavelength for the observer moving toward the source ($\lambda' \to \lambda_0/2$). But you may recall that Einstein's theory of relativity implies that this latter situation is not allowed. The relativistically correct expression does not distinguish between the movement of source or observer; all that is relevant is the relative motion of source and observer toward or away from one another. The relativistically correct expressions are:

Toward One Another	Away from One Another
$\lambda' = \lambda_0 \sqrt{\dfrac{1 - u/c}{1 + u/c}}$	$\lambda' = \lambda_0 \sqrt{\dfrac{1 + u/c}{1 - u/c}}$

(1.17)

The relativistic result is clearly different from the classical result except in the limit $u/c \ll 1$, when they are the same.

For most astrophysical phenomena, $u/c \ll 1$ and so we can safely use the simpler classical expressions (1.16). But for cosmological red shifts, we will have to turn to the relativistic equations (1.17).

1.7.2. How Is It Used?

To use the Doppler effect, we have to know some wavelength that the star has sent out, λ_0, and measure the corresponding wavelength we receive, λ'. The tricky question is this: how do we know what λ_0 is, if it is no longer λ_0 when we receive it? The answer is easy enough: in their proper rest frame all elements, whether here or in the envelope of a star, will produce the same absorption lines (emission lines too, but they are not as important). Since there are only a few elements and their spectra are well known, we can usually figure out from the spectral pattern, even though that pattern has been shifted, what the elements are and therefore deduce the various λ_0s. With the corresponding λ', we have for $u \ll c$, $\lambda' = \lambda_0(1 \mp u/c) = \lambda_0 \mp \lambda_0(u/c)$ and therefore

$$\frac{\lambda' - \lambda_0}{\lambda_0} \equiv \frac{\Delta\lambda}{\lambda_0} = \mp \frac{u}{c}$$

← relative vel.

which is a direct measure of u. But what u?

1.7.2.1. Star Translating Relative to Earth

If the star is moving, then the absorption line will shift to either higher or lower λs and we easily can measure u. One must be careful here; the Doppler shift only measures the component of the velocity directly along

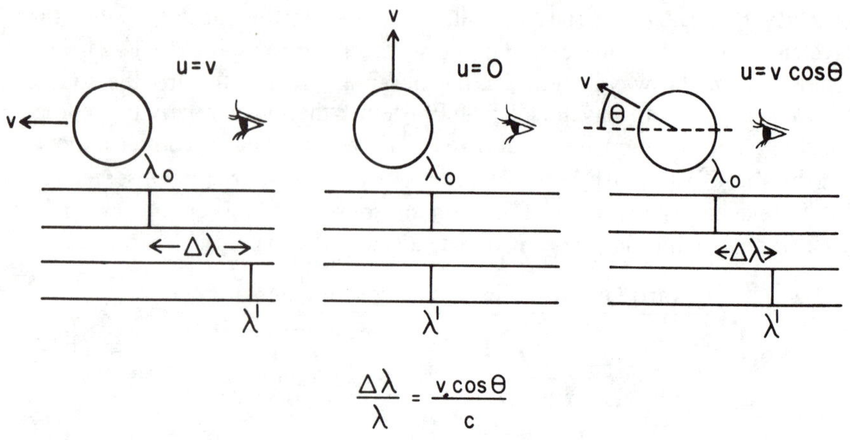

Fig. 1.21. Doppler shift for a source moving in one of three directions relative to observer.

our line of sight as shown in Fig. 1.21. If the star is moving at an angle θ with respect to our line of sight, then all we can measure is the so-called radial component of v, $u = v\cos\theta$. Unless the star is part of a cluster, one cannot determine θ, so that in many instances the Dopper shift can only give us a lower bound on the star's velocity, $v \geqslant u = c(\Delta\lambda/\lambda)$. Even this, however, is a valuable piece of information, and when we can determine θ then $v = u/\cos\theta$.

Another important use of the Doppler effect has been in the study of spectroscopic binaries. Binary stars, two stars revolving around a common center of mass, are placed in several categories. One is visual binaries, those we can see. With the resolving power of telescopes limited to 0.005 seconds of arc, about 6×10^4 visual binaries have been catalogued. In the case of spectroscopic binaries, distance is of no concern since all we are interested in are the stars' absorption spectra. In Fig. 1.22a a small star A revolves around a relatively massive star B. At positions (1) and (3), the velocity of A is at right angles to our line of sight and the measured wavelength is $\lambda' = \lambda_0$. At positions (4) and (2), it is moving towards and away from us, respectively, with the result that λ' is blue-shifted and then red-shifted with respect to λ_0. This moving or time-dependent spectral line is the giveaway that we have a spectroscopic binary. The period of the motion can easily be determined since it is the time it takes λ' to swing between wavelength extremes. In principle, the orbital velocity is given by $v_{\rm orb} = c(\Delta\lambda/\lambda)$, but in practice we are once again foiled because we do not know the angle the orbital plane makes with our line of sight (see Fig. 1.22b). The best we can say is that $v_{\rm orb} \geqslant u$, and from $v_{\rm orb} = r\omega$ that $r \geqslant c(\tau/2\pi)(\Delta\lambda/\lambda)$, where r is the radius of the orbit and τ the period. Around 1000 spectroscopic binaries have been analyzed to date.

$u = c(\Delta\lambda/\lambda)$

$v_{orb} \geq u$

$v_{orb} = r\omega = r2\pi\nu = r(2\pi/T) \Rightarrow r \geq c(T/2\pi)(\Delta\lambda/\lambda)$

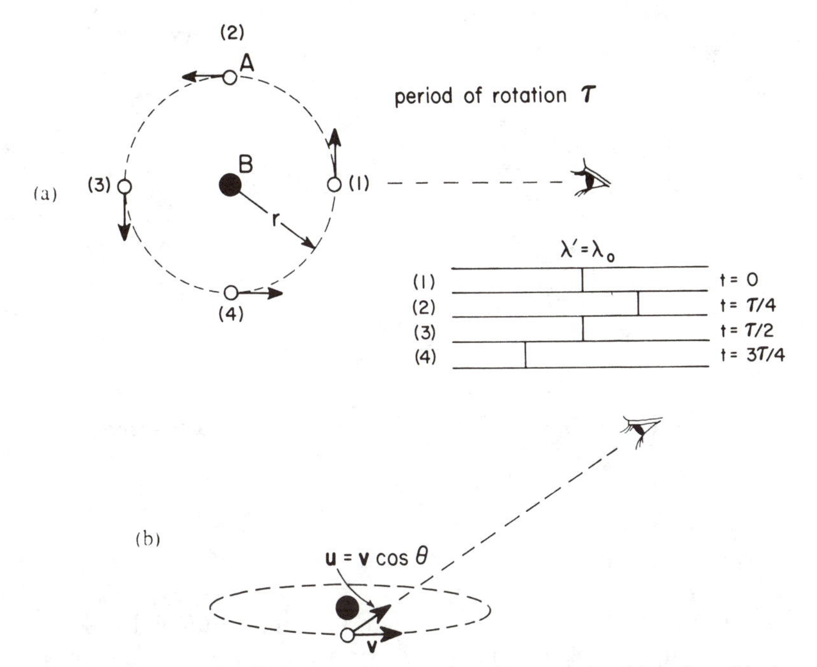

Fig. 1.22. (a) Doppler shift for a rotating star when observer is in orbital plane; (b) observer at angle θ above the orbital plane.

1.7.2.2. Surface Temperature of a Star

A natural spectral line will be very narrow, its width determined only by the uncertainty principle, $\Delta E \Delta t \sim h$ (to be discussed in Chapter 4). Since the decay time is $\Delta t \approx 10^{-8}$ sec, we have as the minimum expected natural line width $\Delta E \sim h/\Delta t \sim 4 \times 10^{-15}/10^{-8} \sim 4 \times 10^{-7}$ eV, where we have taken $h = 4 \times 10^{-15}$ eV-sec. In terms of wavelength, we note that since $\lambda = c/\nu = ch/E_\gamma$, we have

$$\left|\frac{\Delta\lambda}{\lambda}\right| = \left|\frac{\Delta E}{E}\right|,$$

so that for visible photons ($E \sim 2$ eV), $|\Delta\lambda/\lambda| = |\Delta E/E| \sim 4 \times 10^{-7}/2 \sim 2 \times 10^{-7}$. Narrow indeed! But the absorption lines we see are nowhere near that narrow. One reason is thermal broadening. We know that the atoms at the surface of the star are moving about randomly with $\langle \mathrm{KE} \rangle = \frac{1}{2} m v_{\mathrm{rms}}^2 = \frac{3}{2} k T_s$. Thus at any given time, some are moving toward and some away from us, no matter what our line of sight. Even if the star as a whole is not translating, the spectral line would have to be broadened with

$$\frac{\Delta\lambda}{\lambda} \sim \frac{v_{\mathrm{rms}}}{c},$$

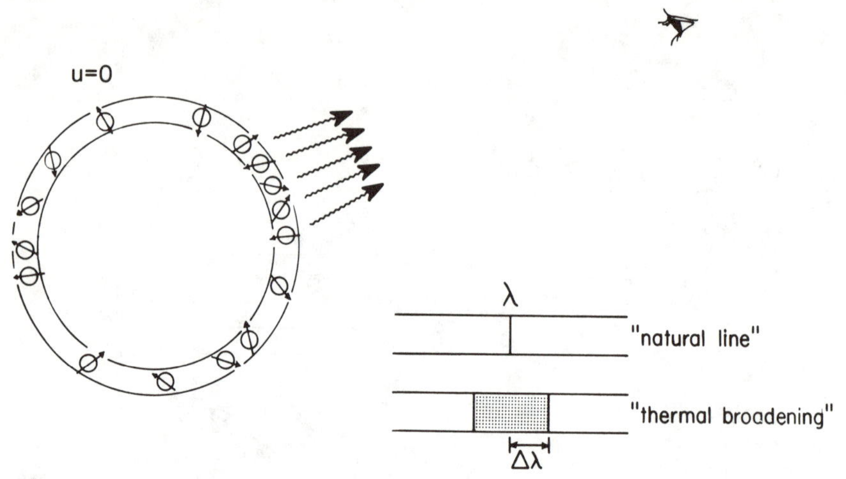

Fig. 1.23. Thermal Doppler-broadening with random motion of charges at surface represented by circles with arrows.

as depicted in Fig. 1.23. For a star whose surface temperature is 10,000 °K, $v_{\text{rms}} = 17 \times 10^3$ m/sec and the thermal Doppler broadening of the visible line would be $(\Delta\lambda/\lambda)_{\text{thermal}} \sim v_{\text{rms}}/c \sim 17 \times 10^3/3 \times 10^8 \sim 5 \times 10^{-5}$, or about 100 times the natural width. This is certainly not the best way to measure the surface temperature of a star since, as we shall see below, other factors contribute to the line width.

1.7.2.3. The Angular Velocity of a Star

If a star is rotating about an axis that is perpendicular to our line of sight, then at any given moment, part of the surface at the equator is moving directly away, part directly toward, and part at right angles to us. The result is a rotationally broadened line (independent of the thermal broadening), with

$$\left(\frac{\Delta\lambda}{\lambda}\right)_{\text{rot}} \sim \frac{v_{\text{edge}}}{c} = \frac{R\omega}{c},$$

where R is the radius of the star and $\omega = 2\pi/\tau$, τ being the star's period of rotation. For the Sun, $\tau \approx 25$ days $\approx 2 \times 10^6$ sec, $R \approx 7 \times 10^8$ m, so that $v_{\text{edge}} \sim 2$ km/sec and $(\Delta\lambda/\lambda)_{\text{rot}} \approx 10^{-5}$, again substantially more than the natural line width. Of course, thermal broadening and rotational broadening overlap, but astronomers are able to separate the two effects and find v_{edge}. If we also know the radius of the star (from L and temperature), we can also find ω. The axis of rotation, of course, may not be perpendicular to our line of sight, in which case $\Delta\lambda/\lambda$ yields a lower limit to v_{edge} and ω, and therefore an upper limit to the period of rotation. Although our sun rotates once every 25 days, some stars have been found that rotate once every 3 or 4 hours! Was our sun once rotating this fast and if so how did it lose its original velocity?

In retrospect, even though we are buried under several hundred kilometers of atmosphere with only a tiny "visible" window to look through, and in spite of the fact that we are many light years from the clouds and stars we wish to study, the laws of atomic physics, the Doppler effect, the black body radiation, the kinetic theory equation all in effect put us "right next door."

"THAT 187,000 MILES PER SECOND MAKES ME A BIT SKEPTICAL ABOUT THE WHOLE THING."

Problems

1. An EM wave has a wavelength of 25 mm.
   a. What is the frequency of the wave?
   b. What is the energy carried by each photon?
   c. In what region (microwave, infrared, etc.) of the EM spectrum is the photon found?
2. Estimate the number of air particles per unit volume ($n = \mathrm{N/V}$) at the surface of the Earth if the pressure at sea level is 1 atm (10^5 N/m^2). What is the rms speed of an air molecule?
3. Formaldehyde (H_2CO) has recently been discovered in interstellar clouds. Calculate the "mean" molecular speed of H_2CO if the cloud's temperature is 50 °K.
4. If a gas of protons is found to have an rms speed of 10^5 m/sec, what is the kinetic temperature of the gas?
5. Say we have 10^6 particles in the lowest energy bin (bins are 1/10 eV wide) of an ideal gas at a temperature of 5000 °K.
   a. In what single energy bin would we expect to find 135,000 particles?
   b. How many particles would have a bin-kinetic energy of 1.29 eV?
6. A container is at a temperature of 10^6 °K.
   a. What is the energy flux of photons exiting a small hole?
   b. At what frequency is the flux the greatest?
   c. What is the energy density of photons in the container?
   d. What is the approximate number density of photons in the container?
   e. What is the photon pressure on the walls of the container?
7. In a container with a fixed number of particles, the energy of the particles is $\propto T$ while the energy of the photons is $\propto T^4$. What special properties of particles and photons accounts for the difference in the power of the temperature? Show that the photon number density goes as $\propto T^3$.
8. What is the wavelength of a photon that makes a transition from an $n = 5$ to $n = 4$ level in the hydrogen atom?
9. What is the speed of an electron with just sufficient energy to ionize by collision a sodium atom ($Z = 11$) in the ground state with ionization energy of 5 eV?
   What would be the speed of a proton capable of ionizing the same atom?
   What would be the corresponding gas temperature if these speeds determined the average KE of the bombarding particles?

10. How much more energy is emitted by a star at 20,000 °K than at 5000 °K?
What is the predominant color of the above two stars?
Is enough information given to determine the luminosity of either star? Explain.
Say we are able to measure the energy per unit area per unit time that reaches the Earth from one of the stars. Would this allow us to determine the luminosity of that star with the information given?
11. Given the fact a 1 sq. meter area on the Earth receives 1300 J/sec from the Sun, infer the surface temperature of the Sun, for $R_{ES} \approx 200\, R_\odot$.
12. At what wavelength does the black body spectrum of the Earth ($T_s \sim 300$ °K) peak?
What is the luminosity of the Earth? ($\equiv L_E$)
If the luminosity of the Sun is $L_\odot \approx 4 \times 10^{26}$ J/sec, what is the energy/sec ($\equiv L_{E,\mathrm{inc}}$) received by the Earth from the Sun?
How do you account for the fact that $L_E > L_{E,\mathrm{inc}}$?
13. If a star is 35 pc away, what is its distance from us in (a) meters and (b) light years?
Why is the technique of trigonometric parallax limited to 650 LY?
14. The black body spectrum of a star is found to peak at $\lambda = 2900$ Å. The star is 300 LY away and its luminosity reaching the Earth, over an area of 10^{-6} m^2, is found to be $L' = 3 \times 10^{-16}$ J/sec.
    a. What is the absolute luminosity of the star?
    b. What is the radius of the star? Express your answer in meters and in terms of the Sun's radius.
15. We saw in the text that the appearance and disappearance of various spectral absorption lines can indicate a star's surface temperature. The ionization energy of sodium is 5 eV. Its 1st excited state is 2 eV above the ground state. The 2nd excited state is 3 eV above the ground state. Between what range of temperatures do we expect to "see" the 1-eV absorption line?
16. It is shown in Appendix C that if the source is moving toward us then

$$\lambda' = \lambda_0(1 - u/c) \tag{A}$$

    a. Show that if the observer is moving toward the source then

$$\lambda' = \lambda_0 \frac{1}{(1 + u/c)} \tag{B}$$

    b. If $u/c = 0.01$, by how much does λ', given by (A), differ from λ' given by (B)? (Carry out the computation far enough to get a few significant figures.)

c. Show that for $u/c \ll 1$ (B) is the same algebraic expression as (A).

d. Comment on how (A) and (B) appear to refute special relativity.

17. At what wavelengths will the following lines be observed?

a. Line emitted at 5000 Å by a star moving toward us at 100 km/sec.

b. The Ca^+ line ($\lambda = 3970$ Å) emitted by a galaxy receding at 90,000 km/sec.

18. What would be the thermal Doppler-broadening of the Na D line ($\lambda = 5800$ Å) for a star whose surface temperature is 40,000 °K? Show that the Doppler-broadening is independent of the star's translational velocity by carefully considering $\lambda' = \lambda_0(1 \pm u/c)$.

19. The $n = 3$ to $n = 2$ Balmer line ($\lambda = 6500$ Å) emitted by a star is found to have a *width* of 0.02 Å while λ' is found to be 6500 Å.

a. What is the velocity of the star relative to the Earth?

b. Assuming the width is due to thermal broadening, what is the spread in random velocities?

c. Assuming the width is due to the rotation of the star and the star's period is 24 hours, what is the size of the star?

CHAPTER 2

The Interstellar Medium and Evidence for Stellar Evolution

Have ever you stood where the silences brood,
And vast the horizons begin,
At the dawn of the day to behold far away
The goal you would strive for and win?
Yet ah! in the night when you gain to the height,
With the vast pool of heaven star-spawned,
Afar and agleam, like a valley of dream,
Still mocks you a Land of Beyond.

Robert Service, *The Land of Beyond*

By the late 19th century, astronomers had developed some ideas about the structure of our corner of the universe:

i. Simply from looking at the sky they had concluded that the stars were distributed in a thin disk—The Milky Way—but that the disk had large holes in it—regions devoid of stars (Fig. 2.1).
ii. The late 18th century–early 19th century astronomer Herschel, counting stars in the sky, had developed a cross section for the Milky Way which (shades of Ptolemy) had the Sun in the center.
iii. In 1875 an emission nebula was detected that for the first time was shown to be a diffuse gas and not a star. Except for this nebula it was assumed that the space between the stars was empty.

All of these ideas began to change due to two dramatic "discoveries," one in 1904, the other in 1915. We shall start with an explanation of the latter.

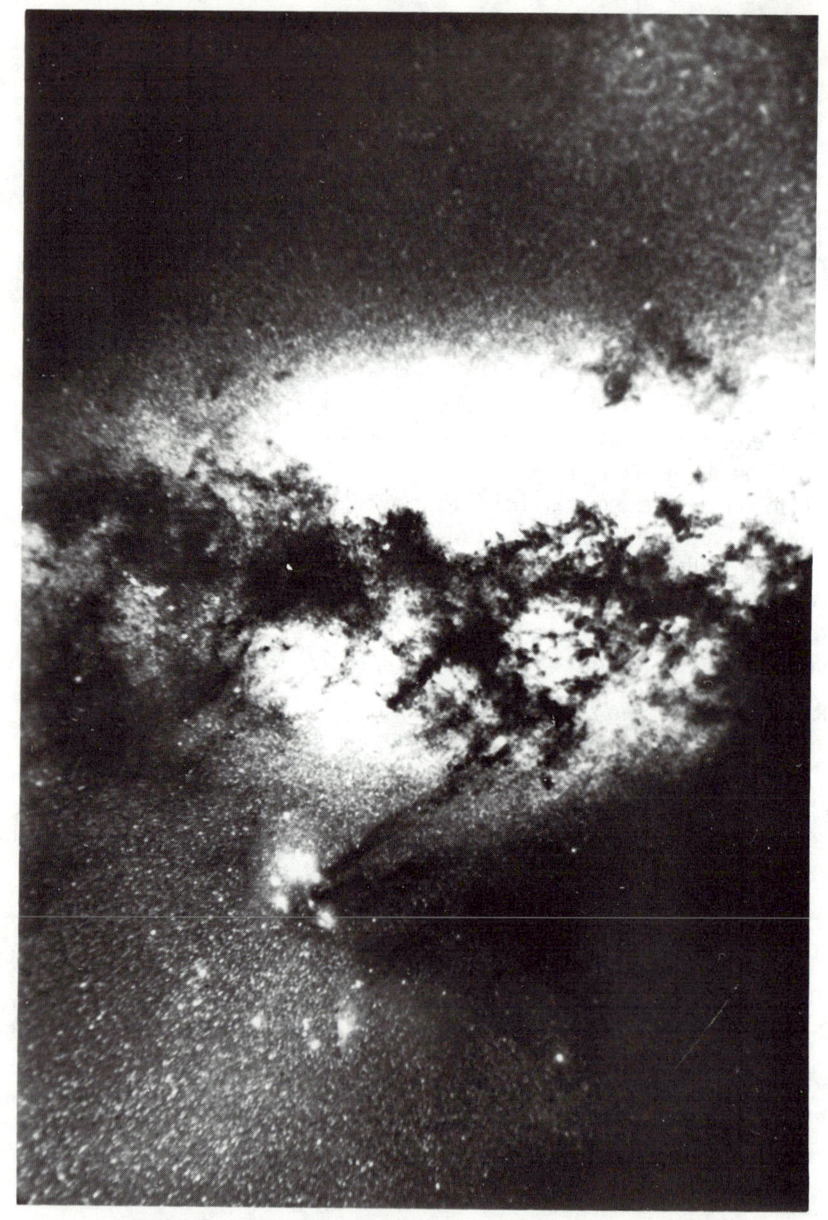

Fig. 2.1. A section of the Milky Way (Hale Observatories).

2.1. 1915—The Sun's Place in the Galaxy

Prior to 1900 astronomers had found large "globs" which they called nebulae but which later and better telescopes had shown to be *globular clusters* of stars as shown in Fig. 2.2. Each cluster contained upwards of 10^4 stars (today we know of $\sim$100 such clusters) and became very visible bench marks in the sky. Unfortunately most were too far away ($\sim$650 LY) for a parallax distance measurement. But in 1915 Shapley reasoned out an ingenious way of determining their distance from the Sun.

Although most stars shine with a constant luminosity, early 20th century astronomers had catalogued a group of stars called Cepheid variables whose luminosities varied with time. (The north star, Polaris, is a Cepheid variable; its luminosity varies by 9% over a 4-day period.) Furthermore, by studying Cepheids in a single star cluster (the Magellanic Clouds) so that all are about the same distance from the Earth, they found that the apparent average luminosity of a Cepheid variable was a function of the period of its oscillation; the greater the period the greater the luminosity. Once the distance to the Magellanic Clouds was known ($\sim$50 kpc $\approx 16 \times 10^4$ LY), one could determine the actual luminosity of the Cepheids from the luminosity-distance relationship we studied in Sec. 1.4.

$$\langle L_0 \rangle = \frac{4\pi d^2 \langle L' \rangle}{\Delta A}, \tag{2.1}$$

where $\langle L' \rangle$ is the average luminosity we receive in a counter of area ΔA and $\langle L_0 \rangle$ is the actual average luminosity of the star. The result is a curve of rising $\langle L_0 \rangle$ vs. the period of oscillation of the luminosity.

When studying globular star clusters, Shapley discovered that they contained Cepheid variables whose period he could easily measure even if he could not resolve their distance in a parallax experiment. What Shapley realized was that he could turn equation (2.1) around and along with the information contained in the 1915 version of the $\langle L_0 \rangle$ vs. T curve, determine the distance to the globular clusters. Thus from a measurement of the period of the Cepheid he could determine $\langle L_0 \rangle$ while experimentally he could measure $\langle L' \rangle$ and ΔA. Plugging these into (2.1) he had d!

What he discovered was astonishing. Plotting the ds, as in Fig. 2.3a, he found the 100 or so globular star clusters are distributed in a large spherical ball (not a thin slab like the rest of the stars) with the Sun apparently near the edge. Now we know the diameter of the ball is $\sim$30,000 pc ($\sim$100,000 LY) with the Sun about 35,000 LY from the center. To assume that the Sun was at the center of the Galaxy would thus require a totally asymmetric distribution of globular clusters (left in Fig. 2.3b), a trick the Creator was

Fig. 2.2. Globular star cluster M3 (Kitt Peak National Observatory).

not likely to play on us. Much more likely it was the globular clusters that were symmetrically distributed with respect to the center of the Galaxy with the Sun near the edge. (Ah, Ptolemy and Copernicus all over again—will we ever learn?) This is what Shapley proposed and indeed this is the way it appears to be.

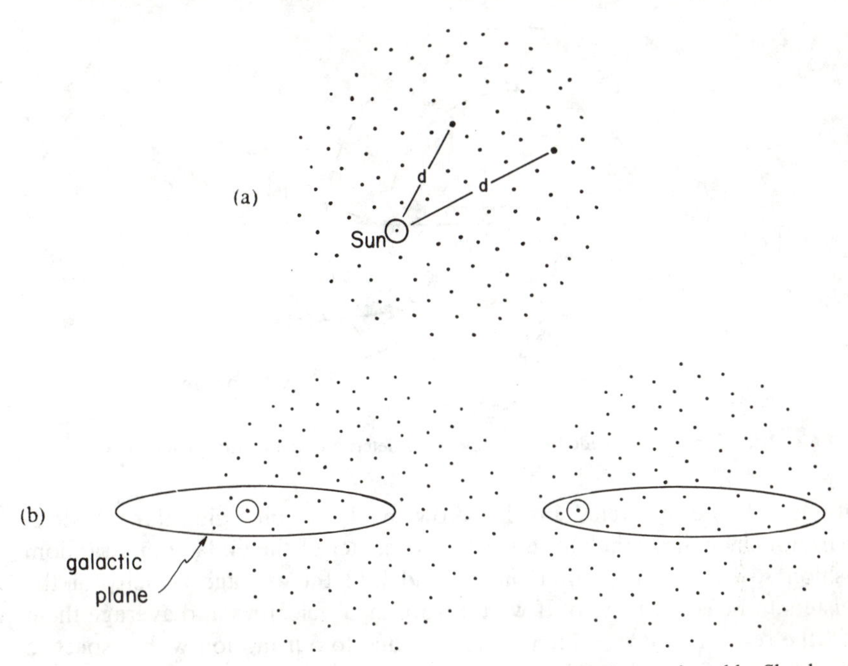

Fig. 2.3. (a) Distribution of globular star clusters relative to the Sun, as found by Shapley; (b) two choices for the position of the Sun relative to the galactic plane.

There are still some complications measuring *distances* to other galaxies using this technique (i.e., Cepheid variables), which, by the way, is effective out to 30 million light years. The main problem is that if we are looking at light produced millions of light years ago and comparing it to the Cepheid curve from the Magellanic Clouds, we have no way of knowing how the $\langle L \rangle$ vs. period curve may have changed over the eons. Within our own galaxy this is not a problem since the Magellanic Clouds are neighbors. But to find the distance to other galaxies requires faith in the constancy of the $\langle L \rangle$ vs. period curve. We shall return to this point when we take up our study of cosmology. But we should not look a gift horse in the mouth. Because globular clusters contain so many stars, we can see them and their Cepheids even if they are in other galaxies and thus we can get a handle on the distances to these island universes. Moreover we can use the globular clusters to measure the mass of our galaxy.

Consider a box filled with randomly moving particles. An observer who is stationary with respect to the box would find that $\langle v \rangle = 0$. However, if the observer is moving relative to the box (say sailing by on roller skates), he would measure the average velocity of the particles in the box not to be zero but instead $\langle v \rangle = v_{\text{obs}}$; his velocity superimposed on the random

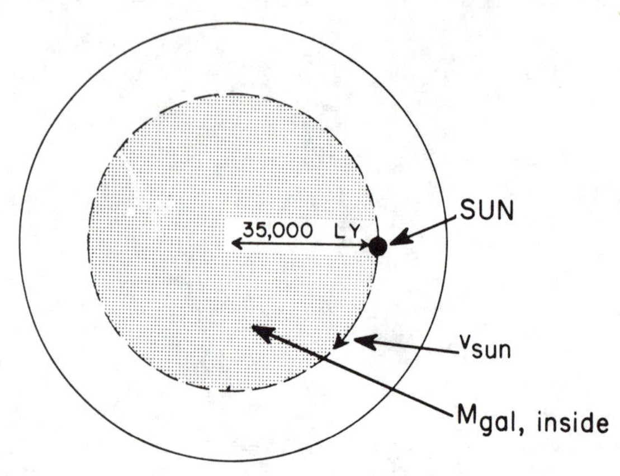

Fig. 2.4. Parameters needed for the dynamic determination of the galactic mass.

motion of the particles inside. Now back to our globular clusters. Astronomers believe the clusters orbit the center of the Galaxy in a random fashion so that at the center one should find the average velocity of the clusters to be equal to zero. If we measure their velocities and average them and the result is not zero then it must be due to our motion with respect to the center of the Galaxy. From such an analysis astronomers find that we are moving tangentially to the galactic center with a velocity of ~250 km/sec. In order to move in a circle with this speed we must experience an acceleration v^2/r and thus the net force $F_{net} = mv^2/r$. What produces this force? It has to be gravity; the Sun's gravitational mass $M_\odot$ interacting with the galactic mass inside its orbit, as shown in Fig. 2.4, giving us $F_{net} = GM_{gal,inside}M_\odot/r^2$. The kinematic expression for the acceleration necessary to produce a circular path coupled with the expression for the force through Newton's second law yields

$$\frac{GM_{gal,inside}M_\odot}{r^2} = \frac{M_\odot v^2}{r}.$$

Therefore, we find

$$M_{gal,inside} = \frac{v^2 r}{G} \approx \frac{(2.5\times10^5)^2\,(3.5\times10^{20})}{6.7\times10^{-11}}$$

$$\approx 3.3\times10^{41}\ \text{kg},$$

where we have taken 1 LY = 10^{16} m.

If we assume all stars have the same mass as the Sun ($M_\odot \approx 2\times10^{30}$ kg) and that all the matter is in the form of stars, then the approximate number of stars acting to keep us in this circular orbit is

$$N_{\text{star,inside}} \approx M_{\text{gal,inside}}/M_{\odot} \approx 3.3 \times 10^{41}/2 \times 10^{30} \sim 10^{11}.$$

We should also note that we make one revolution about the center of the Galaxy in

$$\Delta t = \frac{2\pi r}{v} \approx \frac{2\pi(3.5 \times 10^{20}\text{ m})}{250 \times 10^{3}\text{ m/sec}} = \pi \times 2.8 \times 10^{15}\text{ sec}$$

$$\approx 2.8 \times 10^{8}\text{ yr}$$

where 1 yr $\approx \pi \times 10^{7}$ sec (a rather interesting number to remember).

Once we know the velocity of the Sun, we can use this information and the Doppler-shifted spectra of other stars to determine their velocity with respect to the galactic center. As we shall now show, once we have determined the velocity of other stars, we can piece together a picture of the distribution of mass in the Galaxy. In this regard we write the velocity of a star under the influence of gravity as

$$v = \sqrt{\frac{GM_{\text{gal,inside}}}{r}}\,, \tag{2.2}$$

where r is the radial distance to any star (not just the Sun) and $M_{\text{gal,inside}}$ is the galactic mass inside its orbit.

Is there any hint as to what this distribution might be? If we look at the distribution of stars in the Andromeda galaxy, we might expect ours is similar; a central spherical core containing most of the stars and a disk containing the rest except for our friends the globular clusters which would mill about in a large spherical ball as depicted in Fig. 2.3b. If we consider stars outside the core, then in Eq. (2.2) we can take $M_{\text{gal,inside}}$ to be a constant. The value of v would then decrease as we move to stars further and further from the center of the Galaxy. This is illustrated in Fig. 2.5a. (By the way, this is the shape of the velocity curve for the planets as we move further and further from the Sun.) For stars that are inside the core where the galactic mass density ρ is approximately constant, the mass contained within their orbits would be

$$M_{\text{gal,inside}} = \frac{4\pi}{3}\rho r^{3}.$$

Substituting into (2.2), the expression for v becomes

$$v = \left(\sqrt{\frac{4\pi}{3}G\rho}\right) r.$$

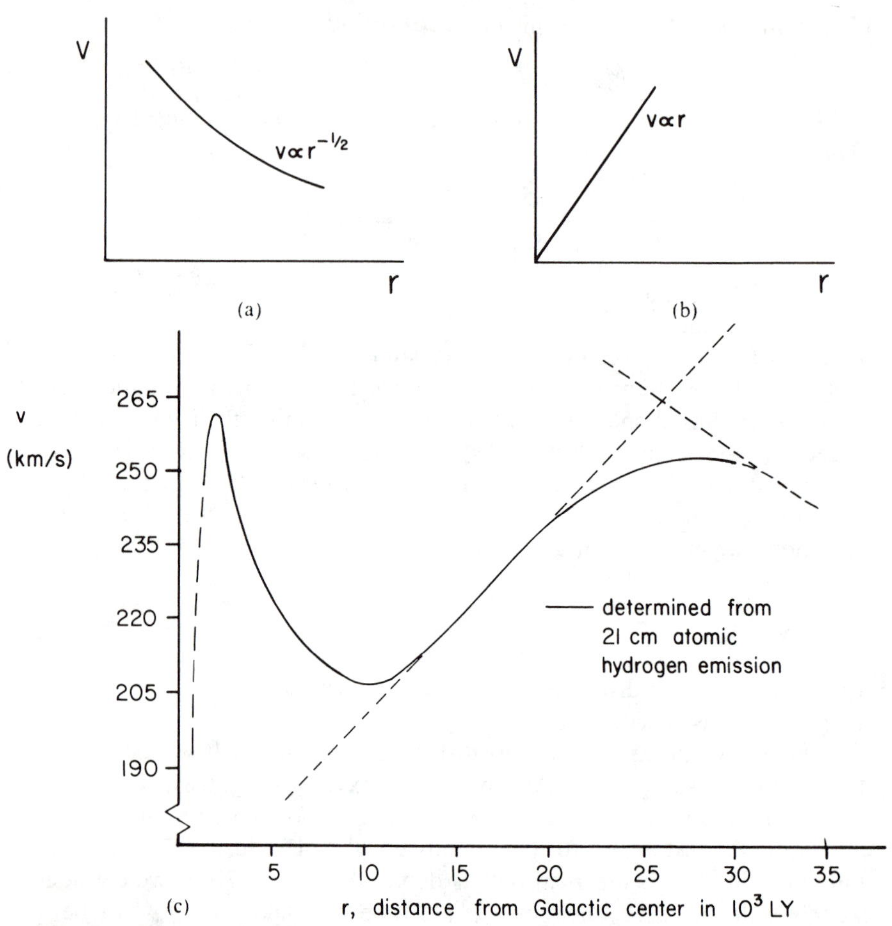

Fig. 2.5. Velocity distribution of stars in a galaxy vs. r: (a) for stars beyond the core; (b) for stars inside a constant density core; (c) for stars throughout the Milky Way.

Under these conditions the velocity of the stars increases linearly as we move out from the center, as illustrated in Fig. 2.5b. (Since a rigid body rotates according to $v = \omega r$, the above expression allows us to draw the surprising analogy between the galactic core and a spinning basketball, for example.) The situation as we find it (based on 21-cm emission from atomic hydrogen) is illustrated in Fig. 2.5c and confirms our expectations; we have a combination of a core of stars, represented by the linear increase in velocity between 10000 and 25000 LY, and a less densely populated disk, represented by the falling velocity curve beyond 25000 LY. Near this point, the peak of the curve crudely delineates the boundary of the bulk of the galactic mass. And what are we to make of the variation in the velocity curve for $r < 10000$ LY? How would you reconstruct the mass distribution in this region? Not a bad haul of physics from just one equation.

2.2. 1904—Hartmann and the Interstellar Medium

If you look at the sky on a clear night there is certainly no way you could possibly tell that there are particles between us and the stars. In fact until 1904 that was exactly what people believed—there was nothing filling the void between the stars. In 1904 Hartmann was studying a star called δ-Orionis, part of a double star system. For reasons we shall discuss later, double stars are very important astrophysical objects. As mentioned in the last chapter, they rotate around a single point (their center of mass, c.m.) under their mutual gravitational attraction. If one star is much more massive than the other, the c.m. is located inside the heavier star and the orbital rotation is confined to the lighter member, in this case δ-Orionis. Because the stars were too far away to resolve, Hartmann decided to use the Doppler shift of the absorption line for the atmosphere around δ-Orionis to find the orbital speed of the smaller star and the periodic shifting of the line to find its period of rotation. Hartmann found more than he had bargained for; the "moving" line was there all right, but so was another absorption line that was stationary and extraordinarily narrow. A similar situation is shown in Fig. 2.6. Hartmann identified this line with—of all things—a line of once-ionized calcium Ca^+. Now a line that over a period of days stays at the same absorption frequency can only mean the absorbing medium is moving at a constant radial velocity with respect to us. If the line is exactly at the frequency associated with resonance lines of Ca^+, then its radial velocity is zero; if the line is shifted, then from that shift we can obtain its radial velocity. In either case the absorbing medium is not moving in a periodic manner. Hence the conclusion that the calcium absorber is detached from δ-Orionis and indeed lies between us and the binary star. The very narrowness of the Ca^+ line confirms this conclusion. For an absorption line to be this narrow, all the factors that contribute to its width must be very small—thermal motion, rotation of the medium, etc. From a detailed analysis of the line width, it was determined that the temperature of the absorbing medium was ~ 100 °K. Since the surface temperature of a star will be at least an order of magnitude higher than this, we again conclude that the medium containing the calcium is not part of the binary system. This cold distribution of Ca^+, drifting at constant velocity through the regions of interstellar space, will henceforth be called a "cloud."

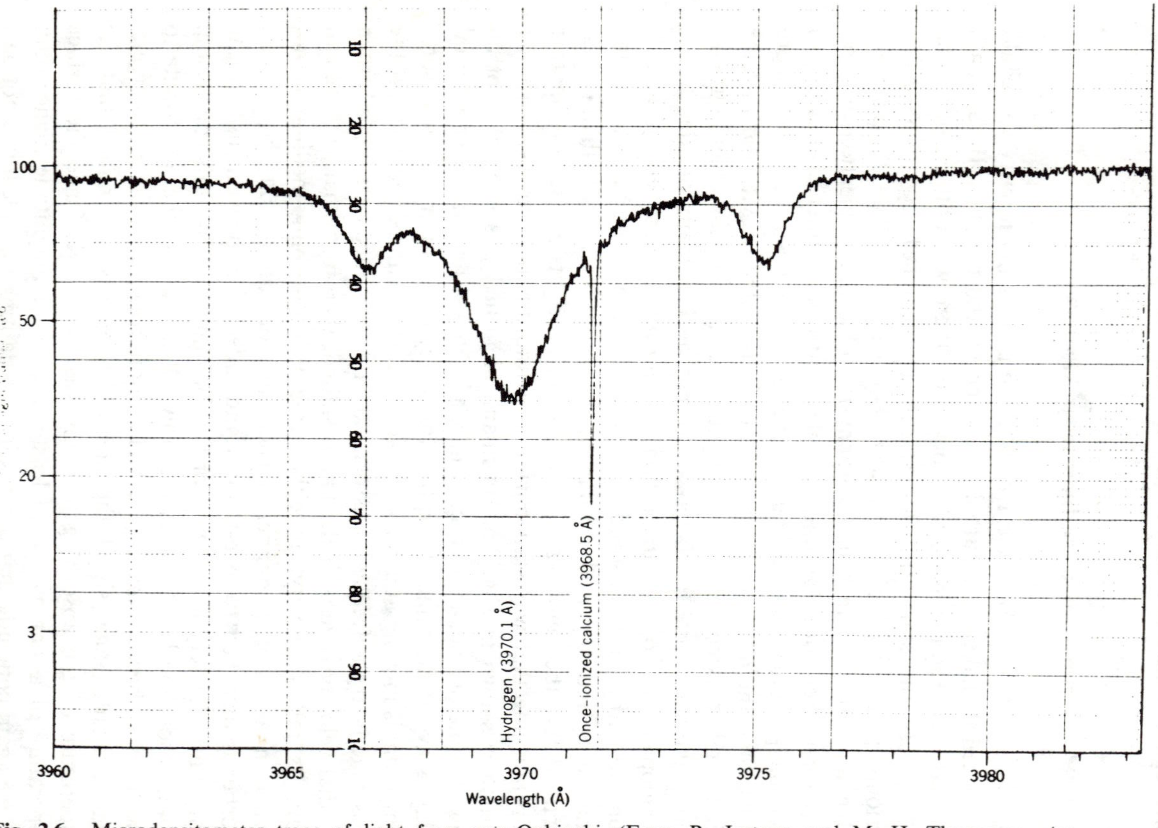

Fig. 2.6. Microdensitometer trace of light from zeta-Ophiuehi. (From R. Jastrow and M. H. Thompson, *Astronomy: Fundamentals and Frontiers,* John Wiley and Sons, New York, 1974).

Is Hartmann's cloud a loner? Hardly. Within 20 years it was established that on the average there were three or four clouds per 1000 LY along the plane of the Milky Way. Often the double star spectrum showed multiple stationary lines indicating several clouds with a range of radial velocities of up to 30 km/sec. Nor was calcium the only element found. Soon lines due to sodium, potassium, titanium, and iron as well as molecules (!) of CN and CH^+ were identified. The interstellar medium was indeed turning out to be an exciting and busy place—hardly a void at all. But where do these clouds come from? Is there anything between them? How are they distributed—and why do they contain heavy elements like Ca? Why not hydrogen? If complex atoms and molecules are constituents of the interstellar gas, where is hydrogen, the most basic and abundant element of all?

Here is where nature proved to be devious. To excite hydrogen from the ground state to the first excited state requires the absorption of an ultraviolet photon. As we showed in chapter 1, our atmosphere blocks out that part of the EM spectrum. There is simply no way a ground-based observer can detect the UV absorption lines of hydrogen. In this regard Hartmann was fortunate; to reach the first excited state of Ca^+ requires the absorption of a "visible" photon. Of course, no one doubted hydrogen was there, but detecting it was a different matter. Now we can go above the atmosphere—in balloons or satellites. But these adventures were in the future.

2.3. 21-cm Radiation and the Structure of the Galaxy

Fortunately nature has been as kind to us as she has been to Hartmann; hydrogen emits radiation right in the middle of another window we have—in the radio frequency range—only no one thought to look there until 1944! The radiation corresponds to the "hyperfine flip" of hydrogenic electrons, giving off photons with a 21-cm wavelength (i.e., $E_\gamma \approx 6 \times 10^{-6}$ eV).

2.3.1. Hyperfine Transitions in Hydrogen

Let us go back for a moment to the Bohr model of the hydrogen atom that we discussed in chapter 1. As you know, there is an electrostatic force between the electron and the proton, and the energy of the electron ($E = \text{KE} + \text{PE}$) is quantized according to $E_n = -13.6 \text{ eV}/n^2$. If this is the only force between the electron and the proton, then the only transitions would be the usual Lyman, Balmer et al., as shown in Fig. 1.17. But there is another force. It is much weaker than the electrostatic force and until 1925 no one had realized it existed. As we shall now discuss, the presence of this force adds another potential energy term to the total energy, generating a 21-cm hyperfine correction to E_n.

In 1925 the Dutch physicists Goudsmit and Uhlenbeck proposed that both the proton and the electron are "spinning," each with $\frac{1}{2}$ unit of angular momentum i.e., $S_e = S_p = \frac{1}{2}\hbar$. The fact that both the electron and proton are spinning means that in addition to the electrostatic force, there will now be a magnetic force as well. This can be seen as follows: In a classical sense the spinning proton is equivalent to a current loop, and a current loop produces a dipole field, as illustrated in Fig. 2.7a. This magnetic field will interact with the spinning electron to produce a magnetic force, $\Delta\mathbf{F}' = dq\mathbf{v} \times \mathbf{B}$ where $\Delta\mathbf{F}'$ is the force on an increment of the electron's charge dq which is moving with a tangential velocity $\mathbf{v}$. We have illustrated this force in Fig. 2.7b for a particular orientation of the electron's spin relative to the direction of $\mathbf{B}$. Using $\Delta\mathbf{F}' = dq\mathbf{v} \times \mathbf{B}$, we see that the force is radially inward, and as a result there are no net forces or torques acting on the electron due to the magnetic interaction (recall that the electron charge is negative).

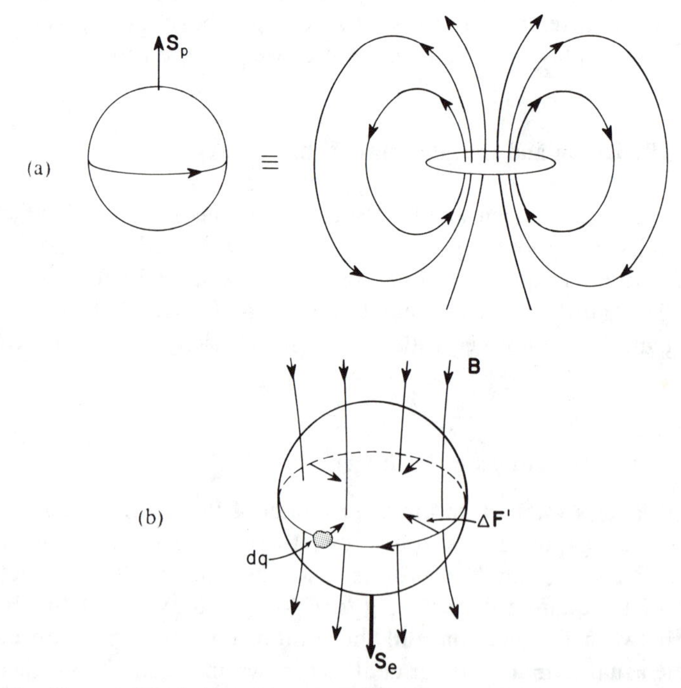

Fig. 2.7. Magnetic dipole interaction: (a) a dipole field produced by a spinning proton; (b) the force on a spinning electron in an external magnetic field.

Since the electron does not lose KE as it circles the proton, the magnetic force is conservative and we may represent it by adding a magnetic potential energy term, PE_m, to the expression for the total energy E. The general term for the total energy thus becomes $E = KE + PE_e + PE_m$. The first two terms, as we already know, are given by the single expression (in units of eV) $-13.6/n^2$, so that we may write $E = -13.6/n^2 + PE_m$. We now must try to understand what the sign of PE_m is and what effect that sign has on the ground-state energy level.

When the spin of the electron and the external field which passes through it are aligned (as shown in Fig. 2.7b), the electron is in an unstable equilibrium position. Equilibrium because the magnetic forces and torques vanish, but unstable because the slightest tilt will cause it to flip over. A point of unstable equilibrium can be thought of as being at a relatively high potential energy, ready to "run downhill" so to speak, at the slightest nudge. As a result, we must add a positive potential energy term to E, $E = -13.6/n^2 + |PE_m|$ which in turn means the ground-state energy level ($n = 1$) will be slightly higher when the electron is in this particular orientation with respect to **B** than it would be if there were no magnetic interaction, i.e., $E_1 = -13.6 + |PE_m|$. Notice in this orientation the spin of the proton and the spin of the electron are opposite (↑↓).

But there is nothing sacred about the orientation of the electron's spin. The atom could well have formed with $\mathbf{S}_e$ pointing up! In this case the spin $\mathbf{S}_e$ and the field **B** are anti-aligned, the magnetic force is radially outward, and the loop is in a *stable* equilibrium position where any tilt will cause it to return to its original orientation. As a result, when the electron is in this orientation with respect to **B**, we must add a *negative* potential energy term to E, $E = -13.6/n^2 - |PE_m|$, which in turn means the ground-state energy level will be slightly lower than it would be if there were no magnetic interaction. Here the spin of the proton and the spin of the electron are aligned (↑↑).

Hence by "turning on" the magnetic interaction we lower the ground state in hydrogen atoms with aligned spins and raise the ground state in hydrogen atoms that are formed with anti-aligned spins. If this was all there was to it we would have a trivial classical derivation of 21-cm radiation, since transitions could take place from the higher to lower ground state. This means we start with an anti-aligned atom and then somehow flip the electron's spin so that the atom ends up in the lower, aligned ground state. The energy involved would be small and hence the emitted photons would have wavelengths where we want them, in the radio window.

But the electron is not a classical beast; it does not obey the laws of classical physics but rather the laws of quantum mechanics. The latter tells us the electron does not move in a simple classical orbit with a discrete path. Instead its path should be thought of as smeared out over a fuzzy ball with a finite probability of being anywhere inside that ball. In the ground state ($n = 1, l = 0$), the probability distribution is spherically symmetric, which means that the probability of finding the electron at a distance r from the proton is the same in any radial direction. The probability varies with r, however, and is a maximum, not surprisingly, at the Bohr radius $r = a_0$. Now in most cases, since the probability does peak at the Bohr radius, one can substitute classical orbits of radius a_0 for the quantum-mechanical smear. But not this time. Since the electron can in effect be all over the place and since the magnetic field of the proton *is* all over the place (as shown in Fig. 2.7a), we see that for a given spin orientation of the electron e, the magnetic field that passes through e may be down (when e is at the equatorial plane of the proton) or it may be up (when e is above one of the poles of the proton). But by reversing **B** we reverse the sign of PE_m! Although the position of the electron is uncertain, its spin direction always remains fixed, either up or down. Thus when we average PE_m over all electron positions for a particular spin orientation of the electron and proton, we see that $\langle \mathrm{PE}_m \rangle = 0$. The magnetic interaction seems to vanish and along with it our 21-cm radiation! A disaster. But what quantum mechanics can take away it can also give back. Quantum mechanics makes another remarkable prediction: the electron has a finite probability of being inside the proton! Moreover this is a very special, singular point (as you can imagine) that is not included in the above average.

Consider then the case when the electron is inside the proton and when $\mathbf{S}_e$ and $\mathbf{S}_p$ are anti-aligned. We have illustrated this in Fig. 2.8a. Since **B** is opposite to $\mathbf{S}_e$, the magnetic force is radially outward, and we are in stable equilibrium where PE_m is a minimum. The ground state energy level thus shifts down according to

$$E_{\uparrow\downarrow} = -13.6\ \mathrm{eV} - |\mathrm{PE}_{m,0}|,$$

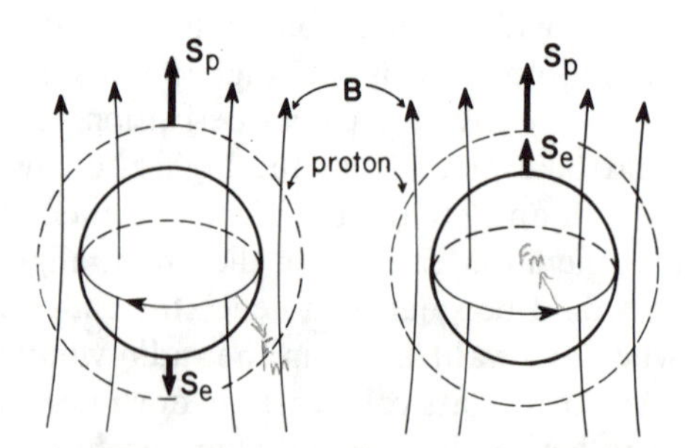

Fig. 2.8. Two orientations of a spinning electron relative to the dipole field "inside" the proton.

where "0" signifies that we are at the center of the proton. But when $\mathbf{S}_p$ and $\mathbf{S}_e$ are aligned, as shown in Fig. 2.8b, **B** is in the same direction as $\mathbf{S}_e$, the magnetic force is radially inward, we are in an unstable equilibrium and the ground-state energy level shifts up according to

$$E_{\uparrow\uparrow} = -13.6 \text{ eV} + |\text{PE}_{m,0}|.$$

So there *is* a ground-state separation between those atoms with $\mathbf{S}_p$ and $\mathbf{S}_e$ aligned and those with $\mathbf{S}_p$ and $\mathbf{S}_e$ anti-aligned but the reason is due to the quantum-mechanical property of the electron. The displacement of the levels from -13.6 eV will depend upon the magnitude of $\text{PE}_{m,0}$ which, in turn, depends upon how fast the electron is spinning, the strength of **B**, and the probability the electron is inside the proton. The latter is very small for states with $n > 1$, or $l > 0$, so only the ground state will show an appreciable magnetic coupling.

In Fig. 2.9 there are three energy-level diagrams comparing the hypothetical $F_m = 0$ case (the Bohr atom) and the two $F_m \neq 0$ cases (where F_m is the magnetic force). We have illustrated the hyperfine transition for an atom that flips its spin, along with the usual non-spin-flip, electrostatic Lyman transition. But there is an interesting and subtle point worth stressing. Classically we found that when the electron was external to the proton, the aligned ground state was lower than the anti-aligned ground state. Quantum mechanics, by placing the electron inside the proton, predicts the *opposite*. This difference is a striking test of quantum mechanics.

Before we discuss how transitions are made between the $\uparrow\uparrow$ ground state and the $\uparrow\downarrow$ ground state, let us go off the deep end for a bit and estimate the strength of the magnetic potential energy term. Consider first a classical current loop whose plane is perpendicular to an external magnetic field (i.e., not its own **B** field). Depending upon the direction of the current, the

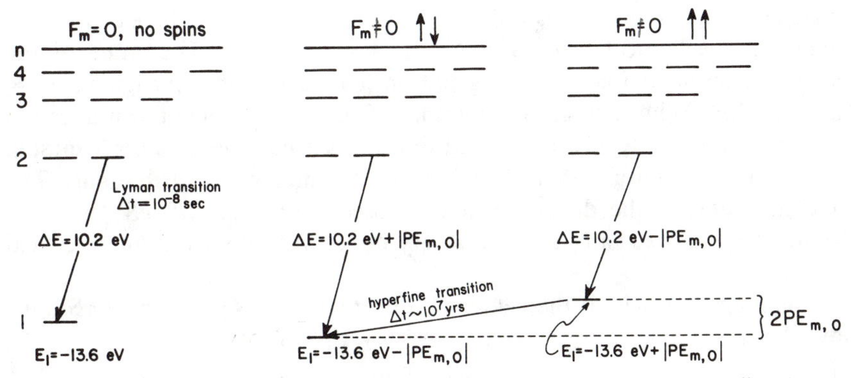

Fig. 2.9. Energy level diagrams of hydrogen without and with magnetic coupling (not to scale).

loop will be in stable or unstable equilibrium, the former corresponding to a minimum potential energy, the latter to a maximum potential energy. Classically the maximum PE_m is given by

$$\mathrm{PE}_{m,\max} = IAB,$$

where I is the loop's current, A its area, and B the external field which threads the loop. This parallels our previous discussion of the hydrogen atom where the field generated by the spinning proton (B) interacts with the current loop of the spinning electron (IA). According to quantum mechanics, we must weight this classical expression by the probability $\mathcal{P}_0$ of finding the electron at the center of the proton,

$$\mathrm{PE}_{m,0} = IAB\mathcal{P}_0.$$

Unfortunately current is not a good variable to use for a spinning electron since all we can measure is its spin angular momentum. We can, however, rewrite the above expression to bring out the intrinsic nature of the angular momentum as follows: we assume the electron is a single loop of current with all its charge at radius b_0. Then $I = e/T = ev/2\pi b_0$ where v is the velocity of the electrons. The orbital angular momentum of this charge is $L = m_e v b_0$. Eliminating v we find that

$$IA = \left(\frac{eL}{m_e 2\pi b_0^2}\right) \times \pi b_0^2 = \frac{eL}{2m_e}.$$

The quantity IA is defined as the magnetic (dipole) moment of the loop, μ. Expressed in terms of μ, the maximum $\mathrm{PE}_{m,0}$ becomes

$$\mathrm{PE}_{m,0} = \mu B \mathcal{P}_0.$$

One might naively assume we can obtain the magnetic moment for a spinning electron by substituting $S_e(=\frac{1}{2}\hbar)$ for L. This turns out not to be the case. The measured value of μ_e is eS_e/m_e; nature in her wisdom has dropped a factor of 2. This only emphasizes the fact that the electron is a quantum mechanical beast and not a classical entity. The value of the magnetic moment for a spinning proton μ_p is even more intriguing, $\mu_p = 2.79\ eS_p/m_p$. Why such an odd number? The reason is the proton does not spend its life merely as a proton; at times it is a neutron plus a π^+ meson, which have different charge and magnetic moment distributions. The average between the dressed and undressed proton is $2.79\ eS_p/m_p$. The value of μ_p, as we shall now see, is very important in determining the field B.

The magnetic field at the center of a current loop is given in MKS units by

$$B_0 = 2\pi \frac{k_0}{c^2} \frac{I}{r_0},$$

where k_0 is the electrostatic force constant, 9×10^9 N-m^2/C^2, c is the velocity of light, and r_0 is the radius of the loop. Multiplying the top and the bottom of the above expression by the area of the loop, πr_0^2, we obtain

$$B_0 = 2\pi \frac{k_0}{c^2} \frac{IA}{\pi r_0^3} = 2 \frac{k_0}{c^2} \frac{\mu}{r_0^3}.$$

If we *assume* the magnetic moment of the proton μ_p replaces the current loop magnetic moment μ, then the field at the center is given by

$$B_0 \approx 2 \frac{k_0}{c^2} \frac{\mu_p}{r_0^3}.$$

This field combined with the magnetic moment of the electron and the probability for finding the electron at the center of the proton gives us the potential energy term we are after,

$$\mathrm{PE}_{m,0} = 2 \frac{k_0}{c^2} \mu_e \mu_p \left(\frac{\mathcal{P}_0}{r_0^3} \right).$$

Since $(4\pi/3) r_0^3$ is equal to a volume of roughly the size of the proton, $\mathcal{P}_0 / r_0^3$ is proportional to a probability density—the probability per unit volume at $r = 0$ of finding the electron. But this is exactly what quantum mechanics is equipped to predict. Indeed quantum mechanics tells us that at $r = 0$

$$\mathcal{P}_0 / r_0^3 = \tfrac{4}{3} a_0^{-3},$$

where a_0 is the Bohr radius. Thus our estimate of the hyperfine energy becomes

$$\begin{aligned} \mathrm{PE}_{m,0} &\sim 2 \frac{k_0}{c^2} \mu_e \mu_p \frac{4}{3 a_0^3} \\ &\sim \frac{8}{3} \frac{k_0}{c^2} \left(\frac{e\hbar}{2m_e} \right) \left(2.8 \frac{e\hbar}{2m_p} \right) \left(\frac{\alpha}{\hbar / m_e c} \right)^3 \\ &\sim \frac{8 \times 2.8}{4 \times 3} \left(\frac{m_e}{m_p} \right) m_e c^2 \alpha^4 \sim 1.5 \times 10^{-6}\ \mathrm{eV}, \end{aligned} \tag{2.3}$$

where we have evaluated $\mathrm{PE}_{m,0}$ using $\alpha = (k_0 e^2 / \hbar c) = (\hbar / m_e c a_0)$.

It turns out that three out of every four electrons that are captured by a proton to form hydrogen are captured into the aligned ground state while only one out of four is captured into the anti-aligned ground state. The aligned ground state is unstable, however, being $2|\mathrm{PE}_{m,0}| \approx 3 \times 10^{-6}$ eV higher in energy than the anti-aligned ground state. Unfortunately, in our semi-classical approach, we have been forced to ignore spin-dependent terms implied by quantum mechanics. These add a factor of 2 to the energy difference between the aligned and anti-aligned ground states so that the final result for the energy difference is 6×10^{-6} eV, as shown in Fig. 2.9. Thus, if the aligned electrons can flip their spin, they may end up in a lower energy configuration, emitting a photon in the process with

$$E_\gamma = E_{\uparrow\uparrow} - E_{\uparrow\downarrow} \approx 6 \times 10^{-6} \text{ eV},$$

or equivalently, using $hc \approx 1.2 \times 10^{-4}$ eV-cm,

$$\lambda_\gamma = hc/E_\gamma = 1.2 \times 10^{-4}/6 \times 10^{-6} \to 21.06 \text{ cm}.$$

It is important to note that these two electron states are characterized by different values of the *total* angular momentum, orbital plus spin. In the aligned state, $l = 0$, $S = \frac{1}{2}\hbar + \frac{1}{2}\hbar = \hbar$ and $L + S = \hbar$, whereas in the anti-aligned state $l = 0$, $S = \frac{1}{2}\hbar - \frac{1}{2}\hbar = 0$ and $L + S = 0$. In the transition it is the *total* angular momentum that must be conserved, where the total is for the whole system of atom plus photon. Since the above shows the angular momentum of the atom changes by one unit of $\hbar$, one unit of angular momentum must be carried off by the photon, which of course is exactly what it does. So energy and angular momentum will be conserved when an electron flips its spin and emits a photon. The trouble is, that will happen spontaneously only once every 10^7 years!!

2.3.2. Transition Amplification by Collision

This slow spontaneous transition rate is speeded up significantly by collisions between atoms. If two atoms collide, there is about a 50% probability that if one of them is in the aligned configuration, then the electron will flip and a 21-cm photon will be emitted. It was the Dutch physicist Van de Hulst who suggested in 1944 that these collisions should be frequent enough for us to detect the 21-cm line emanating from the interstellar gas.

To understand this, we must understand what "collision" means. Possibly a better word would be interaction. In this case, when the atoms are close enough the "collision" is an electrostatic interaction which causes the atoms to scatter so that sometimes the electrons flip their spins. The kind of collisions in general will depend upon the nature of the interacting particles. In the center of a star, collisions between nuclei can lead to nuclear interactions. On the other hand, if one of the particles is a photon, the "collision" can cause Compton scattering or photoelectric annihilation. We can generalize the "collision" and account for all possible interactions by defining an area around one of the particles such that if the other particle enters that area—wham—we have an interaction. The size of the area, or cross section, will depend upon the nature of the interaction; nuclear interactions having small cross sections, atomic interactions larger cross sections. A great deal of effort has gone into experimentally determining

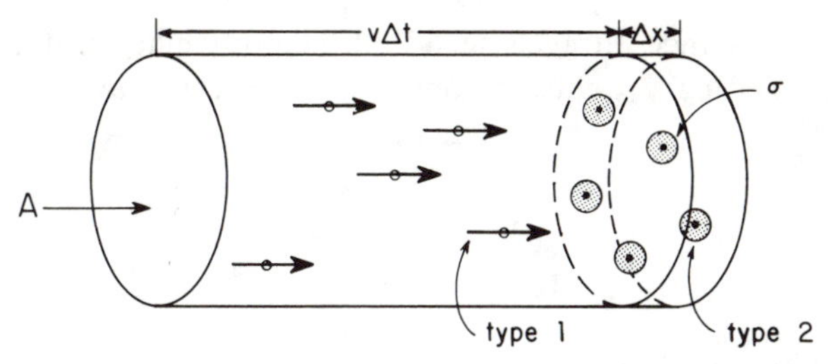

Fig. 2.10. Particles of type 1 in tube $Av\Delta t$ incident upon particles of type 2 in slab $A\Delta x$.

cross sections, not only because we want to study interstellar clouds but because of contemporary interest in fission, fusion, plasma, and many other reactions.

Consider then a slab containing $N_2 = n_2 A\Delta x$ particles of type 2, where n_2 is the number density of particles. Surrounding each particle and depending upon particle type 1 that is incident on the slab, there will be an interaction area σ. If we look at the slab end on, as shown in Fig. 2.10, the total interaction area will be $\sigma \times N_2$. For a single particle of type 1 entering the slab, the probability of a collision between this particle and all the particles in that slab of type 2 will be

$$\begin{aligned}\mathcal{P}_{12} &= \frac{\text{Interaction area}}{\text{Total area}} = \frac{\sigma N_2}{A} \\ &= \frac{\sigma n_2 A\Delta x}{A} \\ &= \sigma n_2 \Delta x.\end{aligned}$$

In time Δt, all the particles of type 1 in the tubular volume $Av\Delta t$ heading toward A will pass through the slab, as shown in Fig. 2.10. This number will be $N_1 = n_1 A v\Delta t$. The total number of interactions these particles undergo will be $N_1 \mathcal{P}_{12}$, with

$$\begin{aligned}N_1\mathcal{P}_{12} &= (n_1 A v\Delta t)\sigma n_2 \Delta x \\ &= n_1 n_2 v\sigma(A\Delta x)\Delta t,\end{aligned}$$

where v is the relative velocity between the particles. Thus the number of interactions per unit volume ($\Delta V = A\Delta x$) per unit time is the rate density,

$$\mathcal{R} = n_1 n_2 v\sigma. \tag{2.4}$$

We now define two other very important but related quantities: the *interaction length* and the *interaction time*. Focusing our attention on a single particle of type 1 incident upon many particles of type 2, the interaction length of particle 1, sometimes called its *mean free path*, is defined as the distance it travels such that the probability of it interacting is almost unity. Setting $\mathcal{P}_{12}$ equal to one and $\Delta x = l$, we have

$$1 = \sigma n_2 l_1,$$

and therefore,

$$l_1 = \frac{1}{\sigma n_2}.$$

The interaction time is simply

$$\Delta\tau = \frac{l_1}{v} = \frac{1}{\sigma n_2 v}.$$

As we shall see later, the interaction length is a very useful quantity for determining the likelihood of an interaction taking place. For example, say we have a container of length L containing particles of type 1 and 2. If the interaction length l is less than L, we have a high probability that 1 will interact before it traverses the "box," but if l is much greater than L, then that probability is very small. We thus have a very clean way of deciding if an interaction will take place within an interstellar cloud, or a star, a galaxy, or even the entire universe.

Returning to the enhancement of the 21-cm radiation in the interstellar clouds, the particles in the cloud are certainly colliding with one another and so we may apply $\mathcal{R} = n^2 v\sigma$ (note that since type 1 and 2 are identical $n_1 n_2 = n^2$). Of course the particles are really moving around randomly and coming at each other from all directions so this rate density $\mathcal{R}$ is only a rough estimate of the enhancement. Nonetheless we can use it to estimate the number of collisions/cc/sec in the clouds that Hartmann found. Since we know the cloud's temperature is $\sim$100 °K, we can estimate the relative velocity of the hydrogen atoms as twice their average thermal speed or $\sim 3 \times 10^3$ m/sec. Van de Hulst estimated the hydrogen density to be about 1 atom/cc ($n = 1$/cc). He also assumed the cross section to be about the area of the atom, say around $(2 \times 10^{-10}\ \text{m})^2$. Putting this all together (in CGS units) we have

$$\mathcal{R} \approx (1/\text{cc})^2 (3 \times 10^5\ \text{cm/sec})(4 \times 10^{-16}\ \text{cm}^2)$$
$$\sim 10^{-10}/\text{cc-sec}.$$

Thus, once every 10^{10} sec, in each cubic cm of the cloud there is an interaction, and therefore the emission of a 21-cm photon. This is once every 300 years, a spectacular improvement over the 10^7 years it would take for the one atom per cubic cm to emit spontaneously a 21-cm photon.

It is one thing to predict the existence of 21-cm radiation; it is quite another to find it. Fortunately we have a window at 21 cm, the T.V.

window we spoke of earlier, which stretches from about 1 cm to 100 cm. Remarkably enough and for reasons that had nothing to do with astrophysics, the techniques for detecting 21-cm radiation were being developed at about the same time van de Hulst made his prediction. During World War II, photons with wavelengths in the cm region were found to be very good for bouncing off ships and planes and thus locating them in the dark (radar). A large effort went into developing the right electronics, senders, receivers, etc. for radar. When van de Hulst made his suggestion the techniques for detecting 21-cm radiation were pretty much ready and waiting. Some refinements were necessary, but in 1951 E. Purcell of Harvard announced the "discovery" of 21-cm radiation. So van de Hulst was right—even at one flip/300 years there was enough atomic hydrogen in those huge clouds to send out a detectable amount of radiation.

You might wonder why all the hydrogen atoms by this time have not ended up in the low-energy anti-aligned ground state where no radiation is possible. The answer is that cosmic rays, X-rays, and ultraviolet radiation continuously ionize the hydrogen atoms, breaking them apart into protons and electrons. When the protons and electrons recombine, three out of four end up in the aligned ground state. Thus there always has been and always will be 21-cm radiation.

It is important to realize that only atoms or molecules with an odd number of electrons can emit spin-flip, microwave radiation (as radiation in the millimeter and cm region is often called). For example, the He atom has two electrons in the ground state, one with spin up, the other with spin down. The Pauli exclusion principle says that no two electrons can be in the same state at the same time. If one should flip, so that both are down or both are up, the Pauli principle would be violated. Since nature does not choose to violate the Pauli principle, such atoms cannot emit microwave radiation. Likewise molecular hydrogen H_2 with two electrons, will not emit 21-cm radiation. Thus, the 21-cm radiation only detects atomic hydrogen; any hydrogen ionized or bound up in molecules will have to be detected in a different manner.

2.3.3. 21-cm Radiation and the Interstellar Gas

What do we expect to see with our "21-cm telescope?" Shapley had shown that the Sun is a fair way out toward the rim of the Galaxy and that we are rotating about its center with a velocity of about 250 km/sec. As we have already shown, we are beyond the galactic core so that our velocity and the velocity of stars and clouds further out will go as $v = \sqrt{GM/r}$; the velocity of the clouds falls off as r increases. If we point our telescope toward the outer rim of the Galaxy in the direction of rotation as shown in Fig. 2.11,

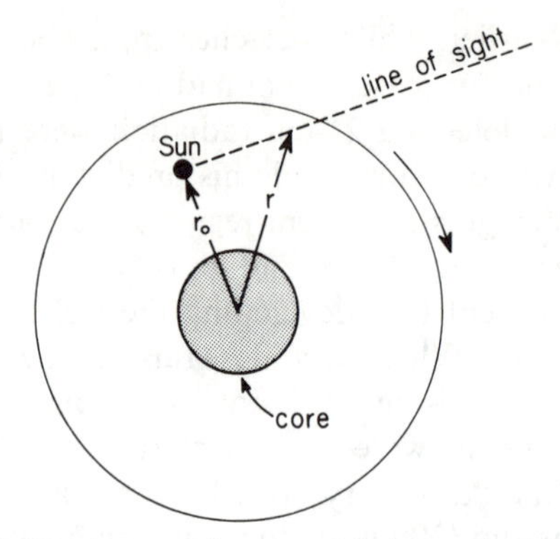

Fig. 2.11. Atom at distance r relative to the Sun at r_0.

all hydrogen atoms beyond r will be moving more slowly than our 250 km/sec. So at this moment we are moving faster than these outlying atoms, and therefore catching up with them. Then all the 21-cm radiation we receive should be blue-shifted, with the radiation from the atoms that are furthest out blue-shifted the most. However, since we receive less radiation from the atoms that are furthest away (reduced by $4 \pi d^2/\Delta A$), we expect the blue-shifted intensity to fall off the further we are from 21 cm. If the hydrogen atoms are uniformly distributed throughout the Galaxy, then the intensity of spin-flip radiation should have the spectrum shown in Fig. 2.12.

What we actually find is a startling modification of this expected effect. We do see blue-shifted 21-cm radiation and the intensity does fall off toward the blue but the distribution is *not* continuous. Instead, as we see in Fig. 2.13, the distribution is in three clumps. Apparently three distinct regions are emitting the 21-cm radiation. Region A, shifted a small amount, is closest. Region B, blue-shifted much more, is therefore further out. Finally, showing the effects of its distance from us, is region C. As we scan

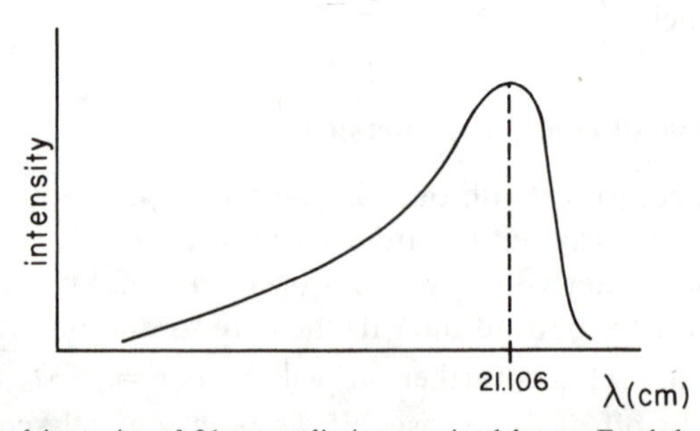

Fig. 2.12. Expected intensity of 21-cm radiation received by an Earth-bound detector for a uniform distribution of hydrogen.

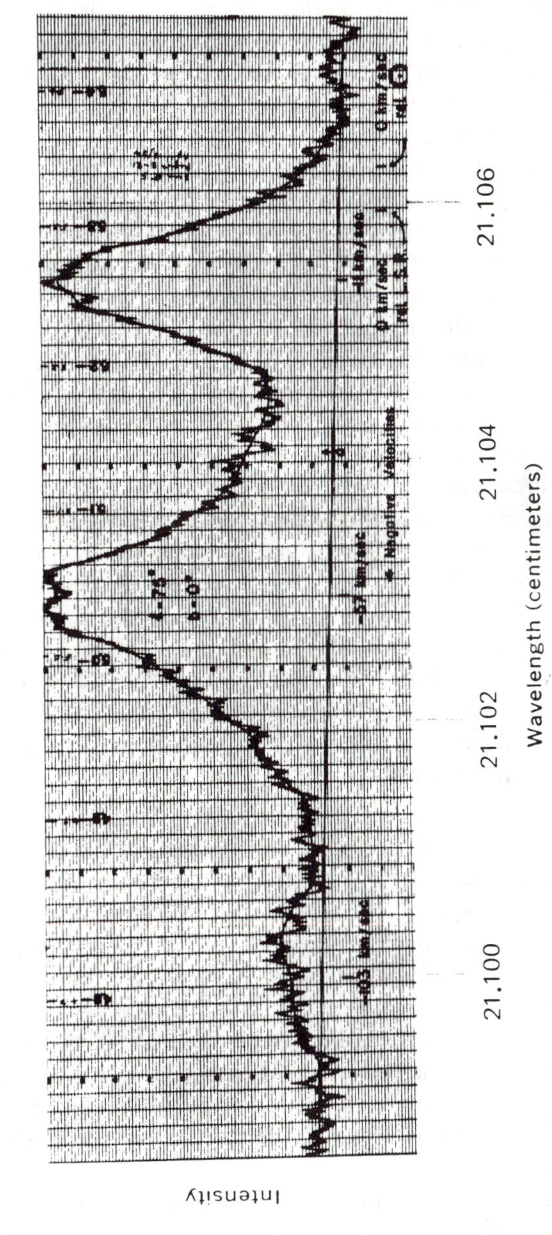

Fig. 2.13. Intensity received by a "21-cm telescope" pointing along the line of sight as depicted in Fig. 2.11. (From R. Jastrow and M. H. Thompson, *Astronomy: Fundamentals and Frontiers*, John Wiley and Sons, New York, 1974).

the galactic longitude, the pattern of Fig. 2.13 is repeated, so that we find the three clumps are really bands, or "arms" as they are often called, that curve about the center of the Galaxy. Arm A is called the local arm, or Cygnus arm. Arm B is the Perseus arm, and arm C the so-called outer arm. If we turn our "21-cm telescope" inward we see three more arms—Sagittarius, Norma, and the 3-kiloparsec arm (because it is 3 kpc from the galactic center). Now these clumps or "arms" are really segments of a larger structure. But what is this larger entity?

A drawing of the structure of the Galaxy put together from 21-cm observations made at Leiden, in the Netherlands, and at Sydney, in Australia, is shown in Fig. 2.14a. One does not see a clearly defined structure but a little imagination and a peek at other galaxies, like the one in Fig. 2.14b, yields the simpler picture shown in Fig. 2.14c. There appears to be two spiral arms that wind continuously outward and contain the Norma, Sagittarius, and Perseus "arms" we mentioned above. About half-way out two more arms appear. Not every astronomer agrees with this 2 + 2 structure but there seems to be little doubt that our galaxy contains at least two continuous arms.

Our sun, by the way, is not in a spiral arm but is between arms and close to a spur that connects the Cygnus and Sagittarius arms (note the position of the small dot in Fig. 2.14a that represents the location of the Sun). Another fact is that material apparently is being fed into the spiral arms from the nucleus of the Galaxy at a rate of one solar mass per year and at a velocity of 45 km/sec. Why? We do not know. Then there is the annoying question of why there is a spiral structure at all. If you think about it for a moment, you will see that after a few revolutions, any potential spiral should wind around itself and disappear. Why? We know that $v \propto \sqrt{1/r}$. Thus particles outside the nucleus at r_1 and $4r_1$ should be moving at v_1 and $v_1/2$, respectively. When particle 1 goes through 90 deg, particle 2, four times further out and traveling at one-half the speed, will have only gone through 1/8th the angle or 11 deg. When 1 is at 180 deg, 2 will be at 22 deg. When 1 has gone full circle, 2 will be at 45 deg. Initially, if one draws a hypothetical line connecting the particles between 1 and 2, the line will bend into a spiral as time evolves as in Fig. 2.15, but as 1 goes round and round and 2 drops further and further behind, the spiral will wind tighter and tighter. Since our galaxy has been turning for $\sim 10^{10}$ yr, we have been through approximately 40 revolutions; it would be hard to imagine a spiral structure containing material particles remaining after that. But apparently it does.

The arms are not a continuous distribution of atomic hydrogen but rather a series of clouds, the ones Hartmann saw in 1904. Called diffuse clouds (to distinguish them from dark and molecular clouds which we shall

Fig. 2.14. (a) An artists impression of the 21-cm map of the Galaxy. The radiation from the northern sky was obtained in Leiden (in the Netherlands) and from the southern sky in Sydney, Australia (Leiden Observatory).

study shortly), they typically have a density ~10 atoms/cc, a kinetic temperature ~100 °K, a diameter ~10 pc ≈ 30 LY, a mass ~100 $M_\odot$, and are separated by about ~100 pc~300 LY. We shall now explore why the temperature is ~100 °K and then take up the question of what might be between the clouds.

diffuse clouds: $\rho \sim 10$ atoms/cm^3, $T = 100$°K
diam = 10 pc ≈ 30 LY, $m = 100\, M_\odot$
separ. dist. = 100 pc ≈ 300 LY

Fig. 2.14. (b) A typical spiral galaxy, M51 (Kitt Peak National Observatory).

2.3.4. Why 100 °K? Fine Structure of Carbon

Clearly something must be operating to keep the interstellar gas at this particular temperature. With a corresponding average KE per particle of

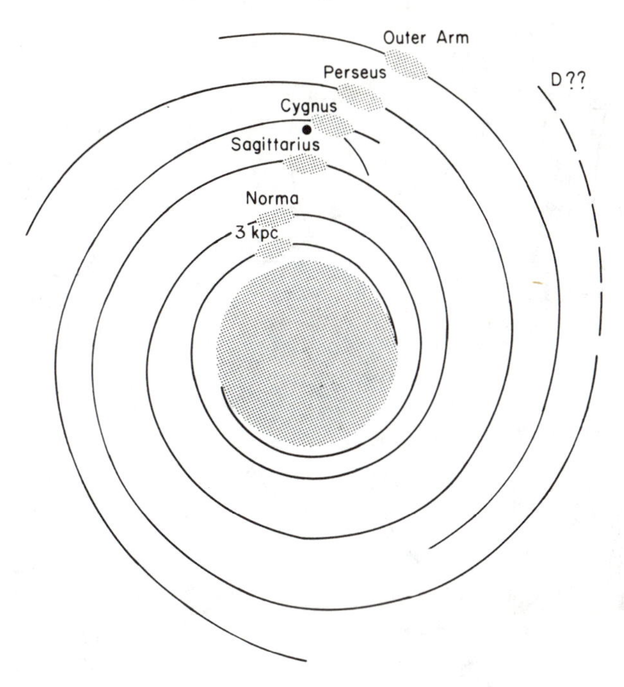

Fig. 2.14. (c) A simplified schematic picture of our galaxy showing clumps or "arms" and possible overall structure.

about 10^{-2} eV, energy has to be absorbed. But a continuous absorption of energy would lead to an infinite temperature. So the gas must have a cooling mechanism as well. The balance between heating and cooling leads to a stable temperature. But what is the source of the outside energy—how is it absorbed and what is the cooling process?

It is now believed that the main source of outside energy is cosmic rays—mostly high-energy protons that many astronomers believe are accelerated in supernova explosions. Cosmic radiation fills the Galaxy. As you read this sentence, as many as ten per second will pass through your body. The rate would be a lot higher if it were not for the Earth's magnetic field which bends the low-energy cosmic rays back into space. But this is not so for the clouds in the spiral arms. As low-energy ($\sim$5 MeV) cosmic rays sweep through the clouds, some have close encounters with the atomic electrons of hydrogen. The resulting Coulomb force on the electron is sufficient to jar it out of the atom and leave the electron with an observed average kinetic energy of about 34 eV. Indeed, the area under the force-vs.-time curve will be the momentum transferred to the electron ($\sum F\Delta t = m\Delta v = \Delta p$) and since $\langle \mathrm{KE} \rangle \sim (\Delta p)^2/2m_e$, the electron gains KE. (What happens

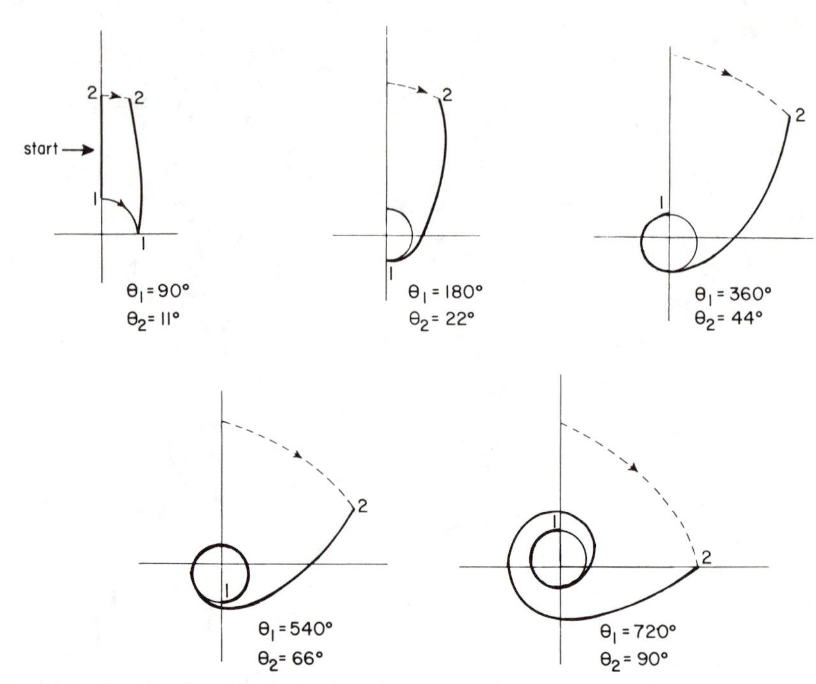

Fig. 2.15. Particles 1 and 2 depicted at the same time as they revolve around the galactic center.

to the energy of the cosmic-ray proton?) It is also possible that ultraviolet radiation from stars and X rays from X-ray sources (we shall study these later when we take up black holes) produce ionization, but we believe cosmic rays are the dominant factor of ionization.

Now a gas of 34 eV electrons in equilibrium would have a kinetic temperature of 4×10^5 °K—much too high! One way to cool the electrons off is by letting them ionize and excite atoms on their own. Each time an electron ionizes hydrogen it loses 13.6 eV. It will continue to do this until its energy falls below 13.6 eV, at which point the electron no longer has sufficient energy to ionize the hydrogen atom. The electron can, however, raise H from an $n = 1$ to an $n = 2$ level and in the process lose 10.2 eV. Assuming the photons leave the cloud, this puts an upper limit on the electron's KE of 10.2 eV, leaving the electrons with roughly an average kinetic energy of 5 eV. If equilibrium with the hydrogen gas is established here, the temperature would be 50,000 °K—still too high. We need another coolant, something with much more closely spaced energy levels. It turns out to be ionized carbon.

We saw in Sec. 2.3.1 that the spinning electron interacts with the magnetic field of the *spinning* proton to add another PE term to the total

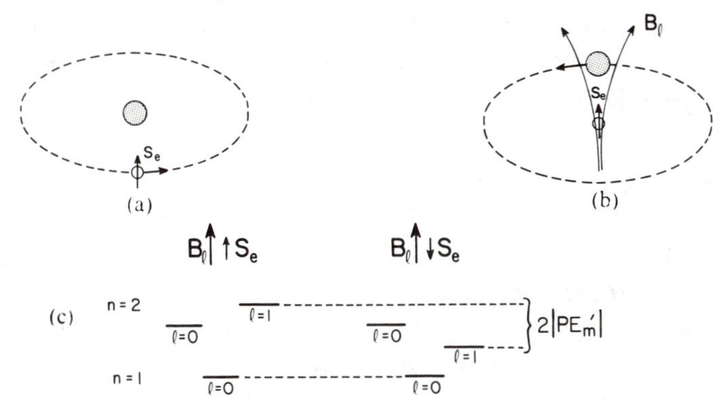

Fig. 2.16. (a) An electron with spin S_e and angular momentum l orbiting a proton. (b) Same situation as (a) but from the point of view of the electron. The electron experiences a magnetic field $\mathbf{B}_l$ due to the orbiting proton. (c) Energy level diagram for the two orientations of $\mathbf{B}_l$ and $\mathbf{S}_e$.

energy E and change the Bohr energy levels. But as far as the electron is concerned, there is still another magnetic field. Recall that in the Bohr atom the electron moves about the proton in a circular orbit (see Fig. 2.16a). From the point of view of the electron, however, it is the proton that moves in a circle, generating a big current loop and producing an orbital dipole magnetic field. Since the electron is at the center of the loop, it experiences a strong magnetic field. As shown in Fig. 2.16b, the direction of **B** is up. As we saw in Sec. 2.3.1, the spinning electron is equivalent to a tiny current loop. The electron feels a force ($\mathbf{F} = q\mathbf{v} \times \mathbf{B}$) which is radially outward if $\mathbf{S}_e$ is opposite to **B** and radially inward if $\mathbf{S}_e$ is in the same direction as **B**. When the spin and the *field* are aligned, the loop is in an unstable situation where the slightest tilt will cause it to flip over; the result is a positive potential energy term added to E, $+|\mathrm{PE}_m|$. When the spin and the field are anti-aligned we have a negative potential energy term added to E, $-|\mathrm{PE}_m|$. Since the electron's spin may have either orientation, we have atoms with either of two possible energy level diagrams, as shown in Fig. 2.16c. The various energy-angular momentum levels will be $2|\mathrm{PE}_m|$ apart. That is, all levels except those with $l = 0$. When $l = 0$ there is no preferred axis with which to define the direction of the electron's spin. For $l \neq 0$ there is such an axis; the direction of the angular momentum vector of the orbiting electron. The effect will thus be strongest in the $n = 2, l = 1$ state of an atom.

We may evaluate PE_m as follows. We can estimate the field at the electron by the classical field at the center of a current loop, $B = k_0 2\pi I / c^2 r_n$. For hydrogen we estimate the current from $I = ev/2\pi r_n = eL/2\pi m_e r_n^2$ where L is the *electron's* orbital angular momentum, $L = mvr_n$ and r_n is the radius of the electron in the nth orbit, $r_n = n^2\hbar/m_e c\alpha$ (recall, $\alpha = k_0 e^2/\hbar c$). The potential energy will be $\mathrm{PE}_m = \mu_e B$ where μ_e is the magnetic moment of the electron, eS_e/m_e. If we make the proper substitutions, we end up with the following expression for PE_m,

$$\mathrm{PE}_m = \frac{1}{n^6} m_e c^2 \alpha^4 \left(\frac{LS_e}{\hbar^2} \right). \tag{2.5}$$

With $L = \hbar$, $n = 2$, and $S_e = \frac{1}{2}\hbar$ we find $\mathrm{PE}_m \sim 10^{-5}$ eV. This energy is called the *fine structure* correction to the Bohr energy of the hydrogen atom (as opposed to the much smaller 21-cm *hyperfine* correction (2.3)).

Hydrogen, of course, rarely has an electron in the $n = 2, l = 1$ state. And when it does, the electron remains there for only approximately 10^{-8} sec. But ionized carbon is different. It has five electrons and since only two can go into each $l = 0$ level, the fifth must end up in the $n = 2$, $l = 1$ state. Thus in its ground state, the first excited state of C^+ is only $2|\mathrm{PE}_m|$ away. Each time one of the free electrons collides with C^+ and causes an upward transition by flipping the bound electron's spin, it costs the colliding electron $2|\mathrm{PE}_m|$ worth of KE. In principle if all the electrons underwent a sufficient number of these collisions, there would be an upper bound on their KE of $\approx 2|\mathrm{PE}_m|$, with an average KE of order $|\mathrm{PE}_m|$. And what is $2|\mathrm{PE}_m|$ for ionized carbon? It is about 0.01 eV, some several orders of magnitude larger than the fine-structure correction for hydrogen, due to the large charge of the carbon nucleus ($Z = 6$, $\mathrm{PE} \propto Z^4$). This corresponds to a possible average KE of $\frac{1}{2} \times 10^{-2}$ eV for these electrons and thus a kinetic temperature of $\sim$50 °K! We are right in the ball park for the temperature of a diffuse cloud. It must be emphasized that for cooling to be complete, the radiation omitted when the atom falls back to the lower $l = 1$ state must escape the cloud. At the densities of a diffuse cloud this is not a problem. The temperature question also brings up an interesting cosmological problem. Where did the carbon come from? Was it created in the "big bang" or in a continuous manner by the stars? If the latter, then there must have been less carbon a long time ago and therefore the gas clouds must have been hotter.

In spite of the importance of carbon as a coolant for the interstellar spiral gas, it makes up only a small fraction of the gas by weight. Indeed, for every 10,000 hydrogen atoms there are only two carbon atoms.

2.3.5. The Two-Component Interstellar Medium

With further refinement in radio astronomy techniques, a second component of neutral hydrogen was found uniformly distributed both between and in the spiral arms. Called the intercloud medium (ICM), its measured density is low with $n_{\mathrm{ICM}} \lesssim 0.1/\mathrm{cc}$ and its temperature around $\sim 10^4$ °K.

Although there are clouds and intercloud hydrogen in the spiral arms, there are no clouds between the arms, just stars and the ICM. Indeed, the clouds occupy by volume only ~1% of the interstellar medium. But the ratio of hydrogen in the clouds to hydrogen in the ICM is about 1:1. The result is an overall average hydrogen density in the spiral arms of about 1/cc.

It may seem surprising that the temperature of the ICM is so high. Of course, coolants like C^+ are almost nonexistent. Still why is $T \sim 10^4$ °K? This can be understood if we assume the clouds and the ICM are in equilibrium. If they are in equilibrium, then at their boundary we expect the pressure they exert on each other to be the same. Since they are both ideal gases satisfying $PV = NkT$, we have

$$P_{\text{cl}} = P_{\text{ICM}}$$
$$(nkT)_{\text{cl}} = (nkT)_{\text{ICM}}$$
$$(nT)_{\text{cl}} = (nT)_{\text{ICM}},$$

where we have written $P = (N/V)kT = nkT$ with $n =$ number density $\equiv N/V$. With $(nT)_{\text{cl}} = 10 \times 100$ we expect $(nT)_{\text{ICM}} = 1000$ °K/cc. With $n_{\text{ICM}} < 0.1$/cc this means

$$T_{\text{ICM}} > 1000/0.1 \sim 10^4 \text{ °K}.$$

A word is in order here on what we mean by equilibrium. If P_{ICM} did not equal P_{cl}, there would be a net force ($F_{\text{net}} = (P_{\text{cl}} - P_{\text{ICM}})\Delta A$) on area ΔA across the boundary that defines the two mediums. Since there is literally no mass across such a boundary (ideally a boundary is of infinitesimal size), a finite force means an infinite acceleration. This is physically absurd and also not seen experimentally. Therefore, the pressures must be equal.

One of the surprising features about the ICM is that between the arms it is moving at about twice the speed of the spiral arms! We shall come back to this point later.

2.4. Other Components of the Spiral Arms

There is more to the structure of the spiral arms than just clouds of low-density hydrogen and the ICM. In this section we shall show that the spiral arms contain dust, dark and molecular clouds, emission nebulae, and associations of hot stars. We shall also explore some of the physics of molecular formation and show one possible way the galactic magnetic field may have originated.

2.4.1. Emission Nebulae and Small Clusters of Stars

The HII regions that we mentioned earlier (Fig. 1.18a) are also located in the spiral arms. The remarkable photographs in Fig. 2.17 of a neighboring

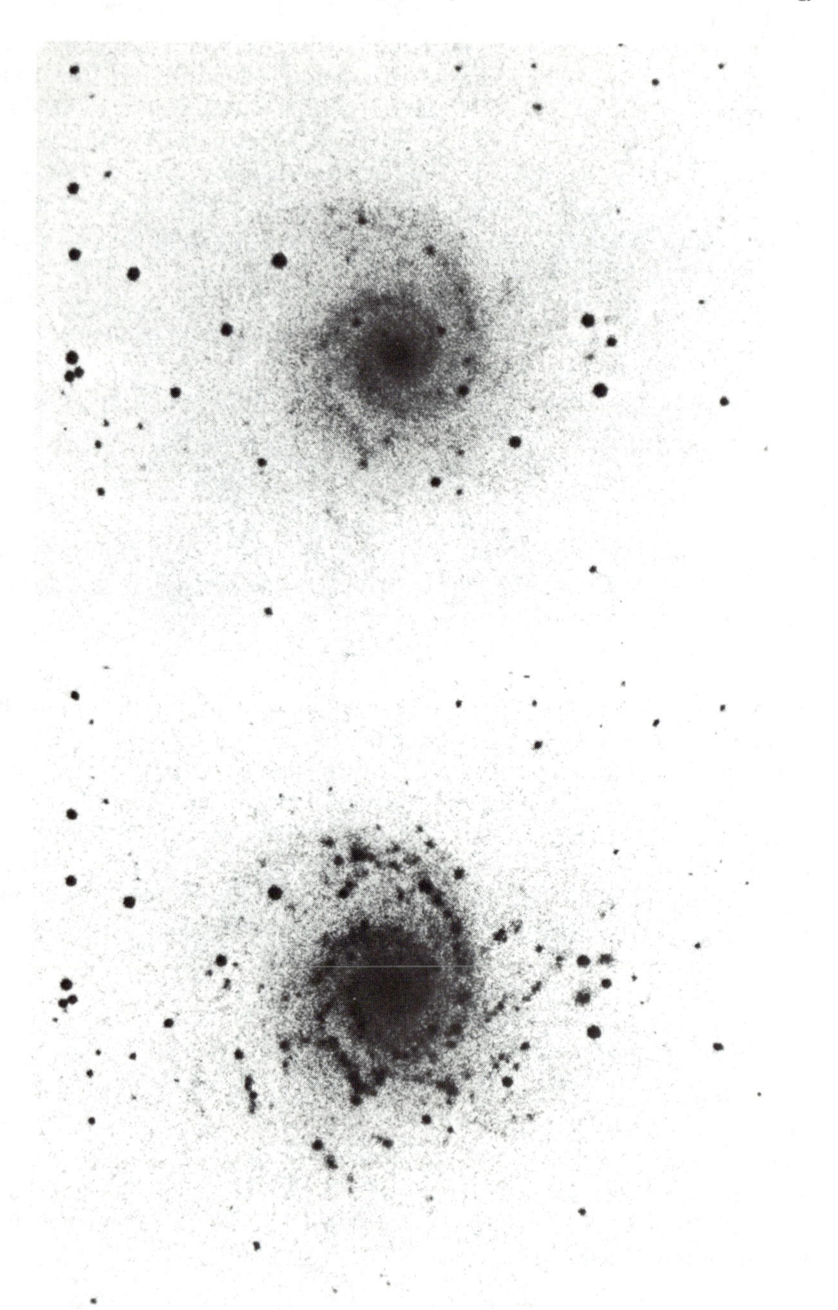

Fig. 2.17. Direct photographs of the galaxy NGC 628. Bottom, using an Hα interference filter. Top, using a yellow transmission filter (courtesy of W. W. Morgan).

galaxy were taken with an H_α filter and a yellow filter, respectively. The emission nebulae clearly show up only along the spiral arms. For the record, the temperature of the emission nebulae is ~10,000 °K and their hydrogen density is ~100 atoms/cc.

Most of the 10^{11} stars that fill the Galaxy are either loners like our sun or in binary pairs (about 50-50 binaries to singles). Peculiar to the spiral arms are something the astronomers call "associations." These are local groups of 10–50 stars, all of them very hot with kinetic surface temperatures exceeding 30,000 °K. A second form of grouping called an "open cluster" contains about 100–1000 stars. Here the kinetic surface temperatures range from 6000 °K to 30,000 °K. Open clusters are most often found in the spiral arms but this is not an exclusive arrangement as it is for associations. Then there are the globular clusters of 10^4–10^5 stars. Their surface temperatures rarely get above 7000 °K and are usually much lower. Globular clusters are never in the spiral arms but range, as we saw earlier, far above the galactic disk. Thus the spiral arms are a peculiar meeting ground for very hot stars in very small groups. Why?

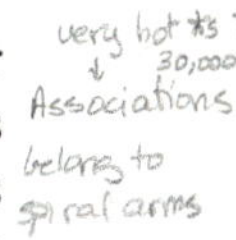

2.4.2. Interstellar Dust and Dark Clouds

There are still many mysteries in astrophysics and one of the most tenacious concerns the makeup of interstellar dust and the role it plays in molecular formation. Like many other phenomena, the realization that our galaxy is filled with dust and the subsequent understanding of its composition was slow to evolve. The early evidence was there—the dark rifts which the 19th century astronomers thought were holes in the Galaxy. In the early 20th century astronomers were puzzled by stars whose spectra indicated that their surfaces were at $T = 30{,}000$ °K (through the presence of He^+), but whose black body curves peaked well into the red. Then in 1930 Robert Trumpler observed the general reddening of open clusters scattered throughout the spiral arms. Finally in 1946 Hall, Hiltner, and McDonald found, quite by accident, that the reddened starlight was also polarized. It was from this potpourri of evidence that interstellar dust was discovered. Let us start with interstellar reddening.

Consider the color of the sky during the day and the color of the Sun at sunset. The sky is, of course, blue but during the day the Sun changes its appearance from yellow to red. The latter is a local form of interstellar reddening. As we shall now show, the blue sky and the red sunset are part and parcel of the same phenomenon, molecular scattering. There are molecules of O_2 and N_2 in our atmosphere and if we imagine them to be bound structures of electrons and nuclei, then we can use classical arguments to explain the colors of the sky and sunset. As we saw in chapter 1, the incident EM radiation from the Sun is composed of various frequencies ω $(=2\pi c/\lambda)$ whose intensities peak near yellow. The variation in intensity over the visible is not very great however, so for what follows we shall consider the intensity in the visible to be independent of ω. The electric field of the incoming radiation, E_{inc}, will cause the electrons in the air molecules to oscillate at the frequencies of the incident radiation (the air molecules will not ionize since the energy of the incident photons is $\sim$2 eV, well below the ionization energy of the outer electrons in oxygen, $\sim$13 eV). The oscillating atomic electrons accelerate and emit an electric field E_{scatt} which is proportional to the acceleration, $E_{scatt} \propto a$. As we shall show below, the scattered radiation depends very strongly upon the frequency of the incident radiation. Indeed, the intensity of the scattered radiation is proportional to ω^4. Thus the number of scattered "blue" photons ($\lambda \sim 4300$ Å) will be five times greater than the number of scattered "red" photons ($\lambda \sim 6500$ Å). When we look at the sky we are looking at scattered radiation —thus the blue sky. When we are enjoying a sunset we are looking at forward radiation. The sunset is red because the radiation has passed through so much material that sufficient blue light has been scattered so as to leave the red coloration.

Let us see if we can be a bit more quantitative. If we imagine the binding force between the electron and the nucleus to be due to a spring, then the atom will have a natural frequency, $\omega_0 = \sqrt{k/m}$, where k is the classical spring constant. (In reality, the natural frequency is the transition energy between energy levels divided by $\hbar$, but a classical model is easier to understand in this situation.) The vector force equation for the electron in the presence of external radiation is

$$\mathbf{F}_{net} = m\mathbf{a} = -e\mathbf{E}_{inc} - k\mathbf{x},$$

with a $(-)$ sign before k because the spring force is opposite to the displacement ($e \equiv 1.6 \times 10^{-19}$Coul). If $\mathbf{E}_{inc}$ is sinusoidal with frequency ω, it will drive the electrons at *that* frequency so that the displacement $\boldsymbol{x}$ varies sinusoidally and the acceleration of the electron is given by $\mathbf{a} = -\omega^2\mathbf{x}$. Plugging $\mathbf{x} = -\mathbf{a}/\omega^2$ and $k = \omega_0^2 m$ into the force equation and solving for $\boldsymbol{a}$, we find

$$\mathbf{a} = \frac{\omega^2}{\omega^2 - \omega_0^2}\left(-\frac{e\mathbf{E}_{inc}}{m}\right).$$

Thus when the incident frequency equals the natural frequency, the atom oscillates wildly with $\boldsymbol{a} \to \infty$ as shown by curve A in Fig. 2.18. Of course this never happens. There is always some damping and $|a|$ vs. ω looks more like curve B. However, since ω_0 for atoms like O_2 and N_2 is in the ultraviolet, we find that the visible frequencies occupy a region far from ω_0

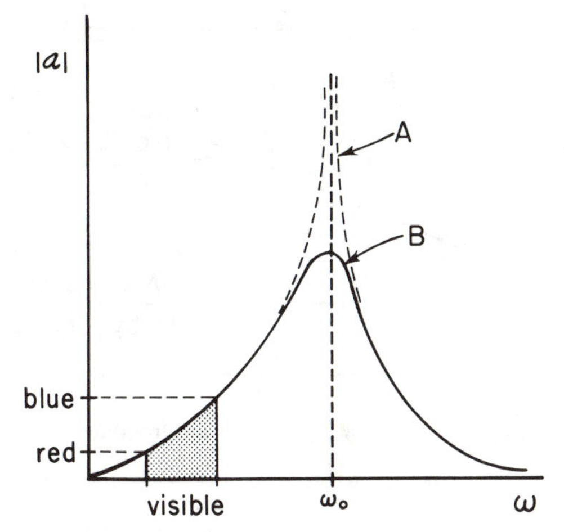

Fig. 2.18. Acceleration vs. frequency for an harmonic oscillator with resonant frequency ω_0 in the ultraviolet.

so that $\omega \ll \omega_0$ and the acceleration becomes

$$|a| \sim \omega^2.$$

Since $E_{scatt} \sim |a|$ and since the intensity (proportional to the number of photons, N_γ) is proportional to $|E_{scatt}|^2$ we have

$$E_{scatt} \propto \omega^2 \qquad \text{and} \qquad N_{\gamma,scatt} \propto \omega^4.$$

Thus the number of photons scattered depends strongly on their frequency. For incident white light, where the number of photons is roughly the same at all frequencies, we have

$$\frac{N_{\gamma,blue}}{N_{\gamma,red}} = \frac{\omega^4_{blue}}{\omega^4_{red}} = \left(\frac{\lambda_{red}}{\lambda_{blue}}\right)^4 \approx \left(\frac{6500}{4300}\right)^4 \sim \left(\frac{3}{2}\right)^4 \sim 5.$$

Thus if we look away from the Sun at the scattered radiation we see blue, but if we look directly at the Sun during a sunset when there is more atmosphere to go through, the blue light has been scattered out and we see red.

It would appear we would not be remiss to suspect something similar is happening in outer space, but alas there is one major difference (wouldn't you know it). Careful measurements show that for the starlight from the interstellar medium (ISM), reddening is not as intense and that, as shown in Fig. 2.19,

$$N_{\gamma,scatt} \propto \omega \qquad \text{(and not } \omega^4\text{)}.$$

Moreover, while light from the setting Sun is not polarized, Hall et al.

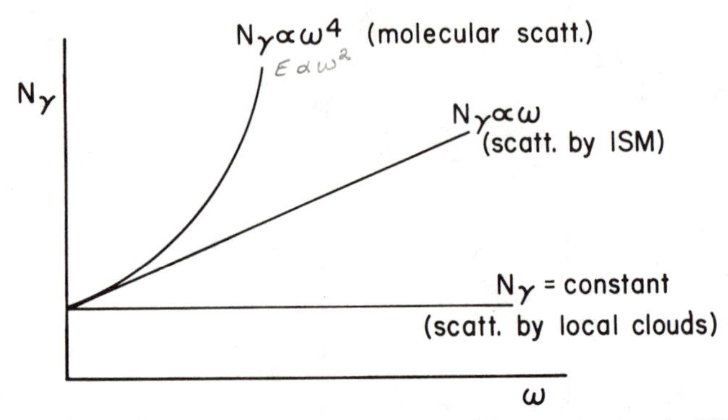

Fig. 2.19. Number of photons vs. frequency as scattered from molecules, the ISM and local clouds.

showed that starlight from space is partially polarized. Clearly something is different about the molecules between us and the Sun vs. those between us and the distant stars. But what? Could it be that in space we are dealing with clumps of molecules? We already see a similar phenomenon in our sky. When molecules condense into large water droplets (due to impurities like dust) and form clouds, the scattered light is not blue, but white! Thus for clumps of molecules of the size of those in the clouds, $N_{\gamma,\text{scatt}}$ is apparently independent of ω as depicted in Fig. 2.19.

To see why, consider a region composed of many closely spaced molecules scattering EM radiation to observer O in Fig. 2.20. All the radiation

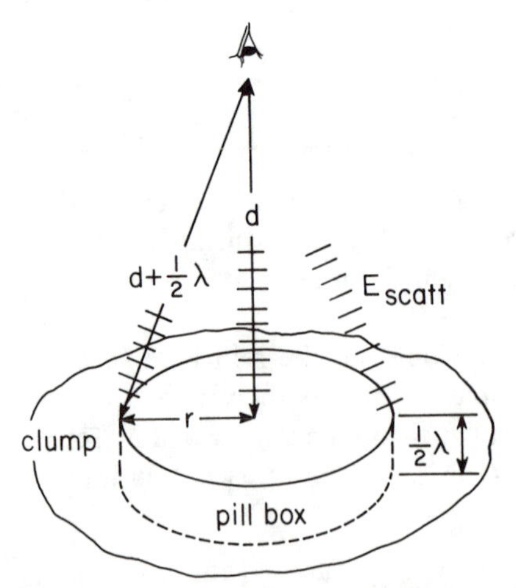

Fig. 2.20. Radiation emitted from a pillbox of radius r and thickness $\lambda/2$ as seen by observer O.

from the surface will be in phase when it reaches O, if it comes from an area such that at the boundary of the area the distance to O is $d + \frac{1}{2}\lambda$. Likewise, if the depth of the region is $\lambda/2$, the radiation between the top and bottom will also be in phase. If all the radiation is in phase, then the electric field vector received by O will be the sum of the electric fields from each molecule,

$$E_{\text{tot}} = E_1 + E_2 + E_3 + \cdots + E_N = NE_{\text{scatt}},$$

where E_{scatt} is the radiation scattered by each molecule, i.e., $E_{\text{scatt}} \propto \omega^2$ and where N is the number of molecules in the region defined by our pillbox diagram. The radius of the pillbox is

$$r^2 = \left(d + \frac{\lambda}{2}\right)^2 - d^2 = d^2 + \lambda d + \frac{\lambda^2}{4} - d^2 = \lambda d + \frac{\lambda^2}{4}\,.$$

Since it is many thousands of feet to a cloud, d is much larger than λ and so $r^2 \cong \lambda d$. The volume of the box is

$$\Delta V = \pi r^2\left(\frac{\lambda}{2}\right) = \pi(\lambda d)\frac{\lambda}{2} \propto \frac{1}{\omega^2}\,.$$

If the density of molecules is constant, we have $N = n\Delta V \propto 1/\omega^2$, so that

$$E_{\text{tot}} = NE_{\text{scatt}} \sim \frac{1}{\omega^2} \times \omega^2 = \text{const},$$

independent of ω!! The radius of the surface, $r = \sqrt{\lambda d}$, is immense compared to λ and since the size of a molecule is only about an angstrom, the pillbox will contain millions upon millions of molecules. So now you know why clouds are white; we are seeing the in-phase radiation from millions of molecules.

Since an individual molecule gives $N_{\gamma,\text{scatt}} \propto \omega^4$ while a large clump of molecules acting coherently gives $N_{\gamma,\text{scatt}} \propto$ constant and since outer space gives $N_{\gamma,\text{scatt}} \propto \omega$, it is reasonable to suspect that it is a clump of molecules that gives rise to the scattering in the latter case, but a clump that is smaller than those necessary to produce scattering from clouds. The calculation for intermediate-sized scatterers is complex but both calculations and experiments show that when the size of the clump is around the wavelength of light i.e., $r_{\text{clump}} \sim 5000$ Å $\sim 5 \times 10^{-7}$ m, we get the proper intensity vs. frequency dependence. Astronomers call this latter scatterer an interstellar grain. Since a molecule is approximately 1 Å, our grain will contain roughly $(5000)^3$ molecules. Such an object is still a small speck—the size of a dust particle—hence the name interstellar dust. But why this size and no bigger or smaller? And how and where did they come from? And what are they made of and what is their shape? Here is where we turn to the work of Hall et al. on the polarization of starlight for some help.

molecule ~ 1Å
interstellar grain: clump of molecules whose size is ~ λ of light, i.e. $r_{\text{clump}} \sim$ 5000 Å ~ 5×10^{-7} m

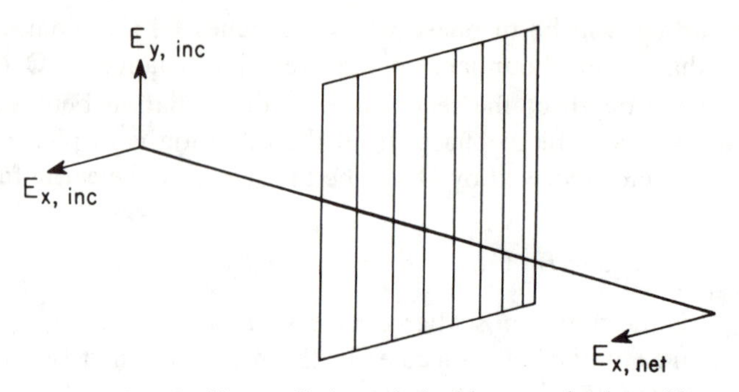

Fig. 2.21. Incident radiation polarized by a conducting grid.

First, a bit of a review. If you have unpolarized EM radiation (E_x and E_y varying randomly in time) incident upon an array of conductors as shown in Fig. 2.21, then only the x component will emerge. Why? Because in the conductor parallel to the y direction the electrons can oscillate freely due to $\boldsymbol{E}_{y,\text{inc}}$ ($\mathbf{F}_y = -e\mathbf{E}_y$), thereby producing a scattered electric vector which is opposite in direction to $\boldsymbol{E}_{y,\text{inc}}$, i.e.,

$$\mathbf{E}_{y,\text{scatt}} \propto \mathbf{a} = \frac{\mathbf{F}_y}{m} = -e\frac{\mathbf{E}_{y,\text{inc}}}{m},$$

where the minus sign means that E_{scatt} is opposite in direction to $\boldsymbol{E}_{\text{inc}}$. The net $\mathbf{E}_y$ vector in the forward direction is

$$\mathbf{E}_{y,\text{net}} = \mathbf{E}_{y,\text{inc}} + \mathbf{E}_{y,\text{scatt}}.$$

Since $\boldsymbol{E}_{y,\text{scatt}}$ is opposite to $\boldsymbol{E}_{y,\text{inc}}$, if there are enough wires, we have complete cancellation. Because the wire has no x dimension, no x component of scattered radiation is produced and therefore, no cancellation of the incident x component is possible. In the case of the above grid the radiation is now polarized in the horizontal plane.

What Hall found was that the radiation coming from the center of the Galaxy was partially polarized in the *plane of the Galaxy* when it reached us. Assuming the radiation to be unpolarized when it left the stars, this means that between us and the stars there must be a grid-like polarizing mechanism. In order for this to happen, there must be elongated conducting particles lined up perpendicular to the galactic plane. To be conducting (we need to produce E_{scatt}), the grains could either be metallic elements like Fe or nonmetallic elements like graphite (i.e., C) or compounds like SiC. Here is where the alignment mechanism gives us some help. Since the grains are moving about randomly due to collisions with electrons, other atoms, and cosmic rays, the only conceivable way to lock them into position would be with a galactic magnetic field $\mathbf{B}_{\text{gal}}$ *providing the grains themselves are magnetic*! Then the magnetic field of the grains will line up with the external field forcing the grains to line up as well.

What comes to mind is that the grains are tiny permanent bar magnets—little pieces of ferromagnetic materials—that line up along the direction of $\mathbf{B}_{gal}$, much like a compass needle. This indeed was the original guess as to what was going on, implying that $\mathbf{B}_{gal}$ was perpendicular to the galactic plane. But is this what actually happens?

Most materials, however, are not ferromagnetic, i.e., do not have permanent fields of their own. One might think that such materials will not line up in $\mathbf{B}_{gal}$, therefore the grains must be made of Fe. However, as Davis and Greenstein pointed out, this is not necessarily the case. There are many materials that are *paramagnetic*. In a paramagnetic material each *atom* has a permanent $\mathbf{B}$ due to unpaired orbiting and spinning electrons. Normally these atoms are randomly oriented throughout the grain so that $\mathbf{B}_{grain} = 0$. But in an external field, the field of each atom will line up along $\mathbf{B}_{ext}$, orienting each current loop perpendicular to $\mathbf{B}_{ext}$. We end up with a weak $\mathbf{B}_{grain}$. However, although $\mathbf{B}_{grain}$ lies along $\mathbf{B}_{gal}$, the grain itself will not. The whole grain is not magnetic, just each atom, so the grain can take on any orientation while the atoms twist around to line up with $\mathbf{B}_{gal}$, as depicted in Fig. 2.22a. Because the grains are spinning wildly, due to collisions with cosmic rays, etc., the atoms inside are forced to twist, trying to line up with $\mathbf{B}_{gal}$. As this happens the grains lose energy and momentarily stop, only to get whacked again, lose energy, stop, get whacked, etc. There is one position (Fig. 2.22b), however, where the grain does not lose energy and that is when the plane of the elongated spinning grain and the plane of the orbits of the electrons are parallel. Then the atoms do not have to twist to stay lined up with $\mathbf{B}_{gal}$ and no energy is lost. Each time a grain stops it has a finite probability of starting up in this orientation. Since it does not lose energy, it will remain in this orientation. Clearly then, after a large number of collisions all grains will be perpendicular to $\mathbf{B}_{gal}$ as shown in Fig. 2.22b. This alignment scheme is called "paramagnetic relaxation."

So which is it? Are the grains ferromagnetic or paramagnetic? By a variety of means we know that in our region of space, $\mathbf{B}_{gal}$ is in the plane of the Galaxy and in fact lies along the spiral arms. The polarization is in the plane of the Galaxy so the grains are perpendicular to the galactic plane. Thus the grains are perpendicular to the galactic magnetic field and therefore *must have paramagnetic properties*. Although the jury is still out on the composition of the grains, the latter are probably graphite or silicates, both tough covalent-bond compounds that have the necessary properties: conductivity and paramagneticity. We would like to stress again that the grains are elongated which of course is necessary if they are to produce a conducting picket fence when aligned. Nonetheless, to make calculations easier we shall treat them as spheres of radius $r = 5 \times 10^{-7}$ m.

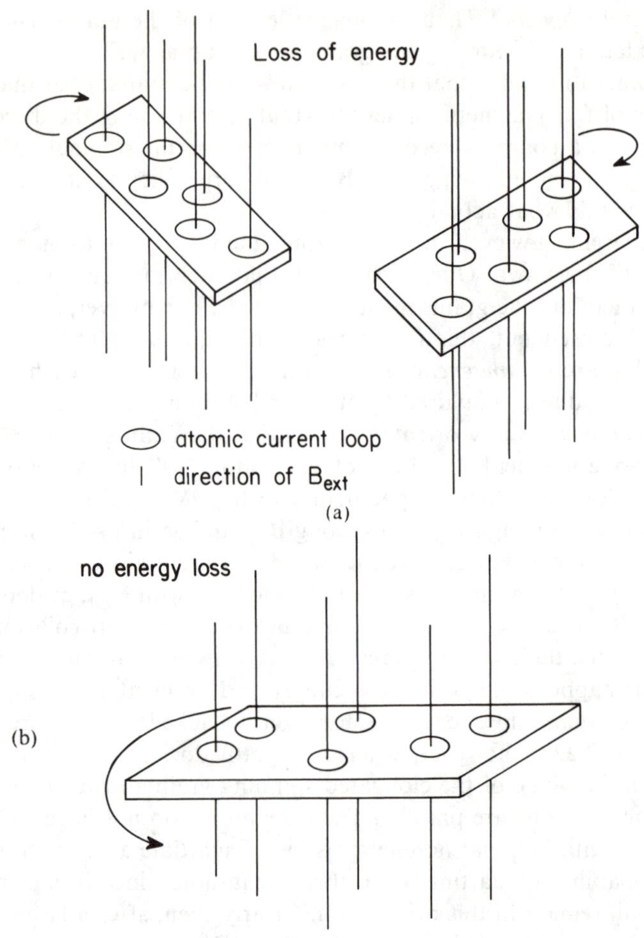

Fig. 2.22. (a) A rotating grain randomly oriented with respect to $\mathbf{B}_{ext}$; (b) A rotating grain perpendicular to $\mathbf{B}_{ext}$.

The interstellar dust is everywhere. But how much is out there and how is it distributed? We can get a handle on these questions by considering another property of dust—its ability to extinguish starlight. Although its preference is to scatter blue light more than red, a dust grain will scatter all components of the EM spectrum. Because the radius of the grain is about the dimension of visible light, however, it is most effective in scattering visible and ultraviolet wavelengths. Atomic hydrogen will not absorb these wavelengths since the first excited state is 10.2 eV above the ground state. Thus the dust in the ISM will be the principal cause of the extinction of visible starlight. If we estimate the cross section for absorbing a photon to be the geometric cross section of the grain, πr^2, we can use the concept of an interaction length to estimate the density of interstellar dust. As you may recall from Sec. 2.3.2, the interaction length of particle 1 with particle 2 is $l = 1/n_2\sigma$. Here we take type 1 as the photon and type 2 as the dust grain. By estimating the surface temperature of a star from the absorption lines of ionized atoms (see Sec. 1.6.2), and by measuring the intensity of radiation that reaches us, astronomers have determined that it takes about 1000 LY of interstellar material to produce one photon interaction length in the interstellar dust. Setting $l = 1000 \times 10^{16}$ m and $\sigma = \pi(5 \times 10^{-7}\ \text{m})^2$ we have

$$n_2 = \frac{1}{l\sigma} = \frac{1}{10^{19} \times \pi \times 25 \times 10^{-14}\ \text{m}^3} \simeq 10^{-7}/m^3 = 10^{-13}/\text{cc}.$$

This is to be compared with an *average* hydrogen density of 1/cc. Because the grains are so large, however, they carry between 0.1 and 1% of the interstellar mass. We can now appreciate the hidden pitfall Herschel faced when mapping out the Milky Way. The Sun is 35,000 LY from the center of the Galaxy. There is no way the light from that distance, even from the brightest stars, can penetrate the ISM dust. Herschel was limited to a myopic view of the Galaxy because of the amount of interstellar dust.

When the visual extinction is almost complete, as it is in Fig. 2.23, we understand this to be caused by a dense cloud of dust. This was not the case at the turn of the century when such rifts were thought to be "holes in the sky" that were totally devoid of stars. Dust clouds are generally divided into three categories, dark clouds, large globules, and small globules. The dark clouds, like the Horsehead nebula, are generally about 10 LY across, roughly the same size as diffuse clouds. But in a dark cloud, the visual extinction over 10 LY (not a thousand light years as for the general ISM) implies a dust density of better than 10^{-11} grains/cc. Indeed, the extinction is so complete that dark clouds are estimated to have dust densities on the order of 10^{-10} grains/cc. If we take as a rule of thumb that the density of hydrogen is 10^{13} times the density of dust, the dark clouds will have a hydrogen density of at least 10^3/cc. The temperatures in the dark clouds will be less than 20 °K. A cold and forbidding place, indeed.

In the 1940s the Dutch-American astronomer Bart J. Bok, looking for objects that might be in the process of gravitational collapse, called attention to smaller, more spherically shaped "holes" which he classified as large globules. Shown in Fig. 2.24a, a typical globule is about 2 LY across, has a mass of about 60 $M_\odot$, a temperature ~10–20 °K, and a hydrogen density of about 10^3/cc. Although one cannot be sure of the geometry of an object from a two-dimensional photograph, the degree of apparent regularity, as well as their smaller size, is sufficient for the large globules to be classified

Fig. 2.23. The "Horsehead" nebula (Kitt Peak National Observatory).

separately from dark clouds. Whether they are in a state of gravitational collapse is an intriguing question to which we shall return. Small globules are tiny but very dense dust balls which can be silhouetted against emission nebulae, as shown in Fig. 2.24b. Indeed, they are so small, usually less than 0.1 LY across, that one could not see them unless they were projected

Fig. 2.24. (a) Barnard 335, a large globule (B. J. Bok, Steward Observatory).

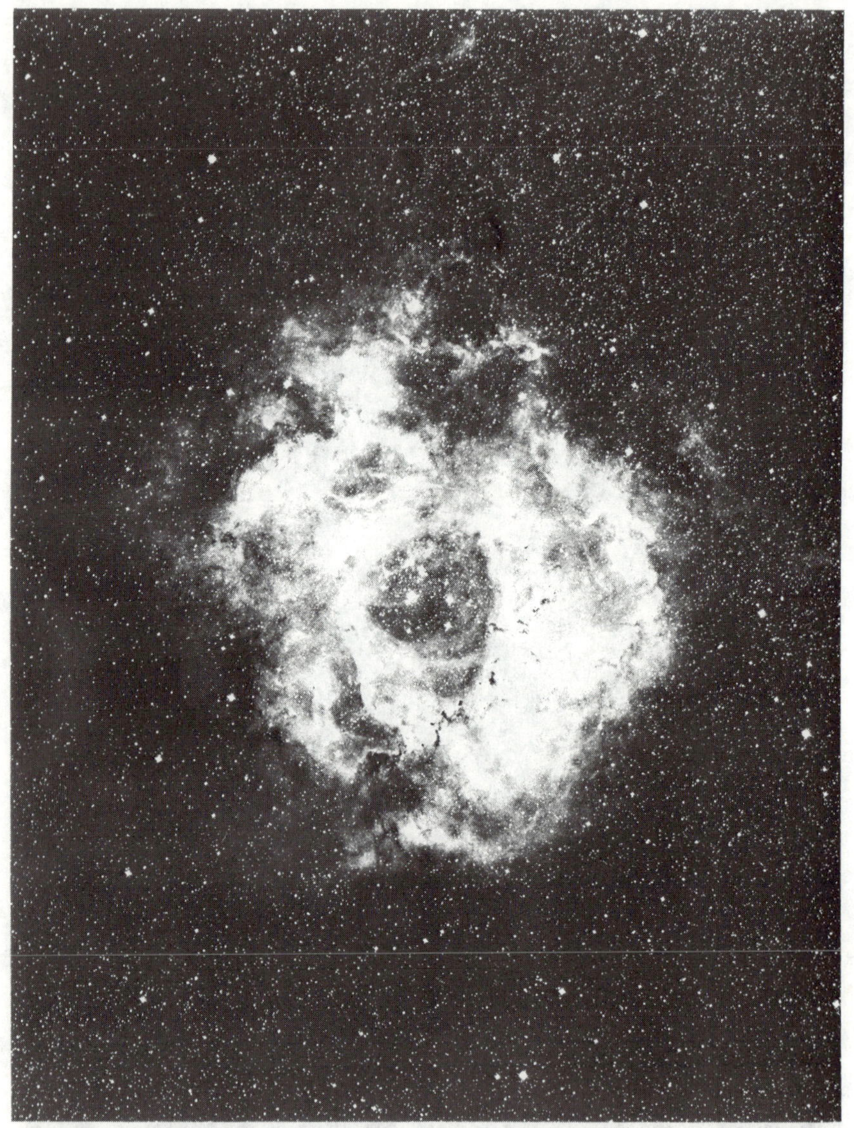

Fig. 2.24. (b) Small globules silhouetted against the Rosette Nebula (Kitt Peak National Observatory).

against a bright background. With diameters ten times smaller than a large globule, their densities are correspondingly greater, with $n_{\text{dust}} \sim 10^{-8}/\text{cc}$ and $n_{\text{H}} \sim 10^{5}/\text{cc}$. Whether the small globules lie between us and the emission nebulae or within the nebulae is not certain. It would appear from the elongated shape of some of the small globules in Fig. 2.24b that they are being torn apart by hot gases and thus must be part of the nebulae and not external to them.

In conclusion then, we see that ~1% of the ISM is locked up in interstellar dust grains. The dust is spread throughout the Galaxy but most particularly is "seen" in opaque, cold regions ranging in size from 10 LY (dark clouds) to less than 0.1 LY (small globules).

2.4.3. Molecules and Molecular Clouds

As we have seen, the spectrum that identifies an atom is associated with the orbital and spin motion of its electrons. If we take as our model of a diatomic molecule two atoms connected by a spring, then there are two more general types of motion possible: the molecules can *rotate*, dumbbell fashion, about their center of mass (c.m.), or they can *vibrate* back and forth along the length of the spring.

Let us consider rotational energy first. The kinetic energy of rotation of a diatomic molecule is given by

$$KE_{rot} = \tfrac{1}{2} I \omega^2,$$

where $I = m_1 r_1^2 + m_2 r_2^2$ is the moment of inertia about the axis of rotation, and ω is the angular velocity. The angular momentum of the molecule about the c.m. is $L = I\omega$. As is usual, we want to write the energy in terms of L, since angular momentum will be the quantized variable, according to quantum mechanics. In terms of L then, we have

$$KE_{rot} = L^2/2I.$$

If there are no external forces acting on the molecules, then $m_1 v_1 = m_2 v_2$, and since $v_1 = r_1 \omega$ and $v_2 = r_2 \omega_1$, we can write

$$I = mr^2,$$

where $m = m_1 m_2/(m_1 + m_2)$ is called the reduced mass, and $r = r_1 + r_2$ is the separation between the atoms. The quantization of L for a rotating system is given by $L^2 = l(l+1)\hbar^2$, where l $(= 0, 1, 2, 3, 4, \ldots)$ defines the rotational energy level. The kinetic energy becomes

$$KE_{rot} = l(l+1)\hbar^2/2I,$$

producing what is often called a "ladder" energy-level diagram, with the spacing between "rungs" increasing with l. The transition energy between an upper level given by l, and the next *lower* one, given by $l - 1$, is

$$\Delta(KE) = l\hbar^2/I,$$

which will also be the energy of the resultant photon. As a quantitative example, consider the molecule carbon monosulfide (CS). Its reduced mass is $m = 12m_p \times 16m_p/(12m_p + 16m_p) \approx 7m_p$ and the separation between atoms is *roughly* 1Å. Thus,

$$I_{CS} = mr^2 \approx 7 \times 1.7 \times 10^{-27} \text{kg } (1 \times 10^{-10} \text{ m})^2 \sim 10^{-46} \text{ kg-m}^2$$

and the transition energy between the 1st excited state and the ground state is

$$\Delta(\text{KE}) \sim (10^{-34})^2/10^{-46} \sim 10^{-22} \text{ J} \sim 10^{-3} \text{ eV},$$

resulting in either an emission or absorption line at $\lambda = hc/E_\gamma \sim 1 \times 10^{-3}$ m ~ 1 mm. Fortunately this wavelength just squeezes through the lower wavelength end of the radio window. We are, therefore, able to use ground-based radio telescopes to detect the rotational transitions of molecules.

Now consider the vibrational energy. Since the molecules vibrate along a straight line, there is no angular momentum to quantize. Instead the vibrational energy is directly quantized in terms of the natural frequency of the oscillator ω_0,

$$E_{\text{vib}} = (n + \tfrac{1}{2})\hbar\omega_0. \qquad n = 0, 1, 2, 3, \ldots$$

Classically $\omega_0 = \sqrt{k/m}$ where k is the effective spring constant and m the reduced mass. Here the energy rungs of the ladder diagram are uniformly spaced and the transition energy between any two *adjacent* levels is

$$\Delta E = \hbar\omega_0.$$

To get some idea of how far apart these levels are and what the photon energies might be, we note that the natural frequency of NaCl is $\nu_0 \approx 10^{13}$ cps, or $\omega_0 = 2\pi\nu_0 \approx 6 \times 10^{13}$ sec^{-1}. Then we have for $\hbar \approx \frac{2}{3} \times 10^{-15}$eV-sec,

$$E_\gamma = \Delta E \approx (\tfrac{2}{3} \times 10^{-15})(6 \times 10^{13}) = 4 \times 10^{-2} \text{ eV},$$

resulting in a wavelength of $\lambda = c/\nu_0 = 300{,}000$ Å which is in the infrared part of the EM spectrum and thus not accessible to ground-based observation. It is interesting to note that quantum mechanics only allows transitions between adjacent levels, so that regardless of the degree of excitation only a single photon energy emerges.

From the above discussion it would appear that we have a wide range of wavelengths with which to look for molecules. We have electron transitions in the visible and ultraviolet; spin-flip transitions at centimeter wavelengths; vibrational transitions in the infrared; and rotational transitions in the millimeter part of the spectrum. In principle these lines can exist either as emission or absorption lines. In spite of this wealth of possibilities, there are still problems: to produce an emission line the molecule must be in an excited state; to produce an absorption line there must be a source of background radiation. This puts us in a difficult situation—a molecule may be effectively hidden by an uncooperative environment, or it may not be there at all. For example, in 1937 CH^+, CH, and CN were found in diffuse clouds through their visible absorption lines at 3950 Å, 4300 Å, and 3875 Å, respectively. In 1971, satellites detected the ultraviolet absorption line of carbon monoxide CO at 1400 Å and the hydroxyl radical OH at 3018 Å. One might suppose that these molecules exist in dark clouds as well. The point is, using these techniques there is no way of knowing, since dark clouds are opaque to the visible and ultraviolet part of the spectrum.

All is not lost, however. Happily the spin-flip and rotational-energy levels are so close to the ground state (10^{-6} eV and 10^{-3} eV, respectively), that even in dark clouds the average kinetic energy of the constituents ($\Delta(\mathrm{KE}) = \frac{3}{2}kT \sim 10^{-4} \times 10$ °K$\sim 10^{-3}$ eV) is sufficient to cause collision excitation. The subsequent radiation emitted when the molecule falls back to the ground state can then be detected. The catch here is that there has to be a sufficient number of collisions to produce a detectable amount of radiation. Since the abundance of heavy elements is *down* by at least a factor of 10^4 with respect to hydrogen, our only hope is that hydrogen is plentiful enough to act as a collision center (recall, $\mathfrak{R} = n_1 n_2 \sigma v$) for molecules like CS. Since there is so much dust in dark clouds, and since the ISM indicates that $n_H / n_g \approx 10^{13}$, one would expect there is ample hydrogen to do the job. It was quite a shock therefore when the 2-mm radiation from CS was found (in 1971 using a millimeter-wave telescope on Kitt Peak in Arizona), but the hydrogen necessary to excite the CS molecule could not be detected. This latter effort centered around probing the CS molecular clouds with the 21-cm radiation being emitted from the diffuse clouds located behind them. There was almost no attenuation of this 21-cm "background" radiation, indicating that *atomic* hydrogen was not present in the molecular cloud. Not giving up easily, astronomers assumed that hydrogen was in *molecular* form. Although it neither absorbs nor emits 21-cm radiation, H_2 does have a very strong absorption line at 1098 Å. But of course such ultraviolet radiation does not penetrate the atmosphere! Frustrating indeed to believe something is out there and yet not be able to detect it. Finally, in 1973 the Copernicus satellite, launched to explore the ultraviolet spectrum, found the expected molecular hydrogen. Interestingly, one can now turn the tables on the molecular emission spectra, and use the intensity of the rotational lines to obtain an estimate of the density of molecular hydrogen. For example, as we have just noted, CS is a strong emitter of 2-mm radiation. Based on the strength of that emission line and the lifetime of the excited rotational state, one can estimate that it would take an H_2 density of 10^5–10^6 H_2/cc to produce the observed intensity. The density of H_2 calculated this way is consistent with the hydrogen-density estimates based on the density of dust particles.

For a long time it was believed that only collisions could account for the molecular excitations, since there was no known source of 10 °K black body radiation to supply the necessary photons. This all changed in 1965 when, as we shall recount in more detail in chapter 6 it was predicted that the remnant radiation of the "big bang" (a catch phrase for the singular space–time point where the Universe may have begun) would have a 3 °K black body spectrum that peaked at $\sim$2 mm. If such radiation exists, one should be able to detect it by looking for molecular excitations in regions where collision excitations are rare. The perfect place to look would be in diffuse clouds where the low hydrogen density (10/cc) would preclude excitation by any mechanism other than photons. Excitingly enough, lines have been found in diffuse clouds for CN at 2.6 mm and for CH and CH^+ at 0.56 and 0.36 mm, respectively.

With the development of radio astronomy, over 45 molecules have been discovered, most of them since 1970. But only a few are found in diffuse clouds. Most are detected in dark clouds and in something called "molecular" clouds. As the name implies, the variety and abundance of molecules are greatest in these, so-called, molecular clouds. Molecular clouds are typically larger than dark clouds ($D > 10$ pc), are more dense ($n_{H_2} \sim 10^4 - 10^6$/cc), and have higher kinetic temperatures (50–100 °K). Most strikingly, they are found near active elements—emission nebulae, as shown in Fig. 2.25, and hot, young stars. Sometimes the juxtaposition of emission nebulae, young stars and molecular clouds can be very complex. Near the Orion nebula, besides the huge cloud of CO, there are small clouds of CS and H_2CO (formaldehyde) and a cloud of hydrogen cyanide (HCN) that is about twice the size of the CS and H_2CO clouds. It is thought these clouds lie in regions not yet ionized by the stars that illuminate the emission nebulae. A molecular cloud found near the center of the Galaxy, Sagittarius B2, is the repository of the largest number and greatest variety of molecules, including eight that are unique to this one region. Sagittarius B2 is also the densest cloud and thus the dustiest, packing $n_{H_2} \approx 10^8$/cc in 20 LY. What are we to make of all of this, the large densities of H_2 molecules, the dust, the complex molecules near emission nebulae and young stars, the uneven distribution of molecules? To gain some insight into what is going on, we turn to some of the physics of molecular formation.

We can understand the juxtaposition of dust clouds and molecular clouds if we assume that the molecules are formed on the surface of dust grains and then later ejected from the surface into the interstellar medium. The first question to answer then is: Are the reaction rates for atoms to

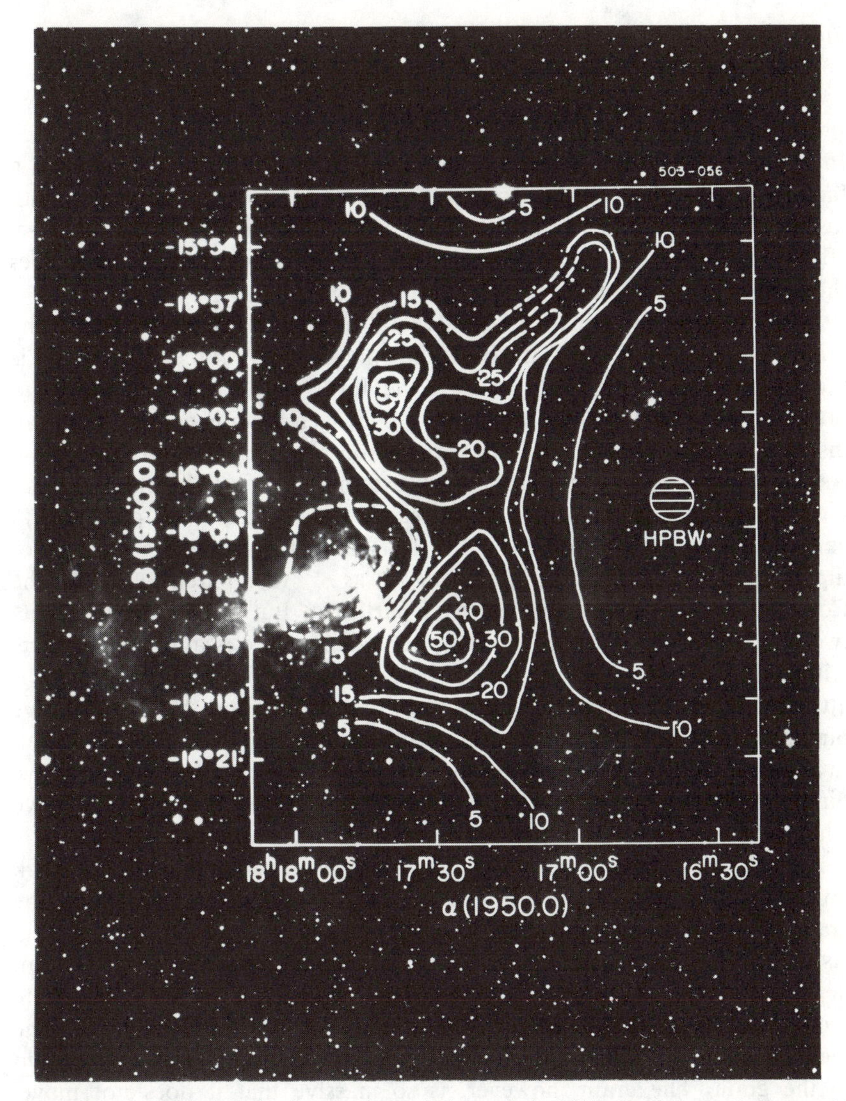

Fig. 2.25. A molecular cloud near the emission nebula M17. Radio map indicates the intensity of emission from CO (courtesy of Charles Lada and *The Astrophysical Journal*).

interact with dust grains favorable? For simplicity we shall consider the case of the formation of the hydrogen molecule, H_2. Then with a grain density in a dark cloud of 10^{-10}/cc, a cross section of $\pi r^2 = \pi(3 \times 10^{-5}\ \text{cm})^2 \cong 3 \times 10^{-9}\ \text{cm}^2$, and with the light hydrogen atom providing all the relative velocity $\langle v \rangle \cong 6 \times 10^4$ cm/sec at 10 °K, the interaction time of our atom with a grain, as defined in Sec. 2.3.2, is

$$\tau = \frac{1}{n_g \sigma_g v} \approx \frac{1}{10^{-10} \times 3 \times 10^{-9} \times 6 \times 10^4} \approx \tfrac{1}{2} \times 10^{14} \text{ sec} \sim 10^6 \text{ yrs}$$

In a molecular cloud, where the grain densities are higher by at least a factor of 1000, the collision time will drop to less than 1000 years. Both numbers are encouraging since they are well within the age of the universe, 10^{10} years. But a chance encounter with a grain is not enough. How does the atom stick to its surface?

In this regard, the worst situation would occur if both the atom and the grain were neutral. If this is the case, the only attractive force that can act is the so-called van der Waal's force, the force between two neutral but polarized objects. Although the incident hydrogen atom is neutral, it has an instantaneous polarization due to the electron revolving around the proton. As a result this atom generates a weak external electric field which *induces* a dipole moment in the atoms on the surface of the grain (see problems at end of chapter 2). The resultant van der Waal's force between the pair of dipoles is extremely short range, falling off as r^{-7} (compared to r^{-2} for the Coulomb force between two charged particles). But unlike the force between two charged particles, the van der Waal's force is always attractive. Although the force between two dipoles is weak, the incoming atom is attracted by many surface atoms at once and it turns out that the effective binding energy of the incident atom to the grain is $\sim$0.1 eV. It goes without saying that if either the incident atom or the grain is charged, the attractive binding energy is much greater. But this is frosting on the cake—we have what we want, an attractive bond between neutral particles.

An attractive force is not enough, however. The atom will not stick unless we get rid of its kinetic energy. This is also taken care of by the grain. Imagine each of the surface atoms hooked to the grain by a spring. As the incident atom approaches the grain, it experiences the attractive van der Waal's force. As it heads into the grain, however, it begins to penetrate the charge cloud of the surface atoms, resulting in a repulsive force. The incident atom will bounce off but not before it has transferred momentum to the grain. The grain, however, is so massive that it does not move. Instead, the surface atoms begin to vibrate and radiate in the infrared. From conservation of energy, the energy carried away by the infrared photons must come at the expense of the incident atom. The incident atom thus bounces away with less KE than when it arrived. Since the average kinetic energy of an atom in a dark cloud is $kT \sim 10^{-3}$ eV and since the van der Waal's binding energy is around 10^{-1} eV, the atom is sucked back onto the grain, bounces a few times, and eventually comes to rest, its energy dissipated in infrared photons. In this manner, one out of every three

Van der Waal's force: force between 2 neutral but polarized objects.
always attractive

hydrogen atoms will stick to the grain. Because they are larger, and hence have larger polarizations, atoms like C, N, and O have close to a 100% probability of sticking to the grain.

Because it is bound to the surface by the van der Waal's force, the hydrogen atom must obtain a series of "kicks" if it is to hop about the grain and meet up with another hydrogen atom. It is thought that the atom needs only $\sim 1/10$ of its binding energy, or about 10^{-2} eV, to "hop" from one surface atom to another. The hydrogen atom picks up this energy randomly from the vibrating surface atoms. The *average* vibrational energy of a surface atom is given by $\langle KE_{vib} \rangle \sim kT_g \sim 10^{-3}$ eV when the typical grain temperature of a dark cloud is 10 °K. It will take about 10,000 collisions for the hydrogen atom to receive a single kick of 10^{-2} eV ($N \propto 1/e^{-E/kT_g} \approx 1/e^{-10} \approx 10^4$). It is estimated, however, that the surface atoms vibrate $\sim 10^{12}$ times a second so that a single "hop" will take place once every 10^{-8} sec. The atom clearly does not waste time looking for a mate. Eventually it will meet up with another atom and form a molecule.

Astronomers suggest that there are three ways to get the molecule off the surface of the dust grains. One is due to the close passage of charged cosmic rays which cause the surface atoms to vibrate. This can increase the temperature of the grain sufficiently so that the atom has a good chance of being jarred loose. A second way is due to the molecular formation process itself which releases kinetic energy that is bound up in the masses of the interacting elements. If our molecule receives enough of this energy, it may "blast off" the grain surface. Finally, the absorption of photons can liberate the molecule from the surface of the grain. Regardless of the circumstances, the initial kinetic energy of the molecule must exceed its binding energy to the grain surface. If it does not, the molecule will remain frozen to the grain. Hence the association of molecular clouds and emission nebulae; juxtaposed to emission nebulae, the grain temperatures in the molecular cloud will be higher and thus the thermal ejection process more efficient.

Having escaped from the grain, the molecule must now survive the onslaught of X-rays, cosmic rays, and ultraviolet radiation. Laboratory studies show that unshielded molecules like CS and NH_3 will dissociate in less than 100 years when exposed to ultraviolet radiation of interstellar intensity. Since even in a molecular cloud it takes 1000 years for a molecule to form, the numbers imply the molecules will dissociate faster than they form. Fortunately the very dust that helps increase the rate of molecular formation also filters out the ultraviolet radiation, so that in a molecular cloud the lifetime of the molecules increases to 10^7 years! This also helps explain why in some instances one can detect different molecules at different levels within the cloud. Molecules close to the surface will be exposed to the most intense ultraviolet radiation and therefore must be very

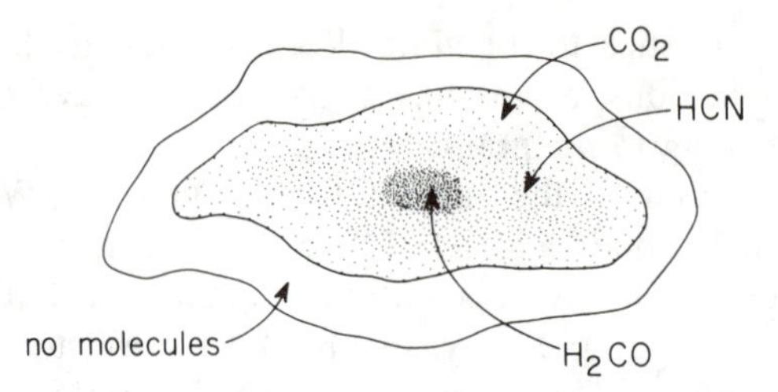

Fig. 2.26. Interior of a molecular cloud showing distribution of various molecules.

tough to survive. An example is CO, both of whose atoms have large polarizabilities. Further in and partially shielded will be molecules like HCN, and finally deep within the bosom of the cloud and well protected against the elements will be the "weak links," like H_2CO. An example of the structure of such a cloud is shown in Fig. 2.26.

The last step is the detection of the molecule, typically by the emission of millimeter radiation. As we discussed earlier, this radiation is produced by collision excitation, with H_2 as the dominant collision center. If we are to detect radiation, there will have to be an appreciable number of molecules in the excited states. Whether there are or not will depend upon two factors: the collision rate and the spontaneous decay rate. The collision time for a molecule with H_2 is given by

$$\tau_c = 1/(n_{H_2}\sigma v),$$

where n_{H_2} is the density of molecular hydrogen, σ the collision cross section, and v the relative velocity. Typically, $\sigma = 10^{-15}$ cm^2 and $v \sim 10^5$ cm/sec so that in CGS units,

$$\tau_c = \left(10^{10}/n_{H_2}\right)\text{sec}.$$

The time for the spontaneous decay τ_D has been determined experimentally for many molecules, a few of which are shown in Table 2.1. The point is, there will only be an appreciable number of excited molecules, and therefore detectable radiation, if $\tau_c < \tau_D$. With this inequality, atoms are excited faster than they decay, thus assuring us of a large population of excited rotational states. We can calculate the critical density of H_2 for which the excitation rate exceeds the decay rate by setting $\tau_c = \tau_D$. For OH, the hydroxyl radical, we have

$$n_{H_2} \sim \frac{10^{10}}{10^{10}} \sim 1/\text{cc},$$

which is well within the reach of most clouds. This probably accounts for the fact that OH is seen in far more regions than any other molecule. For H_2CO and CO, the critical density rises to $n_{H_2} \sim 10^{10}/10^7 \sim 10^3/\text{cc}$, while

TABLE 2.1. SPONTANEOUS DECAY TIME τ_D OF VARIOUS MOLECULES

Molecule	τ_D (sec)
OH	1.3×10^{10}
H_2CO	2.3×10^7
CO	1.3×10^7
NH_3	2.8×10^7
CN	0.7×10^5
HCN	0.4×10^5
CS	0.2×10^5

for CS and HCN, $n_{H_2} \sim 10^{10}/10^5 \sim 10^5$/cc. Not surprisingly, H_2CO and CO rank second and third in terms of the number of regions in which they are found.

Having discussed the physics problems associated with molecular formation, we see that molecular clouds are ideal breeding grounds. The high density of dust leads to the formation of H_2 and other more complex molecules; the dust also protects more complex molecules from destruction by ultraviolet radiation. With large amounts of H_2 in the clouds, collision excitations are frequent enough to produce detectable amounts of radiation as seen spectroscopically.

2.4.4. The Galactic Magnetic Field

One of the most intriguing components of the spiral arm is the galactic magnetic field. Where did it come from and why is it still here 10^{10} years after the formation of the Galaxy?

The second question is easier to answer than the first. As long as there are charged particles in the vicinity of the galactic magnetic field, there is almost no way of getting rid of $\mathbf{B}_{gal}$. It was Faraday who first realized that a changing magnetic flux, $\Phi = BA$, will produce an electric field, $\mathbf{E}'$, which circles $\mathbf{B}$ such that $\Delta\Phi/\Delta t = E'2\pi r$, as shown in Fig. 2.27. If there is ionized matter in the vicinity of the changing flux, $\mathbf{E}'$ will cause these ions to be accelerated along the local direction of E'. These moving charges produce a magnetic field of their own, $\mathbf{B}'$, and therefore a new flux Φ'. What Lenz found was that the new flux always opposes the change in the old. Thus if Φ is trying to *decrease*, Φ' will try to prevent it from decreasing by having $\mathbf{B}'$ *in the same direction* as the original $\mathbf{B}$. What all of this means on a galactic scale is this: $\mathbf{B}_{gal}$ is around 10^{-6} gauss—very small indeed—but it is spread over vast regions of space. Therefore it represents a large flux which, if it tries to change, can produce a substantial $\mathbf{E}'$. Since there is charge around due to ionizing cosmic rays, any decrease in $\mathbf{B}_{gal}$ will induce

magnetic flux Φ will produce elec. field E'

FARADAY'S LAW: $\Phi = BA$ and $\frac{\Delta\Phi}{\Delta t} = E'2\pi r$

LENZ's law: New flux always opposes change in old

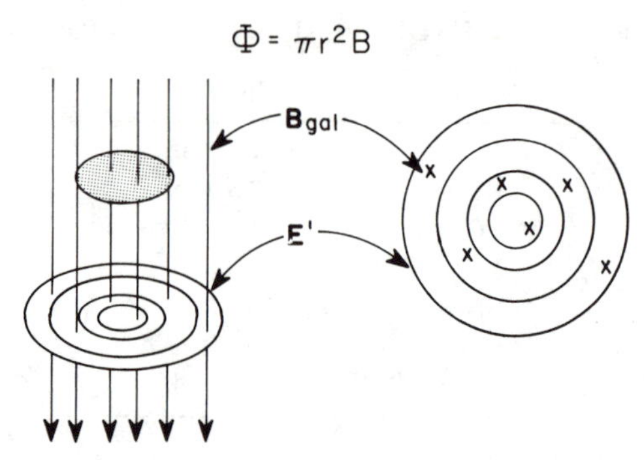

Fig. 2.27. A changing magnetic flux Φ produces an electric field, **E**′.

E′ which produces **B**′, the latter in the same direction as $\mathbf{B}_{\text{gal}}$. Thus the net magnetic field will not decrease quickly—it has its own safeguards in Faraday's and Lenz's laws. Of course, if nature did not behave this way we would not be here. Indeed, as we shall see, the seed of our sun's magnetic field, which is about 1 gauss, is this tiny galactic field.

But where did $\mathbf{B}_{\text{gal}}$ come from? Is its origin cosmological or local? Even now we do not know. It is nonetheless worthwhile to present one argument of how a magnetic field might materialize out of "thin air." Say we have electrons and protons orbiting a star, both at the same velocity, as shown in Fig. 2.28a. This is certainly possible and indeed likely since gravity would supply the radial force necessary to keep them in orbit, while cosmic rays and photons from the star would produce the ions. Each charge is a current loop, but because the charges are opposite in sign and orbit together, the net current is zero. There will, of course, be no magnetic field. As we have just seen, however, the star sends out photons. Sitting on the electron or proton, these photons appear to be coming in at an aberrated angle, $\tan\theta = v/c$, as shown in Fig. 2.28b. (The same aberration principle applies for running at velocity v through vertically falling rain. To "see" the rain you would have to tilt a tube at $\tan\theta = v/v_{\text{rain}}$.) This means that the

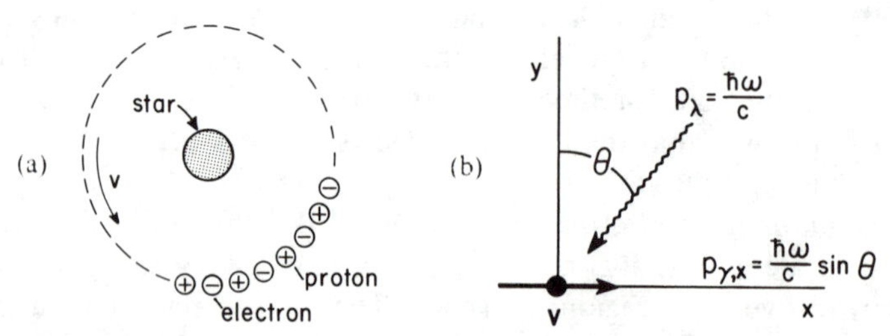

Fig. 2.28. (a) Protons and electrons orbiting a star with zero net current; (b) photon absorbed from the forward direction from such orbiting charged particles.

photon carries a component of momentum opposite to the direction of **v**, $p_{\gamma,x} = -(\hbar\omega/c)\sin\theta$. If the photon is absorbed, conservation of momentum in the x direction gives ($\tan\theta \sim \sin\theta \sim v/c$)

$$P_{i,x} + p_{\gamma,x} = p_{f,x}$$

$$mv_{i,x} - \frac{\hbar\omega}{c}\left(\frac{v_{i,x}}{c}\right) = mv_{f,x}$$

$$\frac{\Delta v_x}{v} = -\frac{\hbar\omega}{mc^2},$$

where x is defined along the initial direction of the particle and where $v_{i,x} = v$. Thus the particle slows up in the tangential direction because it has absorbed the oppositely directed momentum of the photon. When the atom radiates that photon it does so isotropically, i.e., in all directions, so that in the emission process no momentum is gained or lost, on the average. Thus Δv_x is permanently negative. But because of the mass m in the denominator of the expression for Δv_x, the electron loses more velocity than the proton. Whereas initially they were in step, after a while the proton is moving faster than the electron. In terms of the current, $I_{\text{proton}} \neq I_{\text{electron}}$, so that there is now a net current and therefore a dipole magnetic field. So it can be done —we have manufactured a field out of "thin air"—that is all we wanted to show. Does it really happen like this? We do not know.

2.5. Evidence for Stellar Birth and Evolution: Pro and Con

So far we have seen that the Galaxy is composed of stars concentrated in the galactic core, and an interstellar gas composed mainly of hydrogen. About half of the latter is distributed uniformly throughout the Galaxy, while the rest is found in clouds: first, as atomic hydrogen in diffuse clouds ($n \approx 10$/cc) and then as molecular hydrogen in dark clouds ($n \approx 10^3$/cc) and molecular clouds ($n \approx 10^6$/cc). Besides these components there is also dust in the ISM. Did the stars evolve out of this cosmic soup of interstellar matter or not? Here we start by setting you up with some reasons for thinking that stars did evolve out of the interstellar medium and then we shoot you down with physical evidence that implies the contrary. Starting in chapter 3 we shall try to settle the issue of stellar evolution.

2.5.1. Some Circumstantial Evidence—On the Pro Side

If one were to mention stellar evolution at this point, one surely must be drawn to the conclusion that stars evolve out of the interstellar medium. Certainly the step ladder of increasing densities we mentioned above, along

atomic hydrogen is found in diffuse clouds $n \approx 10/cm^3$
molecular " " dark clouds $n \approx 10^3/cm^3$
molecular " $n \approx 10^6/cm^3$

with the more regular shape of the large globules would lead one to believe in a step-by-step evolution from the uniform gas and dust, through dense clouds, to stars. But seeming to be so and being so are two different things. We need evidence based on the laws of physics. Before investigating this pattern we would like to look at evidence that is a bit more circumstantial.

A good lawyer might argue that if the above sequence is true we should be able to find young stars still imbedded in the gas and dust from whence they came. But what constitutes a young star—how would we know one if we saw it? We have already suggested hot stars are young stars because of the rate at which they use up energy. But there is another way to identify young stars that does not rely on dynamics.

Let us say we detect in the Sagittarius arm two members of an association (a group of ~5 to 50 stars spread over a region of ~100 to 500 LY in diameter) that are 30 pc apart along a radial direction from the center of the galaxy. They both move in circles about the galactic center. Since their tangential velocities are given by $v^2 = GM/r$, the star closest to the center of the galaxy will be moving faster and will begin to move ahead of the star further out. After 250 million years the stars will be much further apart than 30 pc and will no longer be in association with one another (see problem 21). Since 250 million years is a mere 2.5% of the age of the universe, we can safely say that those associations we do see are younger than this. And since they are surrounded by gas, dust, and molecular clouds, we have strong circumstantial evidence that these stars evolved out of this material. Along the same line, we can note that the Sun is an old star ($\sim 10^{10}$ years old) and has no gas or dust around it. We also do not find gas or dust clouds amongst the globular clusters. This is reassuring because globular clusters must be very old since they ride well out of the galactic plane and thus are benchmarks of the first stage of *galactic collapse* (we will indicate why they broke away in the next chapter), some 10^{10} years ago.

2.5.2. Some Physical Evidence on the Con Side

All of the above is merely suggestive, but it is not physics. Unfortunately when we begin to look at the physics, our first conclusions will seem to imply that we should not be here!

2.5.2.1. Too Much Kinetic Energy or Not Enough Potential Energy

If a cloud is going to form into a star, there must be some method of collapsing it down to the right size. There are only a few forces available to do this: the gravitational force, the electromagnetic force, and the nuclear force. And just as clearly it is gravity that will have to do the trick; it is attractive, it is long range, extending over infinite reaches of space, and it requires just what every object has—mass. Just because particles in a cloud are under the attractive force of gravity, however, does not mean that the cloud will collapse. On the contrary it may even dissipate! Remember, the particles inside have $\langle v \rangle \sim \sqrt{kT}$ and at some time or another will be moving radially outward. It is possible, therefore, for each and every particle to escape the cloud. In order to do so, the particle must have enough KE to overcome the gravitational pull of the rest of the cloud. If it does not, then it is gravitationally bound and the cloud *may* collapse. How do we know which scenario is correct?

Consider a particle of mass m at a distance r from the center of the cloud. Its total energy with respect to the cloud is

$$E_0 = \tfrac{1}{2} m v_i^2 - G M_r m / r, \tag{2.6}$$

where M_r is the mass inside of r (remember, the mass outside of r does not affect m), and v_i is the rms velocity of the particle. On its outward journey (assuming no collisions), E_0 is conserved. If the particle is to escape to infinity, where $\mathrm{PE}_f = 0$ and $\mathrm{KE}_f \geqslant 0$, then E_0 must be greater than or equal to zero. If E_0 is less than zero (i.e., negative) then the particle cannot escape; it will stop at some finite distance and then return to the cloud. What is true for one particle is true for all. Thus, if we add up the E_0s for all the particles, each of which must be less than zero, then the total, E_{tot}, for the cloud must be less than zero too. To find E_{tot} we need the total KE and the total PE of the cloud: $E_{\mathrm{tot}} = \mathrm{KE}_{\mathrm{tot}} + \mathrm{PE}_{\mathrm{tot}}$. The total KE is easy to find,

$$\begin{aligned} \mathrm{KE}_{\mathrm{tot}} &= \sum \mathrm{KE}_i = \sum \tfrac{1}{2} m v_i^2 = \tfrac{1}{2} m_{\mathrm{H}} N \left(\frac{\sum v_i^2}{N} \right) \\ &= \tfrac{1}{2} M_{\mathrm{cl}} v_{\mathrm{rms}}^2, \end{aligned}$$

where $(\sum v_i^2)/N$ is just our definition of v_{rms}^2, and $M_{\mathrm{cl}} = m_{\mathrm{H}} N$, where we have assumed the cloud is entirely hydrogen. For a diffuse cloud at $T = 100\ {}^\circ\mathrm{K}$, $v_{\mathrm{rms}} \approx 1.5$ km/sec so that

$$\mathrm{KE}_{\mathrm{tot}} \approx 10^6 M_{\mathrm{cl}} (\text{joules}).$$

The total PE is harder to find. Between each pair of particles the PE is $\mathrm{PE}_{ij} = -G m_i m_j / r_{ij}$. The trouble is that we have a huge number of particles. How do we sum over all pairs to get $\mathrm{PE}_{\mathrm{tot}}$? Let us start with just four particles. Then the total PE is

$$\mathrm{PE}_{\mathrm{tot}} = \mathrm{PE}_{12} + \mathrm{PE}_{13} + \mathrm{PE}_{14} + \mathrm{PE}_{23} + \mathrm{PE}_{24} + \mathrm{PE}_{34}.$$

Since only pairs contribute there are a total of six terms. As is fairly easy to show, with N particles there will be $N(N-1)/2$ number of pairs. But each pair does not contribute the same amount to the total PE; some pairs are very close together and some are much further apart. Now we make an

$\langle v \rangle = \sqrt{kT}$ $\quad \frac{3}{2}kT = \langle KE \rangle = \frac{1}{2} m v_{rms}^2 \rightarrow v_{rms} = \left(\frac{3kT}{m}\right)^{1/2}$

approximate guess and take the average separation to be R, the radius of the cloud, so that we have

$$PE_{tot} \cong -\frac{N(N-1)}{2}\frac{Gm_H m_H}{R}.$$

Since N is large, $N-1 \approx N$. Once again $M_{cl} = m_H N$, so that we can write

$$PE_{tot} \approx -\frac{N^2}{2}\frac{Gm_H m_H}{R} = -\tfrac{1}{2}\frac{GM_{cl}^2}{R}. \tag{2.7}$$

Using integral calculus, the full-fledged solution is $PE_{tot} = -\frac{3}{5}GM_{cl}^2/R$, only slightly different from our approximation. What is remarkable about equation (2.7)—and *it is one of our most important equations*—is that we can find the PE of the cloud from its size and mass. For a typical diffuse cloud, $R \sim 15$ LY and $M_{cl} \approx 100\ M_\odot$. Thus its total potential energy is

$$\begin{aligned} PE_{tot} &\approx -\tfrac{3}{5}\frac{6.7\times 10^{-11}(100\times 2\times 10^{30})}{15\times 10^{16}} M_{cl}\ \text{(joules)} \\ &= -0.05\times 10^6 M_{cl}\ \text{(joules)}. \end{aligned}$$

We see that the total energy of the cloud, (2.6), becomes $E_{tot} = 10^6 M_{cl} - 0.05\times 10^6 M_{cl}$, and is greater than zero. A diffuse cloud is not bound and will not collapse. So, already our evolutionary scheme is suspect. Apparently the first step of our ladder is broken. And with this kind of ladder, if you do not have the first step, it is an awfully long way to the second.

What about the large globules that Bok had picked out as likely candidates for collapse? Typically, with $R \approx 3$ LY and $M \approx 60\ M_\odot$, we have $PE_{tot} \approx 1.5\times 10^5 M_{cl}$. It is a bit trickier to calculate the total KE, since it depends upon the temperature of the globule. If we take $T \approx 20$ °K, due to cooling by rotational excitation, then $v_{rms} \sim 0.7\times 10^3$ m/sec and $KE_{tot} \sim 3\times 10^5 M_{cl}$. Although the KE is still a factor of 2 too large, with a little more mass and/or a smaller globule we could manage to get $E < 0$. It is believed that some small globules do have their *thermal* KE less than PE_{tot}. But even in these cases one cannot be certain E is less than zero, since there are other forms of kinetic energy which must be added to KE_{tot}: large scale turbulent motion, rotational KE ($= \frac{1}{2} I_{cl}\omega^2$), and even kinetic energy associated with the magnetic field that threads the globule. This is not the end of our problems, however.

2.5.2.2. Too Much Angular Momentum

You will recall that a particle moving in a circular orbit has an angular momentum $\mathbf{L} = \mathbf{r}\times\mathbf{p}$. Only an external torque can change $\mathbf{L}$ via $\boldsymbol{\tau} = \Delta\mathbf{L}/\Delta t$. A central force like gravity, however, has $\boldsymbol{\tau} = \mathbf{r}\times\mathbf{F}_g = 0$. Thus if gravity is the only force acting, $\mathbf{L}$ is a constant. Consider then a cloud

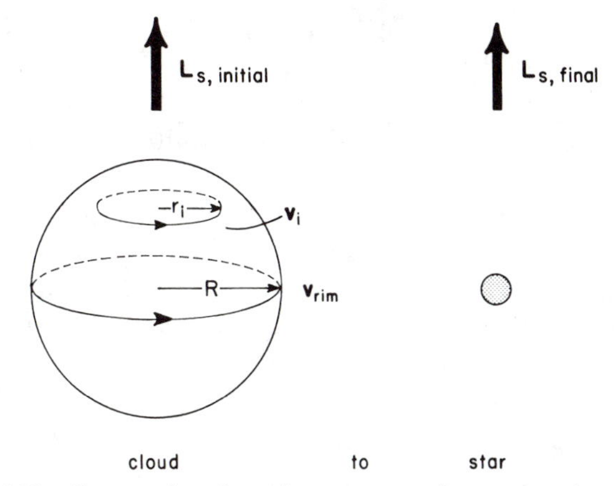

Fig. 2.29. Conservation of angular momentum for a contracting cloud.

which we hope will collapse under gravity. If that cloud is initially rotating, then each atom will have $L_i = |\mathbf{r}_i \times \mathbf{p}_i| = m_\mathrm{H} r_i v_i$ with L_i pointing up for the rotation shown in Fig. 2.29. For the complete cloud the total spin angular momentum, $\mathbf{L}_S$, will be

$$|\mathbf{L}_S| = \sum_i m_\mathrm{H} r_i v_i$$

$$= \omega \sum_i m_\mathrm{H} r_i^2,$$

where we have used $v_i = r_i\omega$. Now the term $\sum m_\mathrm{H} r_i^2$ when summed over all the atoms is just the moment of inertia of the cloud. Assuming a spherical cloud, $I_\mathrm{cl} = \frac{2}{5} M_\mathrm{cl} R^2$, and therefore, we obtain

$$|\mathbf{L}_S| = \tfrac{2}{5} M_\mathrm{cl} R^2 \omega. \tag{2.8}$$

As the cloud collapses, $\mathbf{L}_S$ must be conserved if gravity is the only force acting on the particles. Thus we have from (2.8),

$$|\mathbf{L}_{S,i}| = |\mathbf{L}_{S,f}|$$

$$\tfrac{2}{5} M_\mathrm{cl} R_i^2 \omega_i = \tfrac{2}{5} M_\mathrm{cl} R_f^2 \omega_f.$$

The last equation gives the familiar result, $\omega_f = (R_i^2/R_f^2)\omega_i$, the angular velocity increases as the cloud contracts. In terms of the rim velocity at the equator of the cloud,

$$v_{f,\mathrm{rim}} = R_f \omega_f = R_f \left(\frac{R_i^2}{R_f^2} \right) \left(\frac{v_{i,\mathrm{rim}}}{R_i} \right)$$

$$v_{f,\mathrm{rim}} = \left(\frac{R_i}{R_f} \right) v_{i,\mathrm{rim}}.$$

To collapse from a cloud about 1 LY across to a star the size of the Sun, where $R_j = R_\odot \approx 7 \times 10^8$ m, would mean

$$v_{f,\mathrm{rim}} = \left(\frac{\frac{1}{2} \times 10^{16}\ \mathrm{m}}{7 \times 10^8\ \mathrm{m}} \right) v_{i,\mathrm{rim}} \approx 10^7 v_{i,\mathrm{rim}}.$$

Now a typical dust cloud has an equatorial speed of about 100 m/sec so that if angular momentum is conserved, we expect $v_{f,\mathrm{rim}} \simeq 10^9$ m/sec, which is clearly absurd since it would exceed the speed of light! Indeed, the actual equatorial speed of the Sun is 2×10^3 m/sec. We need a 10^5-fold decrease in $L_{S,i}$ if we are to approach present conditions! But how?

2.5.2.3. Too Much Magnetic Field

For the record let us note one other problem: too much magnetic field. As we shall see in chapter 3, if a star like our sun collapses out of the ISM we expect it to end up with a magnetic field due to $\mathbf{B}_{\mathrm{gal}}$ and conservation of magnetic flux Φ. Unfortunately, naive predictions indicate the Sun should end up with a surface field of around 10^4 gauss, when in reality its surface field is not more than 10^2 gauss. Clearly we have problems on this account too.

2.6. Evidence for Continuing Stellar Evolution: The Hertzprung–Russell Diagram

Solving the problems of energy and angular momentum will only start us down the road of stellar evolution, to be considered in chapter 3. What happens after the star comes into existence is another matter. To help us put the evolutionary scheme in proper perspective, we turn now to the general properties of stars and the Hertzprung–Russell (H–R) diagram.

2.6.1. Stellar Mass

So far we have been able to pin down three properties of stars whose magnitudes we can measure: (1) luminosity; (2) surface temperature; and (3) radius. These are related by the black body equation $L = \sigma T^4 4\pi R^2$. There is one other property we would like to know—the mass of the star. Unfortunately we can not go out and "weigh" a star. While we cannot directly determine the mass of an isolated star, when it is part of a binary system we do have a chance of determining its mass. Since almost 50% of all stars are binary stars, it would appear we are in good shape on this account. As we shall now show, however, nature conspires against us so

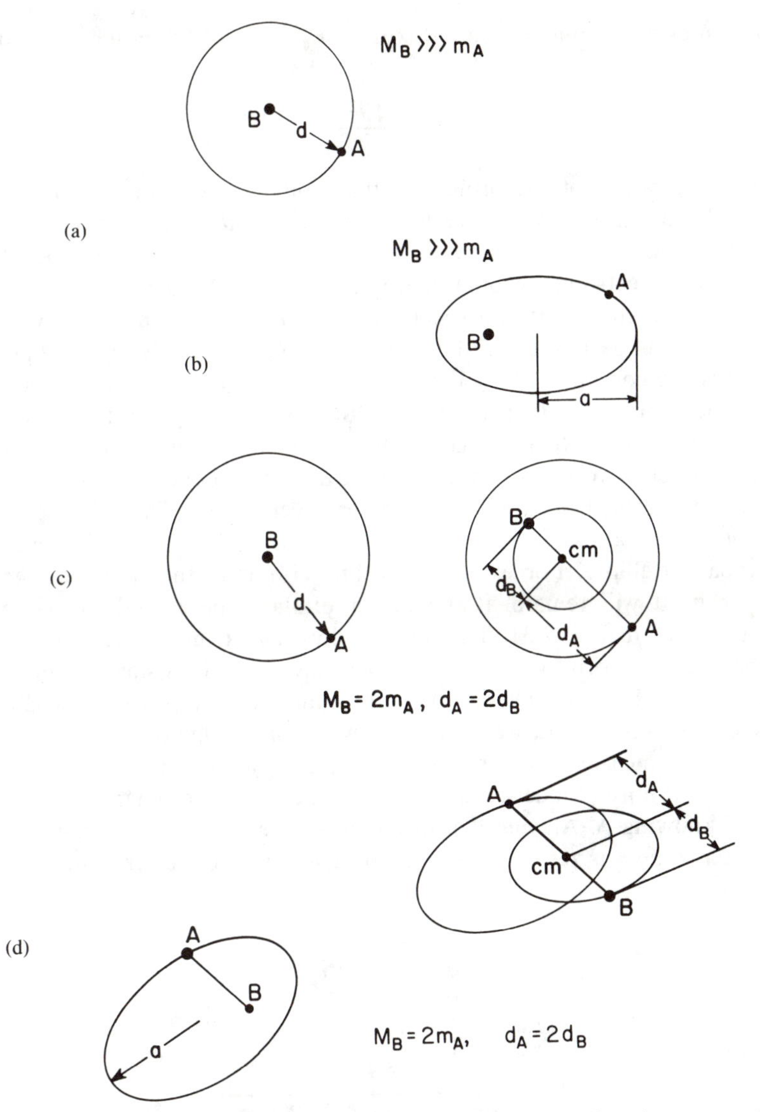

Fig. 2.30. Circular (a) and elliptical (b) orbits of light mass m_A about heavy mass M_B. Two views of approximately equal masses moving in circular orbits (c) and elliptical orbits (d).

that only about 100 or so binary stars can be used to determine stellar masses accurately.

Of the four situations shown in Fig. 2.30, only case (a) is "simple": a star of mass m in a circular orbit about a companion of much greater mass M. In this case the center of mass of the system can be taken as coincident with the center of the massive star B so that $F_A = GmM/d^2 = mv^2/d$ and therefore

$$M = \frac{dv^2}{G} = \frac{4\pi^2 d^3}{G\tau^2}, \tag{2.9}$$

where d is the radius of the orbit, v is the orbital velocity of m, τ is the period of the orbit, and where we have used $v = 2\pi d/\tau$. If it is a visual binary, we can determine d and τ by observing the stars over a period of time. If it is a spectroscopic binary, we can use the Doppler shift of the lighter star's discrete absorption spectrum to find v and the periodicity of the shift to determine τ. We can then obtain d from $d = v\tau/2\pi$. Even these simple cases are complicated by the fact that the orbit may be tilted with respect to our line of sight. In the case of the visual binary, we must view the orbit from above if we are to get a true measure of its shape and size. If it is tilted, a circular orbit will appear elliptical. (And an elliptical orbit will appear as an elliptical orbit of even greater eccentricity.) There is a clue that things may be amiss, however; the massive star will not be at the focus of the apparent ellipse. For example, we know M is at the center of the circular orbit. It will remain at the center of the apparent ellipse when viewed as shown in Fig. 2.31. The mass is thus *not* at the focus, where it should be if the true orbit were elliptical. This giveaway allows astronomers to "untilt" the orbit and calculate its true dimensions. For spectroscopic binaries, on the other hand, we want to view the orbit edge on. If the orbit is tilted we only measure the component of velocity along our line of sight, $u = v\cos\phi$, and not v. Since we cannot see a spectroscopic binary, there is no way of knowing ϕ. All one can do is either guess at ϕ or assume it is zero. If we assume ϕ is zero then all we can determine is a lower bound for

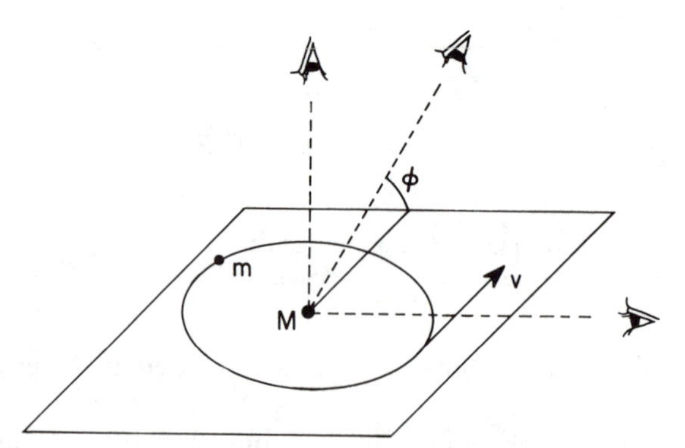

Fig. 2.31. A circular orbit as viewed from directly above (seen as a circle), at an angle (seen as an ellipse) and in the orbiting plane (seen as moving back and forth in a straight line).

M since the measured velocity we plug into Eq. (2.9), is less than the true velocity v.

If the orbit is an ellipse and $M \gg m$, as in case (b) Fig. 2.30, then the expression for M is $4\pi^2 a^3/G\tau^2$. Although this is superficially the same as (2.3), there is one important difference; a is the semi-major axis of the ellipse and not the separation between the stars. With visual binaries, after untilting them, we can determine the distance a directly; with spectroscopic binaries one can deduce the size of the orbit from the variation in the orbital velocity, which is not constant for an elliptical orbit.

If M and m are of comparable size, they will rotate about their common center of mass either in circular or elliptical orbits as shown in Fig. 2.30(c) and (d). If instead of the center of mass, we view the motion from one of the stars, say from M, then the orbit of m will be either a circle of radius d, which is the distance between the stars, as in case (c), or an ellipse of semi-major axis a, as in case (d). As we shall show in chapter 5, when we determine the mass of a black hole, the mass expression for the circular orbit about M is $m + M = 4\pi^2 d^3/G\tau^2$. For the relative elliptical orbit viewed from M we have $m + M = 4\pi^2 a^3/G\tau^2$. In each case the period is the same for the relative orbit about M as it is for the true orbits about the center of mass. Although reducing the problem to a fixed star plus a single orbit allows us to arrive at a simple mass formula, the equation contains *two* unknowns. Unless we find another equation in m and M, we cannot solve for either, just their sum. To get the second equation we have to go back to the pair of orbits about the center of mass (nothing is ever simple) where the distance of each star from the center of mass is related to their masses by $md_2 = Md_1$. If we can find d_1 and d_2 we can couple this information with that of the size (or semi-major axis) of the relative orbit and its period to find m and M. For example, say we plot the orbit of one star as viewed from its companion and find d (or a) is 20 AU $\approx 3 \times 10^{12}$ m and $\tau \simeq 20$ years $\approx 2 \times \pi \times 10^8$ sec. Then

$$m + M \approx \frac{4\pi^2(3 \times 10^{12})^3}{6.7 \times 10^{-11}(4\pi^2 \times 10^{16})} \approx 4 \times 10^{31}\ \text{kg} \approx 20\ M_\odot\,.$$

If we also find $d_2 = 2d_1$ then $2m = M$ and $m + 2m = 20\ M_\odot$, therefore, $m = 6.7\ M_\odot$, $M = 13.3\ M_\odot$. Fig. 2.32 indicates how one finds d_1 and d_2. Because there are no external forces acting on the system, the center of mass moves in a straight line and the stars weave in and out around it. Once the line of motion of the center of mass has been determined, it is easy to measure d_1 and d_2.

So in principle, if we know everything about a binary system, we can determine the mass of its components. In practice things are made difficult

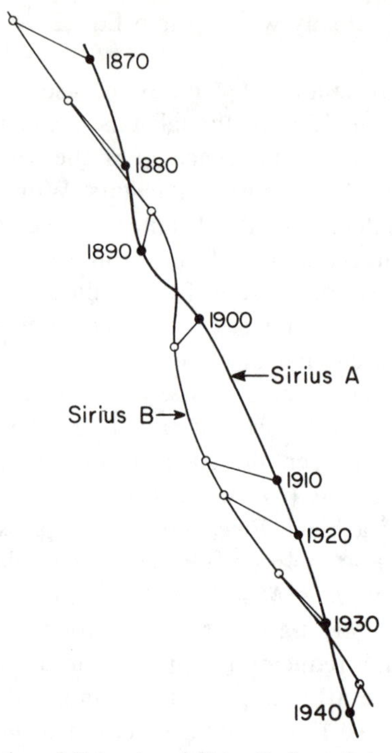

Fig. 2.32. Motion of Sirius A and Sirius B about their center of mass.

because the orbits are tilted, which in the case of visual binaries means that the orbits must be untilted and in the case of spectroscopic binaries only allows for a lower bound on the mass. Clearly our most accurate mass information will come from visual binaries. But here again nature has proved to be rather stubborn. In order to see a visual binary, the orbits must be large, and in fact the further away the binary is, the larger must be the orbit. Since the period τ is proportional to $d^{3/2}$ (see 2.9), most visual binaries have periods that measure in centuries. This makes it hard to get accurate orbital information on all but those few that are comparatively close and have reasonably short periods. Thus with thousands of binaries to choose from, we only have solid information on the masses of about 100 stars.

The result of such analyses is that all observed stars have masses in the range

$$\tfrac{1}{20} M_\odot < M_{\text{star}} < 70\, M_\odot\,, \tag{2.10}$$

with an average being about a solar mass, $M_\odot \approx 2 \times 10^{30}$ kg. The other three stellar parameters range from

$$10^{-4}\, L_\odot < L_{\text{star}} < 10^{6}\, L_\odot$$
$$10^{-2}\, R_\odot < R_{\text{star}} < 10^{3}\, R_\odot \tag{2.11}$$
$$10^{3}\ {}^\circ\text{K} < T_{\text{surface}} < 10^{5}\ {}^\circ\text{K},$$

where $L_\odot \approx 4 \times 10^{26}$ J/sec, $R_\odot \approx 7 \times 10^{8}$ m, and $T_\odot \approx 5800$ °K.

2.6.2. The Hertzprung–Russell Diagram

Although we have four variables to play with, L, R, T, and M, they are not all independent. Because $L = \sigma T^4 4\pi R^2$, we really have only three and as we shall soon see, even L and M are related so that finally only two, say L and T, are independent. What is more, in order to know L we must know d, the distance to the star. Thus the set of stars from which we can obtain useful information is reduced to a much smaller subset—those for which d and T_s can be measured. Indeed, within 10 LY, there are only 12 stars whose distances can be determined accurately. Still we must raise a crucial question. Is there a systematic relationship between L and T, or L and M, for those stars where these quantities can be determined? This was answered in 1911 by the Danish astronomer E. Hertzprung and expanded upon by the American astronomer Henry Norris Russell in 1913 when they developed the now famous H–R diagram, a plot of the star's luminosity vs. surface temperature.

Using a log-log plot, we get the grouping of stars in Fig. 2.33 which are within 20 LY of the Sun. Please note two things about the graph: (i) the luminosity axis is $L/L_\odot$. This normalization cleans things up, making the axis much easier to read and requiring memorization of a simple ratio. Besides the numbers read off the curve, all one needs to know is $L_\odot \approx 4 \times 10^{26}$ J/sec; (ii) the temperature scale—also logarithmic—runs the "wrong" way. Big Ts are to the *left*—small Ts to the right.

What is remarkable about this grouping of local stars is that most are strung out along a line (not straight) instead of being scattered all over the graph in a random fashion. When pairs of variables are related like this, it is natural to look for a physical reason for the correlation. This line has become known as the *main sequence*. Moreover if we locate some binary stars along the main sequence and indicate their masses, we see that a relationship between L and M also exists. For the lower part of the curve, M increases from $0.1\, M_\odot$ to $M_\odot$ as L increases by a factor of 10^3. Thus in this region we may infer

$$L \propto M^3 \left[\text{or } \frac{L}{L_\odot} = \left(\frac{M}{M_\odot} \right)^3 \right]. \tag{2.12}$$

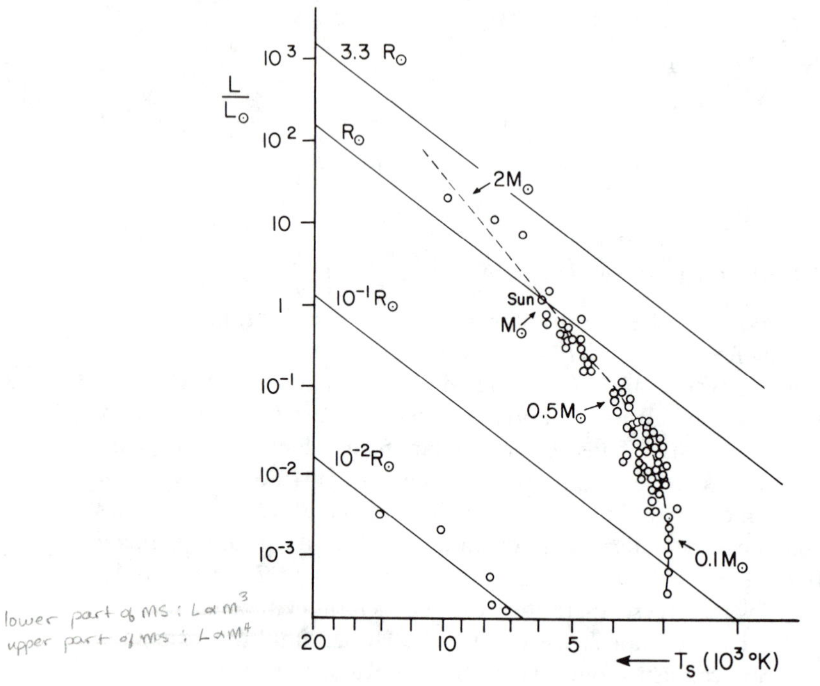

Fig. 2.33. H–R diagram of stars within 20 LY of the Sun.

In the upper region, changing L by a factor of 20 raises M by a factor of 2 so that

$$L \propto M^4 \left[\text{or } \frac{L}{L_\odot} = \left(\frac{M}{M_\odot} \right)^4 \right]. \tag{2.13}$$

Although these equations only hold for the masses of binary stars, we now make the bold assumption (first suggested by Arthur Eddington in 1924) that all stars obey these relationships including single stars whose mass cannot be directly determined. Thus a measurement of L for these isolated stars, which is comparatively easy, suffices to determine M. This expands our mass count by several orders of magnitude. We shall be able to understand the $L \propto M^3$ relation in terms of the ideal gas model for a star (see Sec. 3.3.3). Alternatively, for most main sequence stars, the ratio L/M ($\sim M^2$) varies comparatively little,

$$\frac{1}{20}\left(\frac{L}{M} \right)_\odot \lesssim \left(\frac{L}{M} \right)_{\text{star}} \lesssim 5\left(\frac{L}{M} \right)_\odot, \tag{2.14}$$

where for the Sun,

$$\left(\frac{L}{M} \right)_\odot = \frac{4 \times 10^{26}}{2 \times 10^{30}} = 2 \times 10^{-4}\ \text{J/kg}.$$

In Sec. 4.1.1 we shall see how this ratio enters into the dynamics of nuclear fusion in stellar cores. In that section we shall also uncover the reason for the main sequence curve.

It is important to be able to "read" one more piece of data into the H–R diagram. Since $L \sim R^2T^4$, given L and T, we can find R. Clearly there is a locus of points such that different Ls and different Ts produce the same R. This line can be superimposed on the H–R diagram for a few Rs so that we need not calculate R—just read it off the curve. What Rs do we choose? It is customary to pick fractions or multiples of $R_\odot$. A set of these parametric R lines are shown in Fig. 2.33.

When we add R information to the H–R diagram, we see that the main sequence stars within 20 LY of us run from roughly $\sim(1/10)R_\odot$ to $\sim 3\ R_\odot$. So, we have a wide range of luminosities, temperatures, radii, and masses, yet most stars fall in a regular—albeit peculiar—pattern. But not all stars lie along the main sequence. In the lower left-hand corner of Fig. 2.33 is a second group of stars whose luminosities are very low, whose radii are $\sim 10^{-2}\ R_\odot$, but whose surface temperatures are very high. So high in fact that they have a bluish color although they are known as *white dwarfs*. How did they get to this point on the H–R diagram? Did they evolve from the main sequence?

If one group of stars can deviate from the main sequence, we must see if there are others. One way to check and also look for evidence for evolution is to see how the H–R diagram of a young open cluster compares to that of an old globular cluster of stars.

The young Pleiades open cluster (found in Taurus and named after the seven daughters of Atlas who were transformed into six stars in the cluster while the invisible seventh conceals herself out of shame for having loved a mortal) seems to fall on the main sequence as shown in Fig. 2.34. No surprises here; the Pleiades and the stars in our neighborhood have similar H–R diagrams.

But when we plot the M3 globular cluster in Fig. 2.35, it is a different story. There are very few stars on the main sequence. Instead the stars stretch up and to the *right*—toward larger radii and lower surface temperatures. Since $\lambda_{\text{peak}} \approx 2.9 \times 10^7/T_s$, at $T_s = 4800$ °K, $\lambda_{\text{peak}} \approx 6000$ Å and we have a red star of 10–100 $R_\odot$—a red giant (using a wavelength detector).

The only apparent difference between M3 and the Pleiades is age. Did M3 at one time have a "young" H–R curve? And if so why did its stars evolve away from the main sequence and what evolutionary "track" on the H–R diagram did they follow? Will the Sun move away from the main sequence? The Pleiades are about 1/10th the age of the Sun yet both reside on the main sequence. When does a star begin to leave the main sequence?

M3 is not a unique globular cluster—all globular clusters have similar H–R diagrams. What is more, isolated halo stars—old stars roaming far

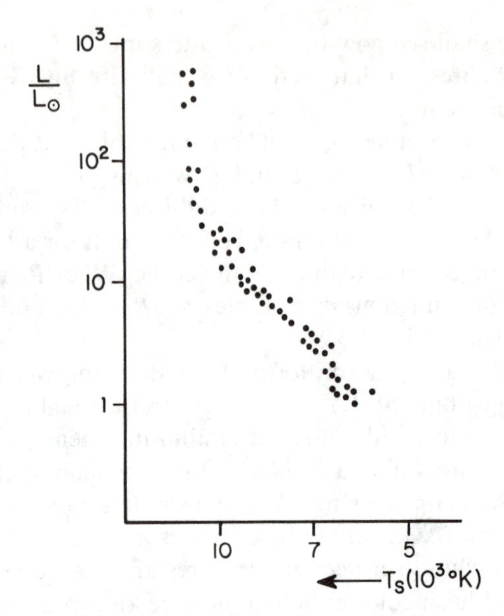

Fig. 2.34. H–R diagram of the Pleiades open cluster of stars.

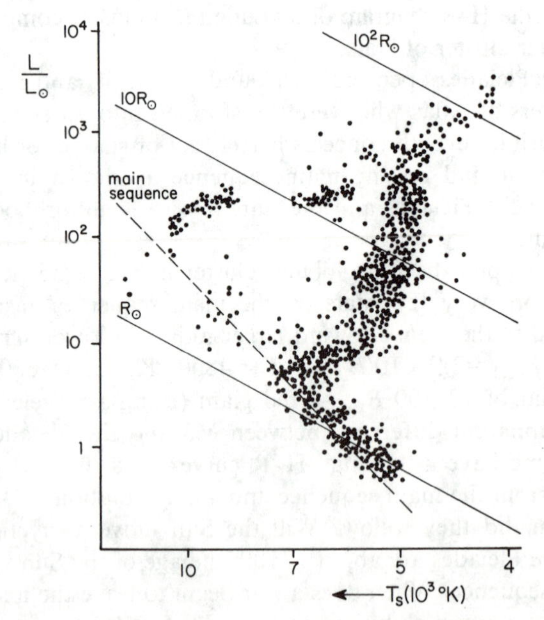

Fig. 2.35. H–R diagram of the M3 globular star cluster.

out of the galactic disk in huge elliptical orbits—have H–R distributions like M3. There seems no doubt about it, some evolutionary scheme is at work here. Because of the striking difference in their H–R diagrams, stars that fall in the Pleiades class of H–R diagrams are called population I stars while the M3 type distributions are called population II stars.

In Sec. 2.1 we mentioned the galactic core. At the center of the core is a compact region of stars that contains 0.1% of the galactic mass yet is only 150 LY across. The stars in this region also follow an M3 type H–R diagram which means they are very old. Similar nuclei, with similar H–R diagrams, are found in many other galaxies (maybe all), indicating that galaxies, unlike stars, may have not evolved or are not presently evolving, but were formed roughly at once.

Finally we must point out one more thing about stars on the H–R diagram and especially those on the main sequence. After a maximum luminosity there are simply no more stars. Expressing this in terms of mass we find there are few stars with $M \geqslant 60\ M_\odot$. Recall that the galactic clouds from which we hope stars evolve have the equivalent of hundreds of solar masses of material, but the H–R diagrams show few stars with M above $60\ M_\odot$.

So at this point we have many questions and few physical answers. In the next chapter we try to remedy this situation.

Problems

1. As we mentioned in the beginning of the chapter, most of the mass of our galaxy is contained in a core region.
   a. Show that the gravitational force on a star inside a core region of approximately constant density is given by

$$F = -\frac{4\pi}{3} G\rho m_{\text{star}} r,$$

   where r is the distance of the star from the center of the core.

b. Assuming the star moves in a circle of radius r, use the above expression to show that its rotational velocity is

$$v_{\text{star}} = \left(\frac{4\pi}{3} G\rho \right)^{1/2} r.$$

c. In Fig. 2.5c the velocity curve appears to peak up at a very small distance from the center before peaking again at about 25,000 LY. From what you have just learned, what can you say about the density of stars in this first region and the mass of the stars located there?

2. Discuss all the properties of the atom that account for 21-cm radiation.
3. Referring to Fig. 2.7b, show that if the current loop is tipped up or down, the spin vector of the electron will flip over. Having flipped over, show that the spin vector is now in a stable orientation with respect to **B**.
4. A particle whose interaction length is 10 cm is in a box 1 cm on a side. Is the probability of interactions between particles closer to 1 or closer to 0? Explain.
5. What would be the rate density for atoms in a box if the temperature of the box were 5000 °K and the number density of atoms were 10^4/cc?
6. What is the mean free path and interaction time of a star if its $v_{\text{rms}} \approx 20$ km/sec and the number density of stars is $0.1/(\text{pc})^3$? (This is typical of our region of the Galaxy.) Take $R = R_\odot$. Does it seem likely we are in for a close encounter with another star?
7. In some 21-cm radiating clouds there are $n \approx 10$ atoms/cc. If the cloud is a sphere of $R = 10$ LY, estimate the number of 21-cm photons emitted from the cloud each second. If the cloud is 1000 pc from the Earth, calculate the number of photons from the cloud that reach a 1 cm^2 counter.
8. For each picture below indicate whether the cloud (C) is receding, approaching, or stationary with respect to the Sun (S) and whether the cloud's 21-cm radiation is blue- or red-shifted.

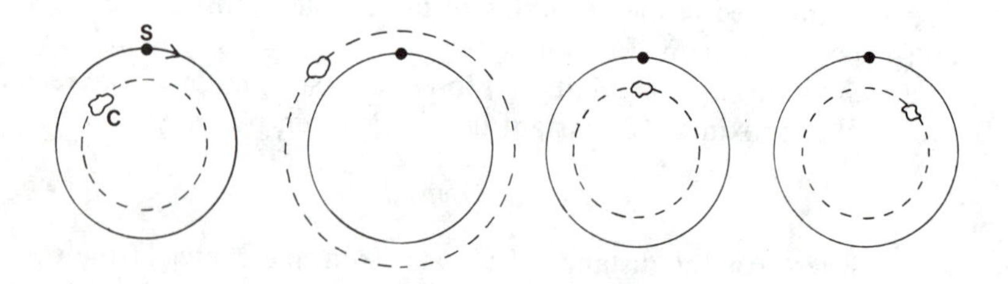

9. Assuming we point our 21-cm telescope in both directions along the line shown, what would the 21-cm spectrum look like?

10. Using the classical Bohr picture, what is the magnitude of the magnetic field at the proton due to an electron in the ground state of hydrogen? What magnetic field does the electron "see"?
11. Derive equation (2.5).
12. In an emission nebula the cloud temperature is always approximately 10,000 °K even though the surface temperature of the stars buried in the nebula may range up to 80,000 °K.
    a. What does this imply about the location of the 1st excited state vis-a-vis the ground state of the coolant in the emission nebula?
    b. Does your answer to (a) mean the energy levels are split by electrostatic forces or by magnetic forces?
    c. Can you suggest an element in the nebula that can act as the coolant?
13. White light is incident on a piece of *red* transparent plastic.
    a. What is the color of the scattered light? Explain.
    b. What is the color of the transmitted light? Explain.
14. Estimate the minimum grain density in an opaque globule that is 2 LY across.
    What is a reasonable estimate of the hydrogen density inside the globule?
    Estimate the mass of the globule.
15. Derive the expression $I=(m_1m_2/(m_1+m_2))r^2$ for a pair of atoms rotating about their center of mass, where I is the moment of inertia of the molecule and r is the separation between the atoms.
16. Find the approximate temperature of a black body radiator that can excite an appreciable number of rotational transitions.
17. In an external electric field the nucleus and the center of the electron cloud of the atom no longer coincide but are separated by a distance a; hence an induced dipole exits.
    a. Show that the effective charge the nucleus sees is given by

$$Q_{\text{eff}} = Ze(a/r_0)^3,$$

    where Z is the atomic number and r_0 is the radius of the atom.

b. Show that the dipole separation a and the external electric field E_{ext} are related according to

$$a = (r_0^3 / k_0 Z e) E_{ext}$$

where $k_0 = 9 \times 10^9$ N-m$^2/C^2$.

c. Evaluate a for a hydrogen atom in a very large electric field, say 10^{10} N/C. (Note for comparison the permanent dipole separation of HCl is $a = 0.2 \times 10^{-10}$ m.)

18. Given the electric dipole configuration indicated below, show that the electric field along the x axis falls off as r^{-3} for r much greater than the separation a_0 between the dipoles.

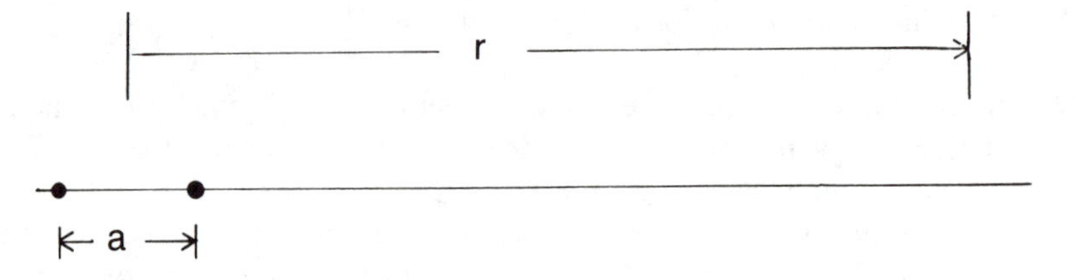

19. Let us use the results of problems 17 and 18 to obtain the results mentioned in the text, that the induced dipole—induced dipole force falls off like r^{-7}.

a. Show that the force between two dipoles lined up as shown below goes like b_0/r^4 for $r \gg b_0$.

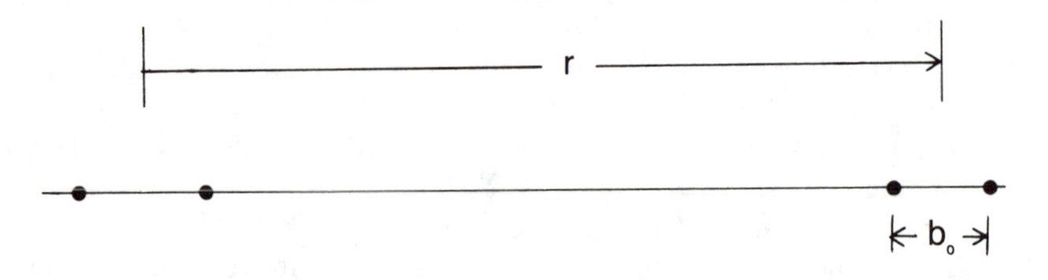

b. Now show that if b_0 is an induced separation, the force between the dipoles falls off as r^{-7}.

20. We mentioned in the text that the van der Waals' binding energy of a hydrogen atom to a grain is 10^{-1} eV. How many collisions with grain atoms will be necessary before a single kick can eject the hydrogen atom from the surface of the grain? Take $T_{grain} \sim 10$ °K.

21. In the 4-kpc arm, we see an "association" whose stars are *radially* separated by ~ 100 pc. How far apart will the stars be in 10^8 years? [Hint: since $\Delta r/r$ is small we can write $(1 + \Delta r/r)^{-1/2} = 1 - \frac{1}{2}(\Delta r/r)$].

Based on your previous answer, what conclusion can you draw about the age of stars we saw in the "association?"

22. What would the temperature of one of Bok's small globules have to be in order that it be gravitationally bound?
23. Derive a single expression for the mass m of a binary pair (m, M) in terms of d_1 and d_2, the distance of m and M from their center of mass, respectively. Assume the orbits about the cm are circular.
24. Show that the period of a small binary star moving in a circle about a much more massive one is proportional to $d^{3/2}$, where d is the separation of the stars.
25. We suspect a binary pair is 200 LY away. Our telescope can resolve angles as small as 0.005″ of arc.
    a. How far apart would the stars have to be in order for us to visually detect them?
    b. What would be their minimum period of rotation? Take $M \sim 1\ M_{\odot}$.
26. In chapter 1, problem 14, the luminosity was calculated for a star that was 300 LY away.
    a. Does that star lie on the main sequence?
    b. What is the approximate mass of the star?
27. If the parallax of a main sequence star is in error by 25%, how far will this star be displaced from the main sequence in an H–R diagram? Use an illustration.
28. Use Fig. 2.33 to determine the mass density ρ of Barnard's star, whose luminosity is $\sim 1/200$ that of the Sun's.

CHAPTER 3

Stellar Evolution I: On the Road to the Main Sequence

Come, let us on the sea-shore stand
And wonder at a grain of sand;
And then into the meadow pass
And marvel at a blade of grass;
Or cast our vision high and far
And thrill with wonder at a star;

Robert Service, *The Wonderer*

3.1. Solving the Initial Problems of Stellar Collapse

The evidence for an evolving stellar sequence seems overwhelming: H–R diagrams, distributed hydrogen, large gas and dust clouds, small gas and dust clouds, etc. But the large clouds are not gravitationally bound and there appears to be too much initial angular momentum. We shall now turn to these latter problems and see how we might solve them.

3.1.1. First Gravity ....

Let us start by looking at the problem of $E > 0$. Since it was energy conservation that got us into this bind, it will have to be energy conservation that gets us out of it. With $E = \frac{1}{2} M_{\mathrm{cl}} v_{\mathrm{rms}}^2 - \frac{3}{5} GM_{\mathrm{cl}}^2/R$, the problem is that the second term is just not large enough. It turns out there are two ways for the $-|\mathrm{PE}|$ term to win out.

3.1.2 The Big Squeeze—Spiral-Density Waves and Supernova Explosions

If we can keep M_{cl} and v_{rms} constant and if we can reduce R, then the PE term will grow in magnitude while the KE term remains the same. But the squeeze cannot be gravitational; it must come from some external source. It is now believed that this squeeze is related to the structure of the spiral arms and to supernova explosions.

We saw earlier that if the spiral arms were always composed of the same stars, then the arms would wind up on themselves and disappear. Clearly then the spiral arms do not always contain the same material. But if they do not always contain the same material, then what are they? Two possible models for galactic arms have been proposed: Spiral-Density Waves and Supernova Percolation. We shall briefly discuss each model in turn.

Spiral-density wave theory, due to C. C. Lin, corresponds to a piling up of the galactic gas due to a slow down of the gas particles. This localized compression then propagates through the Galaxy like a water wave moves in a large body of water. Such a spiral density wave is depicted in Fig. 3.1, where the gas, dust, and stars in the Galaxy are moving faster than the arms. The slower-moving density wave forces the first gas and dust particles that catch the arms to slow down a bit. The gas just behind continues at its original speed. It thus runs into the leading particles, causing a net compressive effect—a pileup of the gas (and dust) into higher density states

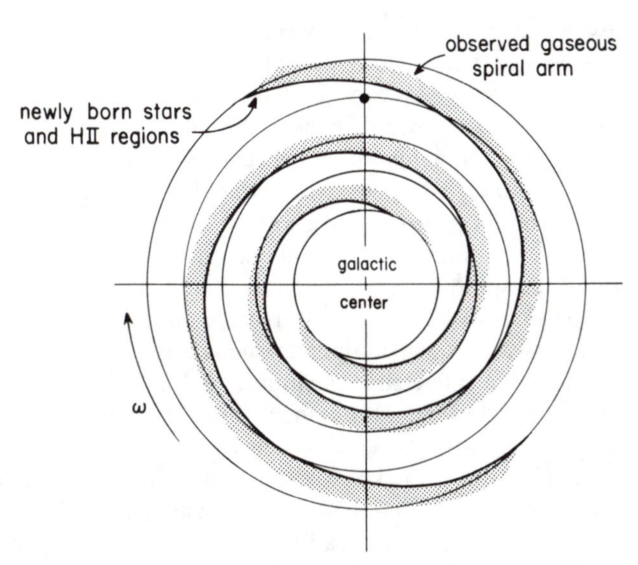

Fig. 3.1. Spiral arms of the Galaxy showing the distribution of newly born stars and HII regions.

—thus feeding the density wave. As long as the incident gas is not uniformly distributed (and it cannot be perfectly uniform since there must be some statistical fluctuations), the compression can lead to,

(a) compression of the $n = 0.1/\text{cc}$ interarm material into $n = 1/\text{cc}$ clouds that are unbound, i.e., $E > 0$;
(b) compression of $n = 0.1/\text{cc}$ interarm material into $n = 1/\text{cc}$ clouds that are bound i.e., $E < 0$; and
(c) compression of the interarm dust into dark clouds where $E < 0$.

In other words, on entering the density wave, the compression of matter could trigger the kinds of collapse that lead to the formation of stars by reducing R such that $\text{KE}_{\text{tot}} < |\text{PE}_{\text{tot}}|$. What is impressive about the observed structure of the spiral arms is that emission nebulae, associations, and dark clouds, are all on the *inner* edge of the arms, where we would expect them to be if they were formed by catching up to the density wave arm, while the unbound $n = 1/\text{cc} = 10^6/\text{m}^3$ clouds are distributed throughout the arms, a result of their inertia carrying them right on through the density wave region. Indeed, eventually all of the gas, dust, and stars continue on through and out of the density wave. The clouds will dissipate and regain their old velocities while the long-lived stars will live out their days in the region between the arms.

The density wave theory raises, however, a couple of sticky questions that as yet have not been answered: (1) How did the density waves originally form? (2) Why haven't they dissipated over the life of the Galaxy? As an aside—some people believe the two extra arms in the outer part of the Galaxy are remnants that separated from a density wave. If this is so, then in a few times 10^8 years these material arms should disappear—unfortunately we cannot wait that long to find out.

There is evidence, however, that stars may form due to a mechanism very similar to that of the spiral-density wave but where the origin of the wave is a supernova explosion. In 1977, young stars were observed on the periphery of an expanding emission nebula. Since the nebula is composed of material that is denser than the interstellar medium, the formation of stars on its leading edge is probably due to a mechanism similar to the possible formation of bound clouds by the spiral-density waves. Here, however, the origin of the "density wave" is known; the expanding nebula is due to a supernova explosion (a topic we shall cover in chapter 5). This particular nebula is 180 LY across and is expanding at the rate of 32 km/sec. An upper limit to the age of the nebula then is

$$\Delta t \sim \frac{90 \times 10^{16}\ \text{m}}{32 \times 10^3\ \text{m/sec}} \sim 3 \times 10^{13}\ \text{sec} \sim 10^6\ \text{yr}.$$

As we shall see at the end of this chapter, this is in good agreement with the expected age of a young star.

A star that is formed by such a supernova explosion may in turn become a supernova and lead to further star formation. Gerola, Seiden, and Schulman have recently shown that such supernova explosions—star formation-supernova explosions, etc.—can in fact "percolate" throughout the Galaxy and not only cause the observed spiral arms but regenerate them as well. This theory assumes that star "ignition" is set off in a *random* manner by a possible supernova explosion of one of the potential star's nearest neighbors, and the theory finds that the probability of star production adjusts itself to lie within a narrow range of values or else the Galaxy will either undergo runaway star formation or cease star production altogether. Such a supernova-percolation theory critically depends upon the differential rotation of the gas in the galactic disk. That is, we must have $\omega = v/r \neq$ constant. (In our galaxy, beyond the galactic core, ω decreases as r increases.) This differential rotation influences the *rate* of star formation because new matter is continually being brought into contact with the supernova region allowing this active region to affect more areas. Indeed, in this model, the faster the rotation of the Galaxy the higher will be the rate of star production. Thus, rapidly rotating Galaxies would be expected to have more old, red, stars than their more slowly rotating counterparts. This appears to be the case.

To contrast the two theories of star formation: For a spiral density wave, the shock wave is created uniformly and simultaneously over the whole wave front by rotating rigid density wave patterns. But in a percolation theory, supernova explosions produce randomly generated shock waves. In either scheme the consequent shock wave provides the "big squeeze" which forces stellar matter to contract into gravitationally bound systems.

3.1.3. Accretion of Matter

There are many astronomers who will accept the density wave hypothesis as being the likely source for compressing the $n \approx 0.1/\text{cc}$ clouds down to $E < 0$ size but who doubt whether this mechanism can do the same for small- and medium-sized dust clouds. There is another way, however, that we can achieve gravitational collapse and which will help us understand how grains are formed. This is accretion of matter.

Accretion of matter is just a fancy way of saying, "picking up mass." Notice that in the energy equation, the KE term goes as M_{cl} while the PE term goes as M_{cl}^2. Thus if we can add mass, the cloud will eventually be able to overcome the effects of excess KE. Of course, if we add mass the cloud grows in size, which means R increases, $1/R$ decreases and this latter decrease in $1/R$ decreases the $|\text{PE}|$. This hurts our quest. That adding mass helps overall is best seen as follows: The mass of the cloud can be written as $M_{\rm cl} = \rho 4\pi R^3/3$. Plugging this into the energy equation we get

$$E = \tfrac{1}{2}\rho \frac{4\pi}{3} R^3 v_{\rm rms}^2 - \tfrac{3}{5} G \left(\frac{4\pi}{3}\right)^2 \rho^2 \left(\frac{R^6}{R}\right).$$

If $\rho =$ constant, the PE goes as R^5 and the KE as R^3 so that eventually we have to win, i.e., the |PE| is bound to become greater than the |KE|. The problem is to determine some sort of time scale. If the time to accrete enough matter is, say, greater than the age of the universe, then we are in deep trouble since we would not have had enough time to form a system with $E < 0$. So let us see if we can develop a simple expression for the time it takes to accrete a given amount of matter onto a cloud that has already formed, say due to the spiral-density wave.

Consider then a gas or dust cloud of density n_1 moving through a medium of density n_2. Then the rate density of interactions per cc per second is given by (2.4)

$$\mathcal{R} = n_1 n_2 \sigma v, \tag{2.4}$$

where σ is the cross section for the interaction between constituents, and v their relative velocity. The rate at which the *cloud* immersed in the gas picks up particles is $\mathcal{R}V$, where V is the volume of the cloud,

$$\mathcal{R}V = (n_1 V) n_2 \sigma v (\text{interactions/sec}).$$

But $n_1 V$ is the number of particles in the cloud and since all have the same cross section σ, we see that $n_1 V\sigma$ is just the total cross section of the cloud, i.e., its geometrical area,

$$n_1 V \sigma = \pi R^2.$$

Assuming the cloud is picking up hydrogen, certainly a fair assumption, the *rate* at which mass is added to the cloud is

$$\frac{\Delta M}{\Delta t} = \mathcal{R} V m_{\rm H} = \pi R^2 m_{\rm H} n_2 v (\text{kg/sec}).$$

For a large globular cluster, moving *through* the $n = 1/\text{cc} = 10^6/\text{m}^3$ region, we have $R \sim 3$ LY, $M \sim 60\ M_\odot$ and a relative velocity of around 10 km/sec (from Doppler-shifted absorption lines), giving

$$\frac{\Delta M}{\Delta t} \approx \pi (3 \times 10^{16}\ \text{m})^2 \times 1.7 \times 10^{-27}\ \text{kg} \times 10^6/\text{m}^3 \times 10^4\ \text{m/sec}$$

$$\approx 5 \times 10^{16}\ \text{kg/sec} \approx 7 \times 10^{-7} (M_\odot/\text{yr}).$$

The time it takes for the mass to double (i.e., $\Delta M = M_f - M_i = 2M_i - M_i = M_i$) is

$$\frac{\Delta M}{\Delta t} = \frac{60 M_\odot}{\Delta t} \approx 7 \times 10^{-7} M_\odot/\text{yr}$$

$$\Delta t \approx \frac{60 M_\odot}{7 \times 10^{-7} M_\odot} \text{yr} \approx 10^8 \text{ yr}.$$

Now what does doubling the mass in 10^8 years accomplish?

(a) Doubling the mass increases R by $2^{1/3} \sim 1.26$ because $M \propto R^3$ implies, if ρ remains constant,

$$\frac{M_1}{M_2} = \left(\frac{R_1}{R_2}\right)^3 = 2; \qquad \frac{R_1}{R_2} = 2^{1/3} = 1.26.$$

Thus in 10^8 yr we increase R by 26%.

(b) This means that in this time the KE doubles ($\text{KE} \propto R^3 \propto (1.26)^3 = 2$).

(c) But then the PE goes up by a factor of 3 ($\text{PE} \propto R^5 \propto (1.26)^5 \propto 3$).

Because the |KE| and |PE| increase by different amounts, each 10^8-yr period will see the |PE| gain on the |KE| by a factor of $\frac{3}{2}$. In the case of a large globule, KE $\approx$ 3 PE (see Sec. 2.5.2) so that the PE must pick up a factor of 3 in order for the cloud to be bound. To pick up a factor of 3 will take $X - 10^8$-yr periods where X is found from

$$\left(\tfrac{3}{2}\right)^X = 3.$$

For this case, X happens to come out around 3 so that it takes 3×10^8 yr to accrete enough matter to bind the original large globule. Actually this time is an upper limit for two reasons. One, ρ will not remain constant but will increase and this will lower the cloud's temperature as C^+ collisions become more prevalent. Secondly, as R grows, the time to double the mass becomes less and less. Thus, compared to the age of our universe or the time it takes for the Galaxy to revolve once, 3×10^8 yr, we have plenty of time to push $E < 0$.

3.1.4. The Formation of Interstellar Grains

Using the accretion model we can investigate how interstellar grains might be formed. If we assume a grain grows by picking up heavy atoms as the grain moves through a cloud, then

$$\frac{\Delta M}{\Delta t} = \pi d^2 m_A n_A v,$$

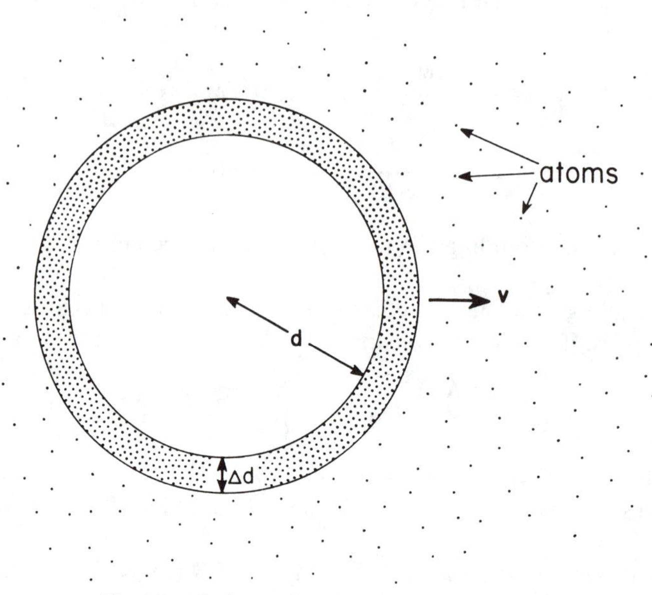

Fig. 3.2. Grain moving through a cloud of atoms.

where d is the radius of the grain, m_A the mass of the heavy atoms, n_A the number density of heavy atoms, and v the relative velocity. If the grain is already a few atoms across, then, as shown in Fig. 3.2, subsequent atoms will be added in a thin spherical shell so that

$$\Delta M = 4\pi d^2 (\Delta d) \rho_g,$$

where Δd is the thickness of the shell and ρ_g the density of the atoms after they are on the grain. Putting both of the above expressions together we can find the rate at which the radius of the grain is increasing,

$$4\pi d^2 \frac{\Delta d}{\Delta t} \rho_g = \pi d^2 m_A n_A v$$

$$\frac{\Delta d}{\Delta t} = \frac{m_A n_A v}{4\rho_g} .$$

Because the numbers on the right are all constants for a given environment (recall $\frac{1}{2} m_A v^2 \sim kT$), the growth rate $(\Delta d / \Delta t)$ is constant. Since we know the final size of the grain is 5×10^{-7} m, the growth time is given by

$$\Delta t_g = \frac{4\rho_g \Delta d}{m_A n_A v} ,$$

where $\Delta d = 5 \times 10^{-7}$ m. For the carbon or silicon grains we have been talking about, we might expect $m_A = 20\ m_H$ and $\rho \simeq 3$ gm/cc $= 3 \times 10^3$ kg/m^3. Thus, in MKS units we find,

$$\Delta t_g \approx \frac{4 \times 3 \times 10^3 \times 5 \times 10^{-7}}{20 \times 1.7 \times 10^{-27} \times n_A v} \approx \frac{2 \times 10^{23}}{n_A v} \text{ sec.}$$

Clearly a great deal will depend upon the values of n_A and the temperature of the region. For example, in a 21-cm cloud, the temperature is 100 °K and therefore the velocity of one of the *heavy atoms* ($m_A \approx 20\ m_H$) will be about 300 m/sec. Since the density of hydrogen is about 10/cc, the abundance ratios tell us that $n_A \sim 10^{-3}$/cc $= 10^3$/m^3. Thus, we find

$$\Delta t_g \approx \frac{2 \times 10^{23}}{10^3 \times 300} \approx 7 \times 10^{17} \text{ sec} \sim 2 \times 10^{10} \text{ yr.}$$

This is much too long, comparable in fact to the age of the universe. In molecular clouds, however, like those near the Orion nebula, n_H may reach 10^5 atoms/cc which decreases Δt_g to 10^6 yr, a quite reasonable number.

3.1.5. ... Then Angular Momentum...

Recall from Sec. 2.5.2.2 that angular-momentum conservation for a closed system of a condensing cloud contracting to the size of a star requires the final rim velocity of a typical dust cloud to be greater than c. Alternatively, inputting the observed rotational velocity of our sun leads to the angular momentum ratio $L_i / L_f \approx 10^5$.

3.1.6. Shedding Matter and Angular Momentum

Let us look a little closer at a rotating nebula. Each part rotates at the same angular velocity ω, in a circle about the spin axis. This means that a piece of the cloud m_x at the nebular equator must feel a force

$$F_{\text{net}} = m_x \left(\frac{v^2}{R} \right),$$

where R is the radius of the nebula and v the speed of m_x as it moves about the spin axis. This equation tells us what the magnitude of the net force must be in order that this segment of the cloud goes in a circle. The *real* force is due to the gravitational attraction between m_x and the mass of the cloud M that lies below it, i.e., $Gm_x M / R^2$. As long as $Gm_x M / R^2 = m_x v^2 / R$, m_x will continue to move in a circle and retain its relative position in the cloud even as the cloud collapses. But as the cloud contracts, conservation of angular momentum demands that v increases. In fact with $L_{cl} = I\omega$ and with the cloud rotating as a rigid body so that $v = R\omega$, we have

$$L_{cl} = I\omega = \tfrac{2}{5}MR^2\left(\frac{v}{R}\right),$$

$$v \propto L_{cl}/R.$$

Since L_{cl} is a constant (there are no external torques), we see that $v \propto 1/R$. Thus the net force needed to keep m_X going in a circle behaves as $mv^2/R \propto 1/R^3$. The gravitational force, however, only goes as $1/R^2$. Therefore the force *needed* to keep the mass moving in a circle increases faster than the force nature can supply. Eventually the cloud reaches a radius where gravity is not strong enough to maintain m_x in a circular orbit; the piece drifts out and is lost to the cloud. Its "memory" of the cloud is recorded in its equilibrium radius R_0, where $mv^2/R_0 = GmM/R_0^2$, or $R_0 = GM/v^2$. While all of this is happening, angular momentum is still conserved but now the system is collectively the cloud, which will continue to collapse, and the piece that has broken away which will circle at radius R_0. Each of them separately will have angular momentum with a sum equal to L_{cl}. Since the angular momentum of the breakaway piece is $L_x = m_x v R_0$, the total angular momentum of the system is $L_{cl} = L_{cl,f} + L_x$ where $L_{cl,f}$ is the angular momentum of the cloud after the piece has broken away. Solving for $L_{cl,f} = L_{cl} - L_x$ we see that the angular momentum of the cloud will decrease as more and more matter breaks away.

Now this theme has many variations: several pieces may break away, or one large piece, or lots of small ones. These are shown in Fig. 3.3 as planets, binary stars and a disk of matter. In each case the total angular momentum is shared, thereby reducing that of the condensing nebula. The fact that over 50% of all stars are binary systems indicates the efficacy of getting rid of angular momentum this way. Indeed, a recent study at Kitt Peak of 123 comparatively nearby stars revealed that 100% had companions of one kind or another! (80% had stellar companions and 20% had only planets—like the Sun.) But do we shed enough angular momentum this way? As the table in Fig. 3.3 shows, the angular momentum of the Sun is a mere 2% of the solar system's total angular momentum. Most of the angular momentum resides in the orbital motion of the planets. In producing the planets, the Sun reduced its original L_{cl} by a factor of 50. Still a factor of 50 loss in angular momentum is only a small part of the $L_f/L_i \sim 10^{-5}$ reduction that we noted in Sec. 2.5. was necessary. Very possibly a disk of matter was thrown off too, and was eventually blown away by the radiation from the newborn Sun.

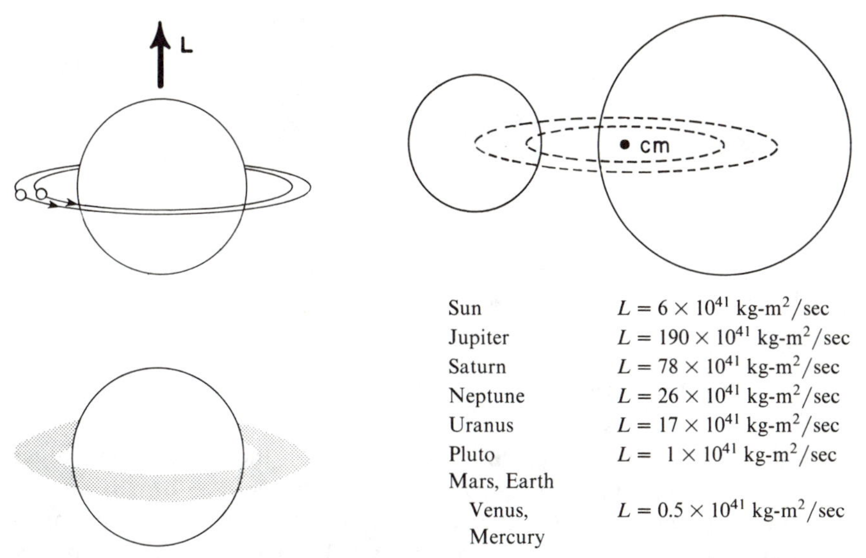

Sun	$L = 6 \times 10^{41}$ kg-m^2/sec
Jupiter	$L = 190 \times 10^{41}$ kg-m^2/sec
Saturn	$L = 78 \times 10^{41}$ kg-m^2/sec
Neptune	$L = 26 \times 10^{41}$ kg-m^2/sec
Uranus	$L = 17 \times 10^{41}$ kg-m^2/sec
Pluto	$L = 1 \times 10^{41}$ kg-m^2/sec
Mars, Earth Venus, Mercury	$L = 0.5 \times 10^{41}$ kg-m^2/sec

Fig. 3.3. Three means of shedding angular momentum: planets, binary stars, disk of matter. The table indicates the angular momentum of the Sun and planets about the solar axis.

3.1.7. Stellar Magnetic Fields

Since planets alone are insufficient to account for all the lost angular momentum, and nonexistent massive disks are a bit speculative, another method of shedding a cloud's angular momentum has been postulated—magnetic drag. Besides introducing an interesting method for losing angular momentum, magnetic drag involves several ideas that we shall meet again when we study neutron stars in chapter 5.

First of all we should ask where the magnetic field of a star comes from. Consider a spinning cloud that is undergoing collapse. What is crucial is that the cloud contains ions. We also need to mix in some galactic magnetic field. Then as the ions are gravitationally attracted inward they feel a *tangential* component of magnetic force $\mathbf{F}_m = e\mathbf{v}_r \times \mathbf{B}_{\text{gal}}$ where $\mathbf{v}_r$ is their radial velocity component and $\mathbf{B}_{\text{gal}}$ the galactic magnetic field. The direction of the force in relation to $\mathbf{v}_r$ is shown in Fig. 3.4. This component of the magnetic force increases the tangential velocity of the ions if the clouds happens to be spinning in the direction shown in Fig. 3.4. The electrons however, experience a force in the opposite direction and hence slow up. Now prior to the application of $\mathbf{B}_{\text{gal}}$ the ions and electrons were moving in tandem with the result that their motion produced no net current. But with their tangential velocities no longer the same, a net current, i_{net}, appears

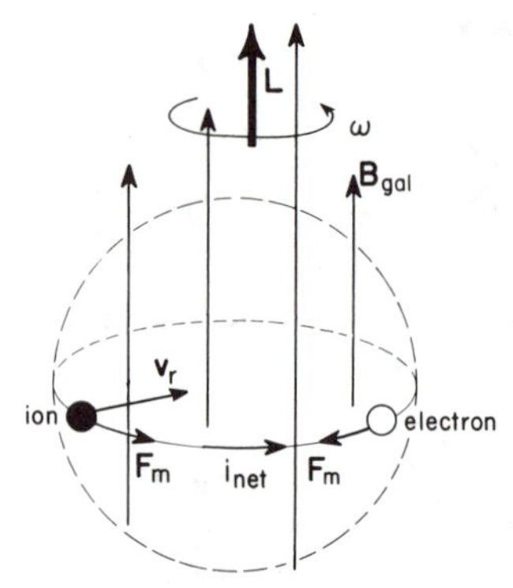

Fig. 3.4. Magnetic forces on charged particles in a rotating cloud.

creating a new magnetic field $\mathbf{B}'$ superimposed on the galactic field. Since $\mathbf{B}'$ is in the same direction as $\mathbf{B}_{gal}$, we have $\mathbf{B}_{net} = \mathbf{B}' + \mathbf{B}_{gal}$. (Can you convince yourself that the direction in which the nebula is spinning is unimportant to the above outcome?) As $\mathbf{B}_{net}$ increases, $\mathbf{F}_m$ increases. Furthermore $\mathbf{F}_m$ also increases because v_r speeds up as the collapse proceeds. The larger magnetic force, $\mathbf{F}_m = e\mathbf{v}_r \times \mathbf{B}_{net}$, drives the ions even faster, producing a larger $\mathbf{B}'$ and thus a larger $\mathbf{B}_{net}$.

The whole process feeds on itself and soon we have a very large $\mathbf{B}_{net}$ which becomes the star's dipole magnetic field. How strong will the final magnetic field be? We can estimate this from the concept of magnetic flux. Since flux is conserved, we have

$$\Phi_i = \Phi_f$$

$$B_i \pi R_i^2 = B_f \pi R_f^2$$

$$B_f = B_i \left(\frac{R_i^2}{R_f^2} \right). \tag{3.1}$$

Thus a collapsing ionized cloud will freeze in the initial field and end up with a much larger dipole field. We believe the *dipole* field of the Sun, with an average surface strength of ~ 1 gauss, was formed this way. (We emphasize dipole because the Sun has intense local fields of over 1000 gauss whose origins are quite different.) If we try to estimate the value of this final field, however, from the initial conditions in the nebula, we run into a bit of a problem. As we shall show in the next section, the cloud from which the Sun was formed started its collapse with $R_i \sim 10^5 \ R_\odot$. The necessary initial ionization was produced by high energy cosmic ray protons. Using the average galactic field of 10^{-6} gauss, we see that (3.1) predicts $B_f = 10^4$ gauss, clearly an unacceptably large value. This is the problem of too much magnetic field we spoke of earlier in Sec. 2.5.2.3. Of course our local value of B_{gal} may have been smaller or the degree of cosmic ray ionization at some stage may have become inadequate. This is still an unresolved problem. Nevertheless the general method of formation of the stellar *dipole* field should be basically as described above.

3.1.8. Magnetic Drag

So, the final field will be a dipole field with the N-S axis along the local direction of $\mathbf{B}_{gal}$. Now an important point: If the cloud has a spin $\mathbf{L}_S$, then $\mathbf{B}_f$ can be in any direction with respect to $\mathbf{L}_S$. All that matters is the original random orientation of $\mathbf{B}_{gal}$ relative to $\mathbf{L}_S$. Two cases are shown in Fig. 3.5. There is a crucial difference between case a, where $\mathbf{L}_S$ and the **B** dipole align and all others (represented by case b), where $\mathbf{L}_S$ and the **B** dipole do not align. In case a, the current *loop* (not the ions) does not move. Thus **B** is steady in all regions of space (see final star blowup in Fig. 3.5a). In all other cases, the current loop must turn since its plane is not perpendicular to $\mathbf{L}_S$. This means that at any point in space, **B** must change with time (see final star blowup in Fig. 3.5b). In effect, **B** is now rotating in space. Thus when $\mathbf{L}_S$ and the **B** dipole do not align, we have what the astrophysicists call a "*corotating magnetic field.*"

Consider then a corotating magnetic field passing through an ionized medium that is initially at rest in the ISM (see Fig. 3.6). There will thus be a relative velocity between the field lines and the ions. Recall from basic physics that it does not matter whether you yank a bar magnet through a stationary loop of wire or pull the loop of wire past a stationary bar magnet. In either case there will be a force on the electrons in the wire, as an ammeter will clearly indicate. The force will be the usual $e\mathbf{v} \times \mathbf{B}$ where $\mathbf{v}$ is the *relative* velocity between the wire and the bar magnet. To find the proper direction of the force, just get in a reference frame moving with the field so that the field appears to be at rest. You can then take the cross product $\mathbf{v} \times \mathbf{B}$ using the usual right-hand rule and you will have the correct direction of the force. For example, in Fig. 3.6 we see that the relative velocity of the ions as referenced to **B** is *out of the paper*, because to an inertial observer the corotating field at the location of the ions is moving *into the paper* (remember, initially the ions were at rest with respect to space). If the ion charge is (+), then the force on it, given by $e\mathbf{v}_{ion} \times \mathbf{B}$ is radially outward, producing an outward component of velocity $\mathbf{v}_0$ (as shown in Fig. 3.6). As this very real current flows outward, it interacts with **B** to produce a second force $\mathbf{F}'$ on the ions where $\mathbf{F}' = e\mathbf{v}_0 \times \mathbf{B}$. For the conditions shown in Fig. 3.6, the right-hand rule shows $\mathbf{F}'$ to be in the direction of the rotating cloud. The ions thus pick up a component of

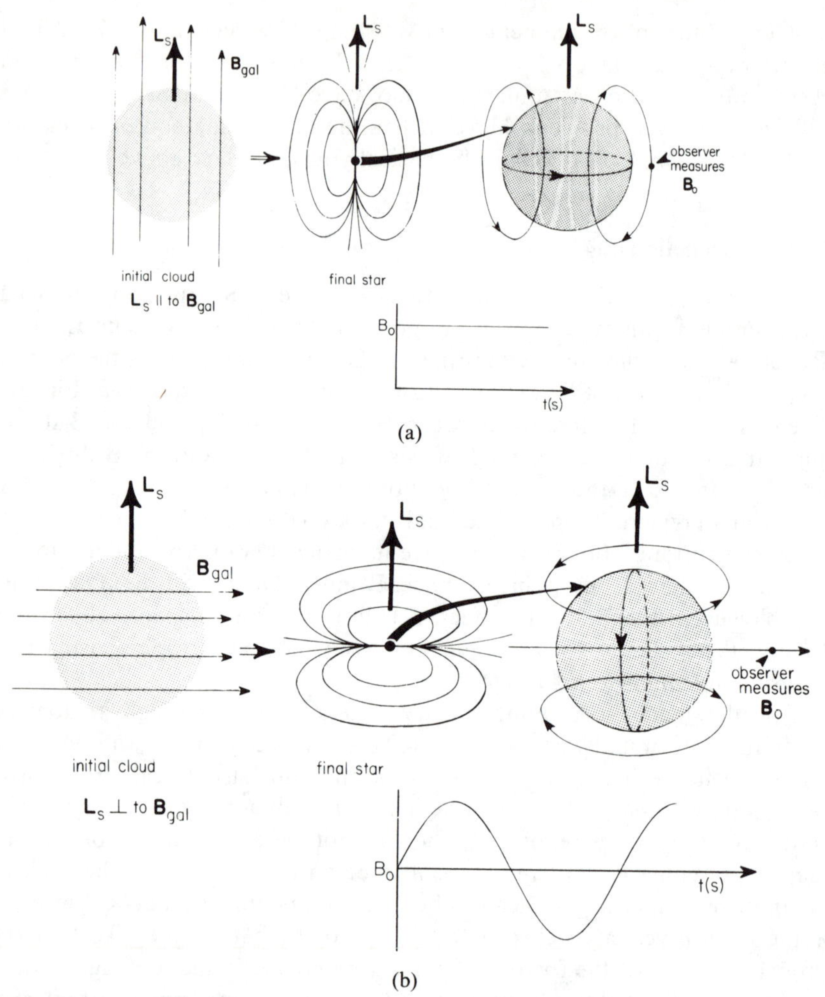

Fig. 3.5. Orientation of angular momentum and galactic-induced stellar magnetic field for (a) $\mathbf{L}_S$ parallel to $\mathbf{B}_{\text{gal}}$; (b) $\mathbf{L}_S$ perpendicular to $\mathbf{B}_{\text{gal}}$. Also noted is the stellar magnetic field $\mathbf{B}_0$ vs. time for the two cases.

acceleration in the direction the cloud is spinning and therefore gain angular momentum $\mathbf{L}_{\text{ion}}$ in the same general direction as the spin vector $\mathbf{L}_S$. The corotating field therefore drags the ions around with it, hence the name magnetic drag.

We can now compare the angular momentum of this object before and after magnetic drag has taken place:

$$\text{Before drag:} \qquad \mathbf{L}_S = \mathbf{L}_{\text{cl},i}$$

$$\text{After drag:} \qquad \mathbf{L}_S = \mathbf{L}_{\text{cl},f} + \mathbf{L}_{\text{ions}}$$

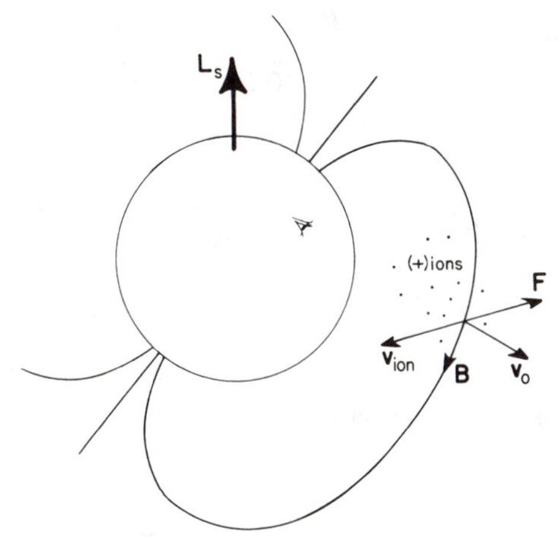

Fig. 3.6. Observer on spinning object sees ions moving out of the paper with velocity v_{ion}.

Since all torques are internal to the ion + cloud system, $\mathbf{L}_S$ is conserved and thus

$$\text{After drag:} \qquad \mathbf{L}_{\text{cl},f} = \mathbf{L}_{\text{cl},i} - \mathbf{L}_{\text{ions}}.$$

The cloud or star is thus effectively braked by magnetic drag. The above arguments are for (+) ions. You should be able to show that (−) charge is also dragged in the spin direction.

(In this era of energy crises, it is interesting to note that the Earth's dipole field does not line up with its spin axis. Thus $\mathbf{B}_{\text{earth}}$ corotates. An entrepreneur of vast imagination has suggested placing a huge copper loop around the Earth. The rotating **B** will then produce a globe-girdling current which we can tap. We will thus have an infinite source of energy! The plan has one serious drawback, however. Can you think of what it is?)

3.1.9. ...And Finally Collapse

So either by compression or accretion, we reach a point where the cloud is ready to collapse and either by shedding material or magnetic drag (or both), its angular momentum will eventually be brought under control. What can we say about the cloud just as it is poised to collapse?

First of all we know its initial radius. At the point of collapse, E_{cl} has just become zero so that

$$E_{\text{cl}} = 0 = \tfrac{1}{2} M_{\text{cl}} v_{\text{rms}}^2 - \tfrac{3}{5} \frac{GM_{\text{cl}}}{R_i}$$

and therefore

$$R_i = \tfrac{6}{5}\,\frac{GM_{\rm cl}}{v_{\rm rms}^2}\,.$$

If we take the cloud to be the progenitor of the Sun, then $M_{\rm cl} \sim 2 \times 10^{30}$ kg and $T_{\rm cl} \sim 50$ °K. Therefore, $v_{\rm rms} \sim 10^3$ m/sec which leads to

$$R_i \approx \tfrac{6}{5}\,\frac{6.7 \times 10^{-11} \times 2 \times 10^{30}}{10^6} \sim 1.4 \times 10^{14}\ \text{m} \sim 2 \times 10^5 R_\odot .$$

This initial radius is only ~ 10 times the radius of our solar system and seems quite reasonable for the onset of stellar collapse. The number density of hydrogen will be

$$n_i = \frac{\rho}{m_{\rm H}} = \frac{M_{\rm cl}}{m_{\rm H}(4\pi R_i^3/3)} \approx \frac{2 \times 10^{30}\ \text{kg}}{(1.7 \times 10^{-27}\ \text{kg})(4\pi/3)(1.4 \times 10^{14})^3}$$

$$\approx 10^{14}/\text{m}^3 = 10^8/\text{cc}.$$

Likewise the initial luminosity will be

$$L_i = 4\pi R_i^2 \sigma T_i^4 \approx 4\pi\,(1.4\times10^{14})^2(5.7\times10^{-8})(50)^4 \sim 10^2 L_\odot .$$

Although it may seem surprising that the initial luminosity of the cloud is greater than the Sun's, you must remember the cloud is much bigger. Also with $T_s = 50$ °K, the radiation will be in the infrared and not seen.

The initial position of the cloud on the H–R diagram is shown in Fig. 3.7. It clearly has a long way to go before it reaches the present position of the Sun.

Let us see if we can figure out what the initial path of the cloud will be. In order to increase T_s, we must somehow increase $v_{\rm rms}$. But $v_{\rm rms}$ is a measure of the random motion of the particles. In free fall, where by definition the particles experience no other interactions but gravity, the particles can only increase their translational velocities and hence $v_{\rm rms}$ must remain constant. Thus initially T_s, which is proportional to $v_{\rm rms}^2$, will remain the same and the cloud's path on the H–R diagram will be straight down. This state of affairs will continue until the cloud's density gets so high that other, nongravitational interactions will take place.

We can get some idea of how rapid the collapse is if we define the "free-fall time," $t_{\rm ff}$, as the time it takes the cloud's initial radius to decrease by 10%, i.e., $\Delta R = (1/10)R_i$. Then an estimate of the free fall time is given by

$$\Delta R = \tfrac{1}{2} a t_{\rm ff}^2 = \tfrac{1}{10} R_i,$$

where we have used the expression $\Delta x = \Delta R = v_{0i} + \frac{1}{2}at^2$ which is only true for a *constant* acceleration. We assume that over 10% of the initial

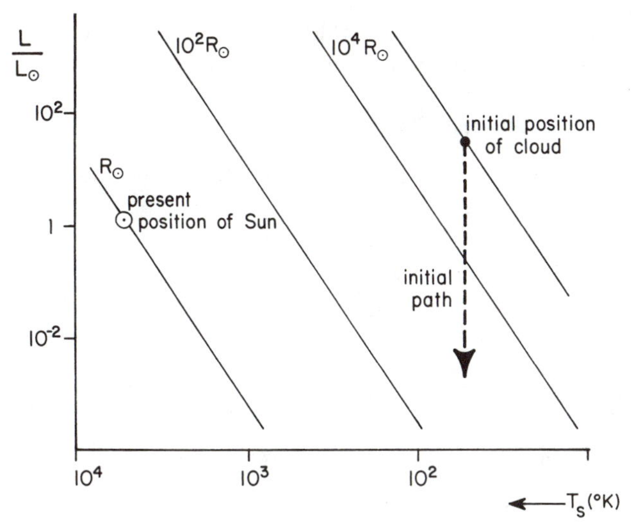

Fig. 3.7. H–R diagram showing initial path of the collapsing stellar cloud.

radius this condition is satisfied. We also take $v_{0i} = 0$. For particles on the edge of the cloud, the acceleration is $a = GM_{\text{cl}}/R_i^2$, so that

$$t_{\text{ff}} = \sqrt{\frac{2 \times \frac{1}{10} R_i}{a}} = \sqrt{\frac{2 \times \frac{1}{10} R_i}{GM_{\text{cl}}/R_1^2}}$$

$$= \sqrt{\frac{1}{5} \frac{R_1^3}{GM_{\text{cl}}}} \ .$$

For $R_i \approx 10^5 R_\odot$, $M_{\text{cl}} \approx 2 \times 10^{30}$ kg, t_{ff} turns out to be ~1000 years! The initial rate of collapse will of course increase because $a \propto 1/R^2$ and since at each subsequent stage the cloud has an "initial" translational velocity. The result, as we shall show in more detail later, is that the cloud can sink to one tenth of its original size in only a few thousand years. Although a few thousand years on a universal time scale is merely a blink, this time is far too long for us to expect to see a cloud collapse while *we* are looking at it, providing we knew where to look with out infrared eyes!

3.2. Quasihydrostatic Equilibrium

Eventually the cloud stops collapsing because forces that oppose the pull of gravity come into play. What these forces are and how they act to stabilize the cloud will be discussed in this and the next section. In this section we shall also discuss some of the general features of a stable cloud or star and in the next section the role radiation plays in achieving that stability.

3.2.1. $F = Ma$ Inside the Cloud

The best way to handle this problem is to look inside the star and "see" what is happening. We shall take as our prototype system a spherical distribution of particles. Inside the sphere we shall focus our attention on an imaginary, thin spherical shell as shown in Fig. 3.8. The shell has an inside radius of r_2 and an outside radius of r_1 (variable, of course, depending on where we imagine the shell to be) and thickness $\Delta r = r_1 - r_2$. The volume of the shell is

$$\Delta V_r = 4\pi r^2 \Delta r$$

and the mass contained within the shell is

$$\Delta M_r = \rho \Delta V = \rho 4\pi r^2 \Delta r.$$

The density involved is ρ at the center of the thin shell. The density may be different in different parts of the cloud or star, or it may be uniform.

The shell feels a pull from all the gravitational mass inside of r_2. We shall call this mass M_r, the mass inside the cross-hatched area in Fig. 3.8. Thus the force of gravity on the shell is

$$F_{g,\text{shell}} = -GM_r \Delta M_r / r_2^2,$$

where the minus sign means an inward force. If this were the only force acting, the shell would collapse into free fall like a deflated balloon. But at the inside boundary of the shell is a force due to the pressure exerted by the

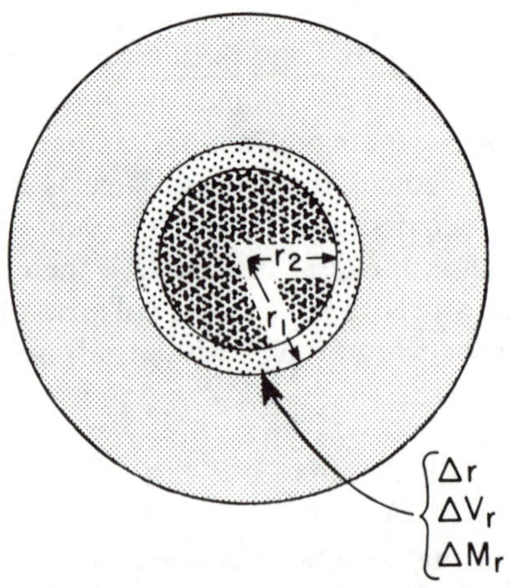

Fig. 3.8. Spherical cloud indicating the parameters for a thin inner spherical shell.

ideal gas particles (or alternatively, photons as we shall soon see). This force, F_2, pushes *outward* with magnitude

$$F_2 = +P_2 4\pi r_2^2,$$

where P_2 is the outward pressure on the inside of the shell. On the outside of the shell is an *inward* pressure and therefore an *inward* force

$$F_1 = -P_1 4\pi r_1^2,$$

where P_1 is the *inward* pressure on the outside of the shell. The net force then on the shell is

$$\Delta F_{\text{net,shell}} = P_2 4\pi r_2^2 - P_1 4\pi r_1^2 - GM_r \Delta M_r / r_2^2. \tag{3.2}$$

The net force will cause the shell to accelerate according to the relationship

$$a_{\text{shell}} \Delta M_r = \Delta F_{\text{net,shell}}. \tag{3.3}$$

Equating (3.2) and (3.3), substituting for ΔM and setting $r_1, r_2 \approx r$ (since Δr is small), we have

$$a_{\text{shell}}(\rho 4\pi r^2 \Delta r) = (P_2 - P_1) 4\pi r^2 - GM_r(\rho 4\pi r^2 \Delta r)/r^2,$$

and solving for a_{shell} we finally get

$$a_{\text{shell}} = \frac{1}{\rho} \frac{(P_2 - P_1)}{\Delta r} - GM_r/r^2.$$

In terms of the Δ symbol, which means outer minus inner here, $\Delta P = P_1 - P_2$. Thus $P_2 - P_1 = -\Delta P$ and so

$$a_{\text{shell}} = -\frac{1}{\rho}\left(\frac{\Delta P}{\Delta r}\right) - GM_r/r^2. \tag{3.4}$$

The quantity $\Delta P/\Delta r$ is simply the slope of the P vs. r curve and is called the pressure gradient. Its value and its sign are crucial to the evolution of the star.

Consider the possible solutions of (3.4):

(i) $a = 0$ — We have perfect equilibrium. Thus $(1/\rho)\Delta P/\Delta r = -GM_r/r^2$. Notice the gradient comes out negative—pressure increases as we go inward—which is to be expected.

(ii) $a \cong 0$ — Here we allow a very small acceleration—so small that in our time frame no motion is noticeable. This is called *quasihydrostatic equilibrium* (QHE) and will still satisfy $(1/\rho)\Delta P/\Delta r \cong -GM_r/r^2$.

(iii) $a > 0$ — The star or cloud expands outward. The pressure on the inner surface of the shell overcomes both the pull of gravity and the pressure on the outer surface of the shell. Since the overall sign of (3.4) must be positive, the gradient will still have to be negative but $|(1/\rho)\Delta P/\Delta r| > |GM_r/r^2|$.

(iv) $a < 0$ This can be achieved two ways. The first is automatic when $\Delta P/\Delta r$ is positive. We shall not consider this case further since it does not correspond to any of the physical situations we will encounter. The second way is important, however. The gradient may still be negative but will have to be less than that for equilibrium, i.e., $|(1/\rho)\Delta P/\Delta r| < |GM_r/r^2|$. The worst case is free fall where the pressure gradient goes to zero and $a = -GM_r/r^2$.

Each situation vis-a-vis the pressure gradient is illustrated in Fig. 3.9. To show how delicate equilibrium is, let us consider the case where $(1/\rho)\cdot \Delta P/\Delta r = -0.9999\ GM_r/r^2$. One might think that this must correspond to case (ii) $a \cong 0$, since the two terms are so close in magnitude. However,

$$\begin{aligned} a_{\text{shell}} &= +0.9999\ GM_r/r^2 - GM_r/r^2 \\ &= -0.0001\ GM_r/r^2, \end{aligned}$$

which yields a $\Delta r = R/10$ collapse time of

$$t_{\text{ff}} = \sqrt{\frac{2 \times \frac{1}{10}R}{a}} = \sqrt{\frac{2 \times \frac{1}{10}R}{0.0001\ GM/R^2}}\ ,$$

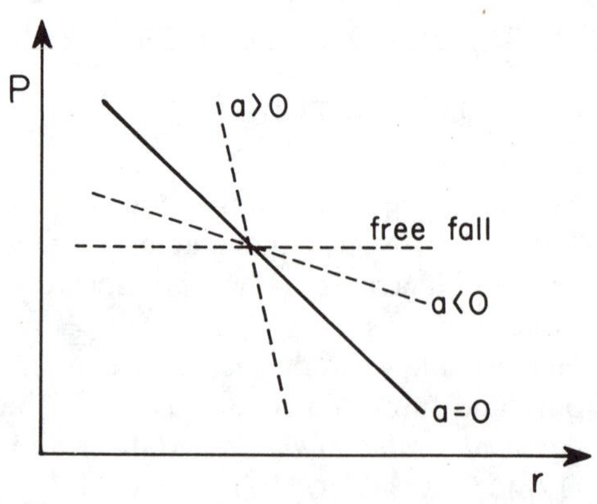

Fig. 3.9. Pressure vs. r curve for four values of the shell acceleration.

where R is the outer radius of the object. For a star like our sun this gives

$$t_{ff} \approx \sqrt{\frac{10^3 \times 2 \times (7 \times 10^8)^3}{7 \times 10^{-11} \times 2 \times 10^{30}}} \sim 7 \times 10^4 \text{ sec}$$

$$\sim 20 \text{ hr!} \sim 1 \text{ day.}$$

This is an incredibly fast collapse. Since the stars we see do not appear to be collapsing at all, they must satisfy condition (ii) to a very high degree of accuracy. Thus we should focus on QHE as the state that determines the end point of the collapse and all subsequent conditions inside the star. We therefore arrive at the end point of the collapse when

$$\frac{1}{\rho}\frac{\Delta P}{\Delta r} = -\frac{GM_r}{r^2}. \tag{3.5}$$

← hydrostatic equil.

$\frac{dP}{dr} = \rho g$

3.2.2. Reformulating the Ideal Gas Law

In order to deal with the expression for the pressure gradient, we must put the equation for an ideal gas into a form more suitable for doing astrophysics. Three things are inconvenient with the equation in its original form (see Sec. 1.2). First, most stars are made up of more than one kind of particle. There is no indication of this kind of division in $PV = NkT$. Secondly, the gas may or may not be ionized. If ionized, there are lots of electrons besides the original atoms. How do we indicate this? Finally, we rarely use volumes directly in studying interstellar clouds and stars. It is better and more general to use the mass density ρ. How do we incorporate this?

We start by assuming, quite reasonably, that the star or cloud is composed of three types of particles: hydrogen, helium, and all heavier elements which we group under a single atomic number Z and a single atomic weight A. Then we define x = fraction by weight of hydrogen $= M_H/M$; y = fraction by weight of helium $= M_{He}/M$; and z = fraction by weight of heavy elements $= M_Z/M$; where M is the total mass of the object in question and M_H, M_{He}, and M_Z are the masses (total) of each constituent. The number of each type of *atom* will be (taking the mass of an atom to be A times the mass of a hydrogen atom m_H),

$m_A = Am_H$

$$N_H = M_H/m_H = xM/m_H$$

$x = M_H/M$ so $M_H = xM$, $A = 1$

$$N_{He} = \frac{M_{He}}{m_{He}} = \frac{M_{He}}{4m_H} = \frac{yM}{4m_H}$$

$M_{He} = yM$, $A = 4$

$$N_Z = \frac{M_Z}{m_A} = \frac{M_Z}{Am_H} = \frac{zM}{Am_H}.$$

$z = M_z/M$ so $M_z = zM$

$N_{TOT} = \left(x + \frac{1}{4}y + \frac{1}{A}z\right)\frac{M}{m_H}$

Thus the total number of particles in an un-ionized gas is

$$N_{\text{tot}} = \left(x + \frac{1}{4}y + \frac{1}{A}z\right)\left(\frac{M}{m_{\text{H}}}\right) \equiv \left(\frac{1}{\mu}\right)_{\text{un-ion}} \times \frac{M}{m_{\text{H}}},$$

where μ is the mean molecular weight of the object measured in units of proton masses.

If the gas is completely ionized, each atom also contributes Z electrons to the number count so that Z + 1 particles will be present for each atom. If N now represents the total number of particles contributed by that atom, we have

$$N_{\text{H}} = 2\frac{M_{\text{H}}}{m_{\text{H}}} = 2x\frac{M}{m_{\text{H}}}$$

$$N_{\text{He}} = 3\frac{M_{\text{He}}}{m_{\text{He}}} = 3\frac{M_{\text{H}}}{4m_{\text{H}}} = \tfrac{3}{4}y\frac{M}{m_{\text{H}}}$$

$$N_{\text{Z}} = (\text{Z}+1)\frac{M_{\text{Z}}}{m_A} = \left(\frac{\text{Z}+1}{A}\right)\frac{M_{\text{Z}}}{m_{\text{H}}} \cong \tfrac{1}{2}z\frac{M}{m_{\text{H}}}$$

and

$$N_{\text{tot}} = (2x + \tfrac{3}{4}y + \tfrac{1}{2}z)\frac{M}{m_{\text{H}}} = \left(\frac{1}{\mu}\right)_{\text{ion}}\frac{M}{m_{\text{H}}},$$

where we have used the approximation $(\text{Z}+1)/A \cong \text{Z}/A \cong \frac{1}{2}$ for elements with $A \geqslant 20$. The ideal gas law, $PV = NkT$, now becomes, for *all* cases

$$PV = \left(\frac{1}{\mu}\right)\frac{M}{m_{\text{H}}}kT.$$

Since $M/V = \langle\rho\rangle$ is the average gas mass density, we finally write

$$P = \frac{\langle\rho\rangle kT}{\mu m_{\text{H}}}, \qquad (3.6)$$

where $1/\mu = (x + y/4 + z/A)$ if un-ionized and $1/\mu = (2x + 3y/4 + z/2)$ if ionized. Although quite disguised, this is the astrophysicist's form of the ideal gas law. By varying x, y, and $\langle\rho\rangle$ (not z because you can see from our definitions that $x + y + z = 1$), one can adjust the pressure to take into account most situations found in stars.

With this relationship (3.6) for the pressure, we see there are 3 possible causes of a pressure gradient required in (3.5):

$\Delta P/\Delta r \sim \Delta T/\Delta r$	a temperature gradient
$\Delta P/\Delta r \sim \Delta\rho/\Delta r$	a density gradient
$\Delta P/\Delta r \sim \Delta\left(\frac{1}{\mu}\right)/\Delta r$	a "type of particle" gradient.

The last two will be important when we take up the red giant phase of the evolution of the Sun, but for the early stages of stellar evolution it is the temperature gradient that will determine the pressure gradient.

3.2.3. *P* and *T* Inside a QHE Cloud or Star

We consider now in greater detail what can be learned about a cloud or star in QHE. We assume the object to be spherically symmetric with constant density ρ. The latter is not a good approximation because in most clouds and stars ρ will vary significantly, falling off as r increases. Unfortunately, this makes calculations difficult so we shall stick with $\rho = \text{constant}$ to make the general point. The condition for QHE is

$$\frac{\Delta P}{\Delta r} = -\frac{GM_r\rho}{r^2}.$$

If we imagine the object divided up into a series of thin concentric shells of thickness Δr, as shown in Fig. 3.10, then the pressure difference across each

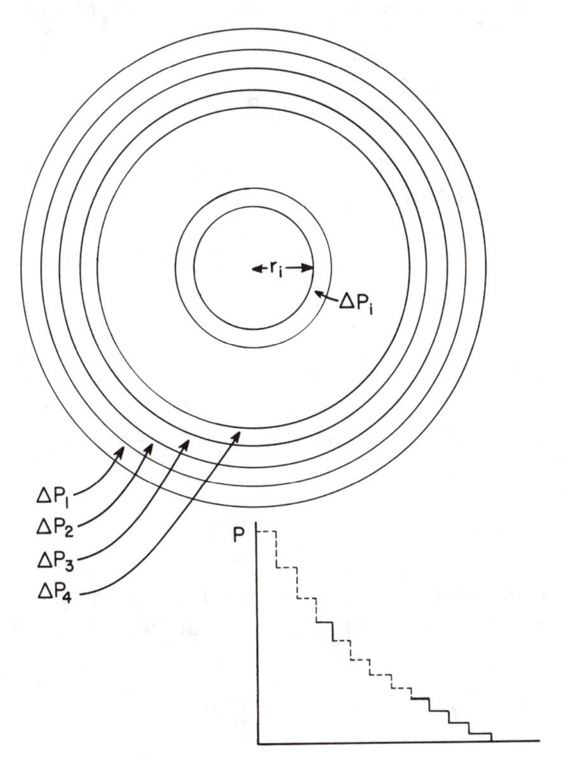

Fig. 3.10. Pressure distribution for incremental values of r.

is ($M_{r_i} = \rho 4\pi r_i^3/3$)

$$\Delta P_i = -\frac{G M_{r_i} \rho \Delta r}{r_i^2} = -\frac{G\left[(4\pi/3)\rho r_i^3\right]\rho \Delta r}{r_i^2}$$

$$= -\frac{G 4\pi}{3} \rho^2 r_i \Delta r.$$

If we know the pressure somewhere—say at the surface where $P_R = 0$—then we can find P anywhere by adding up the changes in pressure across each shell until we go from the surface to the shell of interest, i.e.,

$$\Delta P = P_R - P_r = \sum_{P_r}^{P_R} \Delta P_i,$$

so that

$$P_r = G \frac{4\pi}{3} \rho^2 \sum_{r}^{R} r_i \Delta r.$$

The sum can be evaluated quite easily by a variety of mathematical techniques, such as the area of a triangle is $\frac{1}{2}$ that of the corresponding rectangle:

$$\sum_{r}^{R} r_i \Delta r = \tfrac{1}{2}(R^2 - r^2),$$

where the sum runs from the surface of the star with $r_i = R$ to the radius of interest with $r_i = r$. Substituting for the sum we get

$$P_r = G \frac{4\pi}{3} \rho^2 \left(\frac{R^2 - r^2}{2} \right) \qquad \text{expressed in terms of } \rho$$

or

$$P_r = G \frac{3}{8\pi} M^2 \left(\frac{R^2 - r^2}{R^6} \right) \qquad \text{expressed in terms of the mass of the star.}$$

We can immediately find the pressure at the center (P_c) of the QHE gas by setting $r = 0$:

$$P_c = \frac{3G}{8\pi} \frac{M^2}{R^4}.$$

Since the cloud or star is an ideal gas, we have $P = \rho kT/\mu m_{\mathrm{H}}$. Solving for T ($T_r = (\mu m_{\mathrm{H}}/k)P_r$) and plugging in the value of P_r given above, we find the temperature at any radius inside the cloud to be

$$T_r = \frac{\mu m_{\mathrm{H}}}{k} \left(G \frac{4\pi}{3} \rho \right) \left(\frac{R^2 - r^2}{2} \right).$$

At the center of the star, where again we set $r = 0$, we have

$$T_c = \tfrac{1}{2} \frac{\mu m_H}{k} \frac{GM}{R},$$

where we have set $(4\pi/3)\rho R^2 = (M/R)$. For example, at the center of our sun, these formulae indicate that $P_c \sim 10^9 P_E$ where P_E is an Earth atmosphere (14.7 lb/sq. in $\approx 10^5$ N/m^2). Likewise the central temperature of the Sun, with $\mu = 1$, is.

$$T_c \approx \frac{1.7 \times 10^{-27} \times 6.7 \times 10^{-11} \times 2 \times 10^{30}}{2 \times 1.4 \times 10^{-23} \times 7 \times 10^8} \approx 10^7 \text{ °K}.$$

This is 2000 times hotter than the surface temperature, $T_s \approx 5800$ °K.

3.2.4. The Virial Theorem

It should be clear from the above discussion that the temperature and pressure are not constant throughout the star. Thus it is no longer a trivial problem to convert the kinetic energy per particle, $\langle \text{KE} \rangle = \tfrac{3}{2} kT$, to KE_{tot} for a whole star, where each particle is at a temperature varying according to its distance from the star's center. Still, if we again count the number of particles at distance r, $\Delta M_r / \mu m_H$, that are in the spherical shell of volume $\Delta V_r = 4\pi r^2 \Delta r$, as in Fig. 3.8, then the total KE for the entire star must be

$$\text{KE}_{\text{tot}} = \sum \langle \text{KE} \rangle_r \left(\frac{\Delta M_r}{\mu m_H} \right) = \sum \langle \text{KE} \rangle_r \left(\frac{\rho_r \Delta V_r}{\mu m_H} \right)$$

$$= \tfrac{3}{2} \sum \frac{k T_r \rho_r \Delta V_r}{\mu m_H} = \tfrac{3}{2} \sum P_r \Delta V_r, \qquad (3.7)$$

where we have used $\Delta M_r = \rho_r \Delta V_r$ and the ideal gas law, $P_r = k T_r \rho_r / \mu m_H$ to obtain the final version of this relation.

On the other hand, we may also compute the magnitude of the total gravitational potential energy for the star in general by first considering the potential energy $G M_r \Delta M_r / r$ for the spherical shell, ΔM_r, and the mass M_r. Then, summing over the entire star we obtain

$$-\text{PE}_{\text{tot}} = \sum \frac{G M_r}{r} \Delta M_r = \sum \frac{G M_r}{r} (\rho_r \Delta V_T)$$

$$= 3 \sum \frac{G M_r}{r^2} \rho_r V_r \Delta r,$$

where $\Delta V_r = 4\pi r^2 \Delta r = (4\pi r^3/3)(3\Delta r/r) = V_r 3\Delta r/r$ and M_r is the mass in the sphere of radius r. For a star in QHE, we may replace the factor

$GM_r\rho_r r^{-2}\Delta r$ in the above equation by $-\Delta P_r$, according to (3.5), i.e.,

$$\mathrm{PE}_{\mathrm{tot}} = 3\sum_r V_r \Delta P_r. \tag{3.8}$$

The quantities $\sum P_r \Delta V_r$ in (3.7) and $\sum V_r \Delta P_r$ in (3.8) both measure the work done in moving all the particles from infinity to within the spherical volume of the star and as such are equal in magnitude to one another. This can be verified explicitly for the special case of constant density and ΔP_r given in the last section .(In general it follows from the calculus technique of integration by parts.) Thus, combining (3.7) and (3.8), we learn that in general the magnitude of the total KE and PE for a star in QHE are related by

$$\mathrm{KE}_{\mathrm{tot}} = \tfrac{1}{2}|\mathrm{PE}_{\mathrm{tot}}|$$

or since PE is negative,

$$2\,\mathrm{KE}_{\mathrm{tot}} + \mathrm{PE}_{\mathrm{tot}} = 0. \tag{3.9}$$

Equation (3.9) is called the *virial theorem*. It is not to be confused with conservation of energy which says that $E = \mathrm{KE}_{\mathrm{tot}} + \mathrm{PE}_{\mathrm{tot}}$. The virial theorem is a separate, special equation for an ideal gas in QHE. Conservation of energy is more general and holds for a system whether in QHE or not as long as conservative forces act. For an ideal gas in QHE, both the virial theorem and energy conservation can be employed with profit.

Two variations of the virial theorem are worth noting. The factor of 2 in (3.9) is a statement of nonrelativistic particles moving under the influence of an inverse-square-law force. It is also valid for electrons revolving around a positively charged nucleus, now an inverse-square-law electrostatic force. For a circularly orbiting planet, Newton's second law $GMm/r^2 = mv^2/r$ implies $\mathrm{KE} = \frac{1}{2}mv^2 = \frac{1}{2}GMm/r$ and the virial theorem (3.9) is automatically satisfied. The electrostatic version is implicit in the Bohr energy level formula of Sec. 1.4.3.

For relativistic particles, however, the form of the virial theorem changes. A black body photon "star" with energy density $E_\gamma/V = aT^4$ and pressure $P_\gamma = \frac{1}{3}aT^4$ still obeys (3.8), but (3.7) becomes

$$\mathrm{KE}_\gamma = \sum aT_r^4 \Delta V_r = 3\sum_r P_r \Delta V_r. \tag{3.10}$$

The factor of 2 difference between (3.10) and (3.7) means that the extreme relativistic limit of the virial theorem is not (3.9) but instead

$$\mathrm{KE}_{\mathrm{tot}} + \mathrm{PE}_{\mathrm{tot}} = 0. \tag{3.11}$$

Recall, however, that the left-hand side of (3.11) is just the total energy of the star. Thus $E = 0$. But we know that $E < 0$ if the star is stable, so (3.11) is never achieved for any viable star. Nevertheless (3.11) will be useful to set upper limits for masses of stars, as we shall see at the end of this chapter and also in chapter 5.

As a final point, note that we now have two ways of looking at QHE:

$$\text{(i)} \qquad \frac{\Delta P}{\Delta r} + \frac{GM_r \rho_r}{r^2} = 0$$

$$\text{(ii)} \qquad 2\,\text{KE}_{\text{tot}} + \text{PE}_{\text{tot}} = 0.$$

The first is a pressure difference relationship satisfied at every point in the star if it is in QHE. If not, then as we saw in 3.2.1, the star will either expand or contract. The second relationship, in effect, integrates (sums) the pressure difference equation throughout the star. It too must be satisfied if the star is in QHE. If we should find that $\text{KE}_{\text{tot}} > \frac{1}{2}|\text{PE}_{\text{tot}}|$, then the star must expand; it has too much KE, or alternatively, too much outward pressure. If instead we find that $\text{KE}_{\text{tot}} < \frac{1}{2}|\text{PE}_{\text{tot}}|$, the star must contract; the gravitational pull or inward pressure is too great. Whether one employs (i) or (ii) above to explore the equilibrium dynamics of a star depends upon the circumstances of the problem.

3.2.5. The Onset of Quasihydrostatic Equilibrium

As we saw in Sec. 3.2.2 our constant-density cloud must have a temperature gradient if it is to remain in QHE. In Sec. 3.2.3, we found an expression for T_c, the temperature at the center of the cloud. The surface temperature T_s will be very small compared to T_c—hence the gradient. A reasonable question to ask then is: What is the average temperature of the cloud or star? The average temperature $\langle T \rangle$ will clearly lie between T_c and T_s. This is not much consolation; we should be a bit more specific. In this regard it is reasonable to assume an average temperature such that

$$\text{KE}_{\text{tot}} = \tfrac{3}{2} Nk\langle T \rangle,$$

where N is the total number of particles, $N = M/\mu m_{\text{H}}$. If we accept this definition then we can use the virial theorem to find $\langle T \rangle$, i.e.,

$$2\,\text{KE}_{\text{tot}} + \text{PE}_{\text{tot}} = 0$$

$$2(\tfrac{3}{2} Nk\langle T \rangle) - \tfrac{3}{5} GM^2/\text{R} = 0$$

$$\langle T \rangle = \tfrac{1}{5} \frac{m_{\text{H}}\, \mu GM}{kR}.$$

Comparing this with the central temperature T_c we see that for a cloud with constant density

$$\langle T \rangle = \tfrac{2}{5} T_c.$$

We are now ready to make an attempt to locate the onset of QHE. We argue as follows: Eventually the motion of the infalling gas must be randomized through the absorption of energy. As we shall see in the next section, this will happen when the densities get large enough so that there are an appreciable number of collisions which produce EM radiation, the photons of which are absorbed by the particles. In this way gravitational potential energy, which goes as $1/R$, can be turned into random KE and therefore into the necessary temperature gradient. There are, however, two events which foil the transfer of energy from PE to $\langle \mathrm{KE} \rangle$. The first happens when the collision energy goes into separating the molecular hydrogen into atoms, i.e., $H_2 \rightarrow H + H$. The reaction requires ~ 2 eV of energy which would otherwise go into thermal kinetic energy. (An analogy: When water boils, energy is used up to break bonds between water molecules producing water vapor and so the temperature remains constant.) Secondly, PE is used up in ionizing hydrogen, $H \rightarrow H^+ + e^-$. This takes 13.6 eV and will be a steady drain on the star's PE until most of the hydrogen is ionized. In both cases the diversion of energy prevents the magnitude of the temperature gradient from increasing and thus the pressure gradient will be inadequate to prevent collapse. When all the original H_2 has been converted to H^+, the transfer of energy from PE to thermal kinetic energy can proceed until $\Delta T/\Delta r$ reaches the point where $\Delta P/\Delta r = -\rho_r G M_r / r^2$. It turns out that the onset of QHE is quite rapid when full dissociation has been completed. Thus with the aid of the virial theorem and energy conservation, we can calculate the radius of the cloud at which QHE begins.

Initially the total energy of the cloud is zero. It will remain approximately zero (some energy is radiated away but this is unimportant) until dissociation begins. By the time dissociation is over, the cloud will have *lost* all the energy that went into breaking up molecular and atomic hydrogen. This energy will be (recall, $m_{\mathrm{H}_2} = 2m_{\mathrm{H}}$)

$$E_d = 13.6 \frac{M}{m_{\mathrm{H}}} + 2 \frac{M}{2m_{\mathrm{H}}} = 14.6 \frac{M}{m_{\mathrm{H}}} \text{ eV}.$$

When dissociation is finished, we have therefore

$$\mathrm{KE}_{\mathrm{tot}} + \mathrm{PE}_{\mathrm{tot}} = 0 - E_d,$$

the minus sign because the cloud has lost this energy. Since we have assumed the cloud has also reached QHE, the virial theorem, $2\,\mathrm{KE}_{\mathrm{tot}} + \mathrm{PE}_{\mathrm{tot}} = 0$ is also valid. Plugging this into the above equation, we get

$$-\frac{\mathrm{PE}_{\mathrm{tot}}}{2} + \mathrm{PE}_{\mathrm{tot}} = -E_d.$$

Therefore, we find

$$+\frac{\mathrm{PE}_{\mathrm{tot}}}{2} = -E_d,$$

and since $PE_{tot} = -\frac{3}{5}GM^2/R$ we have (converting to MKS units),

$$-\frac{PE_{tot}}{2} = \frac{3}{5}\frac{GM^2}{2R} = (14.6 \times 1.6 \times 10^{-19})\frac{M}{m_H} \text{ (joules).}$$

Assuming $M \sim M_\odot \approx 2 \times 10^{30}$ kg we can solve explicitly for R,

$$R \approx \frac{3}{10}\frac{MGm_H}{14.6 \times 1.6 \times 10^{-19}} \approx \frac{3}{10}\frac{2 \times 10^{30} \times 6.7 \times 10^{-11} \times 1.7 \times 10^{-27}}{14.6 \times 1.6 \times 10^{-19}}$$

$$\approx 29 \times 10^9 \text{ m} \approx 40\ R_\odot. \tag{3.12}$$

This is about $\frac{1}{2}$ the distance from the Sun to the planet Mercury. Of course it is not the final size of the star. The star will continue to collapse—under the influence of gravity—but it will do so ever so slowly since it is now in QHE. The average temperature of the cloud at 40 $R_\odot$ is

$$\langle T \rangle = \frac{1}{5}\frac{\mu m_H G}{kR}M \approx \frac{1}{5}\frac{1.7 \times 10^{-27} \times \frac{1}{2} \times 6.7 \times 10^{-11} \times 2 \times 10^{30}}{1.4 \times 10^{-23} \times 40 \times 7 \times 10^8}$$

$$\approx 6 \times 10^4\ {}^\circ\text{K},$$

where $1/\mu = 2$ for 100% ionized hydrogen. We can see from (3.12) that $R \propto M$ while the above equation indicates that $\langle T \rangle \propto M/R$. This leads to the interesting conclusion that when QHE is reached, $\langle T \rangle$ will be independent of M and R; $\langle T \rangle$ will be about 10^4–10^5 °K whether M is 10 $M_\odot$ or $M_\odot$!

At 40 $R_\odot$ the surface temperature of the cloud will be about 50 °K (we shall see why in the next section) and its luminosity $L = 4\pi R^2 \sigma T^4 \approx 10^{-4}\ L_\odot$. The mass and number density will rise to

$$\rho = M/V \approx 2 \times 10^{30} / \frac{4\pi}{3}(2.5 \times 10^{10})^3 \sim 2 \times 10^{-2} \text{ kg/m}^3$$

$$n = \rho/m_H \sim 10^{25}/\text{m}^3 \sim 10^{19}/\text{cc}.$$

Our cloud has now plunged straight down the H–R diagram of Fig. 3.11, seemingly no closer to the present position of the Sun than when it started.

Finally, since we know $R_i \sim 10^5\ R_\odot$ and $R_f. \sim 40\ R_\odot$, we can estimate the total time spent in free fall. To do this it is best to concentrate on a single particle of mass m_0, at the edge of the cloud. That particle will "see" a mass $M_\odot$ attracting it. Since the particle's acceleration will not be constant, we must be a bit careful how we go about an analysis over such long distances. The correct approach is integral calculus, but what we shall try to do is find the velocity of the infalling particle at several intermediate radii and then estimate the time to go from one radius to the next from $\Delta r = \langle v \rangle \Delta t$ using $\langle v \rangle = (v_2 + v_1)/2$, where v_2 and v_1 are the velocities of m_0 at the pair of consecutive radii under consideration. We can easily find the velocities

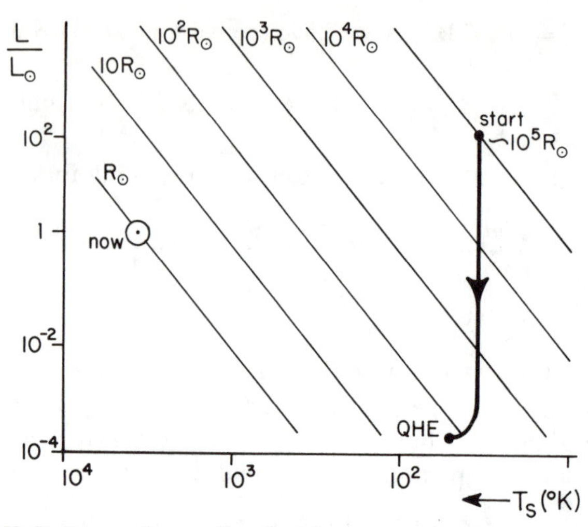

Fig. 3.11. H–R diagram for a stellar cloud from onset of collapse to onset of QHE.

from energy conservation applied to m_0. Taking $v_{i,\mathrm{ff}} = 0$, we have

$$0 - \frac{GMm}{R_i} = \tfrac{1}{2}mv_r^2 - \frac{GMm}{r} .$$

Solving for v_r we have

$$v_r = \sqrt{\frac{2GM}{R_i r}(R_i - r)} . \tag{3.13}$$

As time progresses, r becomes smaller. If we take $r = \chi R_i$, where χ is some fraction of $R_i = 10^5\ R_\odot$, we obtain,

$$v_\chi = \sqrt{\frac{2GM}{R_i}\left(\frac{1-\chi}{\chi}\right)} \approx 2\times 10^3\sqrt{\frac{1-\chi}{\chi}}\ (\mathrm{m/sec}).$$

Although the expression $(V_2 + v_1)/2$ is true for a linear increase in velocity, it is also approximately valid for (3.13) when r is broken down into the following four intervals: $0.9\ R_i < r < R_i$, $0.5\ R_i < r < 0.9\ R_i$, $0.1\ R_i < r < 0.5\ R_i$, $0.1\ R_i < r < 0.01\ R_i$. We can thus estimate the collapse time Δt as $\Delta t_1 + \Delta t_2 + \Delta t_3 + \Delta t_4$ as shown in Fig. 3.12. The time to reach $0.01\ R_i$ $(= 10^3\ R_4 1_\odot)$ is

$$\Delta t = \Delta t_1 + \Delta t_2 + \cdots = 4.75 \times 10^{10}\ \mathrm{sec} \sim 1500\ \mathrm{yr}.$$

As the t vs. R plot of Fig. 3.12 shows, the time to reach $40\ R_\odot$ will not be much more than 1500 years. This is only a rough estimate but most astronomers agree that this phase of collapse does not last more than 5000

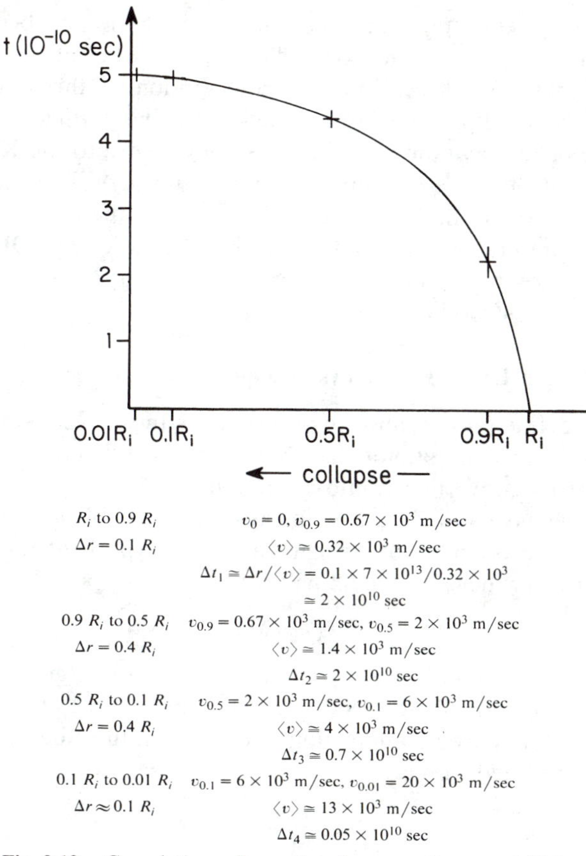

Fig. 3.12. Cumulative collapse time from $r = R_i$ to $r = 0.01 R_i$.

years which is extremely short relative to the time scales thus far encountered.

3.3. Energy Transfer by Photons

If we are to achieve QHE, we must find a mechanism of converting the gravitational PE into thermal motion. We have suggested earlier that "collisions" will accomplish this. Now here we must be very careful. A collision implies a close, but not necessarily a "billiard ball," encounter. In the sense that we use the word, this means that the particles approach close enough to experience a deflection due to their mutual Coulomb interaction (ion–ion, ion–dipole, etc.). These collisions inevitably lead to the emission of electromagnetic radiation, either because the particles are charged in which case they emit photons when they accelerate, or because they become excited and emit transition radiation. By virtue of the deflection, this kind of collision will certainly begin to randomize the motion of the particles, but because photons are radiated away, the particles should lose, not gain, KE. The gravitational PE is thus converted into the KE of the γ rays and not into the KE of thermal motion. And here is the rub—unless these γs are absorbed, the cloud cannot heat up and the requisite temperature and pressure gradients cannot be produced that lead to QHE. So our first job is to study how photons are absorbed and our second job is to see how this absorption produces QHE.

3.3.1. Absorption Cross Sections vs. Temperature

Astronomers define photon absorption events as those events in which the identity of the incident photon has been altered either by changing its wavelength or by converting it into a different particle.

In a sense we have already solved the absorption problem. We showed in Sec. 2.3 that the probability of a single particle of type 1 interacting with a large number of particles of type 2 will be

$$\mathcal{P} = n_2 \sigma \Delta x.$$

If type 1 are photons and type 2 massive particles (dust grains, hydrogen atoms, electrons), then all we need to do is understand the absorption cross section for photon–particle interactions. In the literature the above expression is often modified to read

$$\mathcal{P} = n_2 m_{\mathrm{H}} \frac{\sigma}{m_{\mathrm{H}}} \Delta x = \rho \kappa \Delta x,$$

where $\kappa \equiv \sigma / m_{\mathrm{H}}$ is called the absorption coefficient (vs. σ, the absorption cross section). It will depend upon the specific absorption process of which there are only a few (see Fig. 1.1):

(a) *Photoelectric Ionization*: The photon is annihilated and an electron is ejected from the atom with KE = E_γ − BE. For this process to be effective the photons must have at least 13.6 eV kinetic energy.
(b) *Compton Scattering*: The photon is scattered and loses energy to the electron, corresponding to an increase in the photon's wavelength. The process is most likely to occur in the range $E_\gamma \sim 0.01$ to 1 MeV.
(c) *Pair Production*: A photon, passing close to an atom, annihilates into an $e^+ - e^-$ pair. This requires a minimum of 1 MeV in order to create the rest mass of the electron–positron system.
(d) *Inverse Bremsstrahlung*: A photon hits a *free* electron and is absorbed, thereby increasing the KE of the electron ($\mathrm{KE}_{e,f} = E_\gamma +$

$KE_{e,i}$). (To conserve momentum there must be a heavy nucleus nearby which absorbs momentum but little KE.) Of course there must be a supply of free electrons for this process to be effective.

Because these processes are energy-dependent they are temperature-dependent as well. For example, if $T < 10^4$ °K then process (a) is ruled out because photons from black body radiation ($E_\gamma \sim kT$) will not have enough energy to cause ionization, while processes (b) and (c) are ruled out because there will be few photons energetic enough to initiate the reaction. Process (d) is also ruled out because $KE_e \sim kT$ is too low to produce the collision ionization necessary to supply free electrons. If $T > 10^6$ °K then processes (a) and (d) will now have low cross sections because of the small wavelength of the photon. Process (b) and (c) will now be energetic enough but are not efficient, again due to the small wavelength of the photon.

The result is the curve κ vs. T shown in Fig. 3.13 which is also parametrized for various ρs. It is satisfying to note that based on the previous calculation for the onset of QHE, where we found $\rho \sim 10^{-2}$ kg/m^3 and $\langle T \rangle \sim 6 \times 10^4$ °K, the value of κ falls at the peak of the curve. In a similar fashion we can see why the surface temperature remained ~50 °K during the collapse from 10^5 $R_\odot$ to 40 $R_\odot$; both ρ and κ were too small

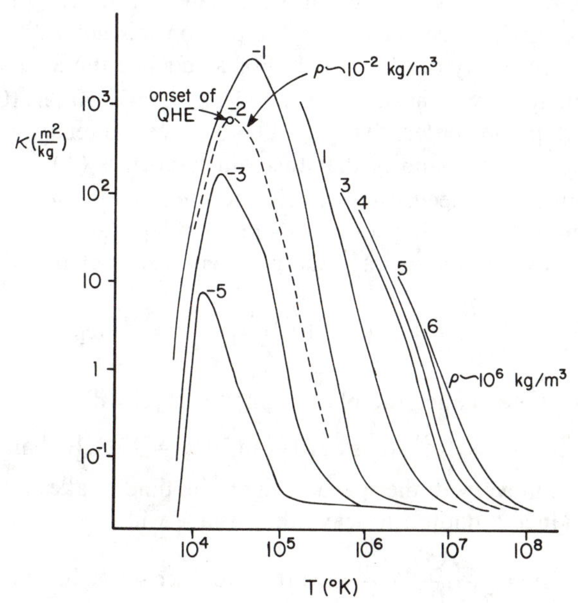

Fig. 3.13. Typical absorption coefficient vs. temperature curve parametrized according to the mass density exponent η ($\rho = 10^\eta$ kg/m^3).

for their product to be large enough to produce appreciable absorption ($\mathcal{P} = \rho\kappa\Delta x$).

To get a better feel for what the product $\rho\kappa$ means, we should note that $1/\rho\kappa$ is the mean free path—the distance in which the photon has a 63% probability of interacting. It is interesting to study $1/\rho\kappa$ for the Sun. Because T goes from 10^7 °K at the center of the Sun to 10^4 °K at its surface, κ roughly increases as r increases. But ρ falls off as we move away from the center. The result is that $1/\rho\kappa$ remains approximately constant inside the Sun. For example, at the center, $\kappa_c \sim 10^{-1}$ m^2/kg and $\rho_c \sim 10^5$ kg/m^3, whereas in the interior, $\kappa_{\text{int}} \sim 10$ m^2/kg and $\rho_{\text{int}} \sim 10^3$ kg/m^3 (the density of water). In either case the mean free path is $1/\rho\kappa \sim 10^{-4}$ m. This means that a photon starting from anywhere in the interior travels 10^{-4} m, before being absorbed. An observer standing, say, at $R_\odot/2$ could only see 10^{-4} m in any direction!

This small but constant value of $1/\rho\kappa$ has interesting consequences vis-a-vis the time it takes a photon starting from the center of the Sun to appear at the surface. Let us follow a photon originating at $r = 0$. After traveling 10^{-4} m, it is absorbed. But there is now the possibility of reemission. The new photon may go forward or backward; let us say with a 50% probability. If it goes forward, it gets 10^{-4} m further out and is then reabsorbed. Again there is a 50% probability for forward emission and the chance to move another 10^{-4} m outward. This is a horribly slow *random-walk* process. After three such steps the photon has gone 3×10^{-4} m but with only a probability of $(\frac{1}{2})^3 = 1/8$. For a star like the Sun with a radius of 7×10^8 m, there will have to be $R \, . \, (1/\rho\kappa)^{-1} = 7 \times 10^8 \text{ m}/10^{-4} \text{ m} = 7 \times 10^{12}$ such steps in order that the photon goes directly out with no backtracking. The probability that this will happen is $(\frac{1}{2})^{7 \times 10^{12}}$ which is a very small number indeed. More likely the photon will work its way back and forth, in and out, before one of its progeny appears at the surface and is radiated away. On the average the number of such steps will be

$$\left(\frac{R_\odot}{1/\rho\kappa}\right)^2 = (7 \times 10^{12})^2 = 49 \times 10^{24} \text{ steps}$$

and therefore the distance the photon has to travel will be

$$d = 49 \times 10^{24} \text{ steps} \times 10^{-4} \text{ m/step} = 49 \times 10^{20} \text{ m}.$$

Since photons move with the speed of light, the time it takes for the original photon to induce a photon to leave the surface will be d/c or

$$\Delta t = \frac{49 \times 10^{20} \text{ m}}{3 \times 10^8 \text{ m/sec}} \approx 16 \times 10^{12} \text{ sec} \sim 5 \times 10^5 \text{ yr}.$$

This is a rather remarkable result. It says that for our sun it takes $\sim 10^6$

years for a photon to "work" its way to the surface. Thus the light we see from the Sun began its journey to the surface 10^6 years ago! If the center of the Sun went dead *now* we would not know about it for 10^6 years. We will get back to this and the solar neutrino problem in the next chapter when we study nuclear processes inside the Sun. Still the question remains—how do we know what is going on in the solar interior now if all our information is 10^6 years old?

At this point, however, we are less worried about the Sun than about the collapsing cloud. We need a temperature gradient and we shall now see that the absorption of photons will deliver that gradient.

3.3.2. Temperature Gradients and Radiative Transfer

So, as the cloud collapses, its PE $(= -\frac{3}{5}GM^2/R)$ decreases, i.e., becomes more negative. This lost PE is converted into photons which, through their absorption, affect the transfer of PE to thermal KE and the formation of a temperature gradient.

Consider then a thin shell of matter with photons incident on the inner side of the shell as shown in Fig. 3.14. Although there are photons flying about in all directions, the net flow is outward. This is because the star is hotter the further in you go. Thus if an observer stands at some arbitrary point in the star and looks toward the center he will see more photons per second coming at him from the hot interior than he would see if he turned around and looked toward the cooler surface. We shall continue to mean, therefore, when we speak of the photon flux the net outward flow of photons implicit in the rising internal temperature. The number of photons absorbed in the shell will be

$$N_{abs} = \rho\kappa\Delta r N_{inc},$$

where N_{inc} is the number of γs incident on the inside of the shell, and Δr

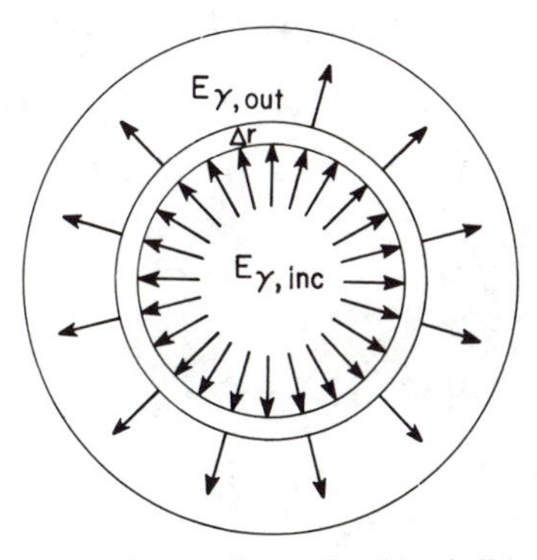

Fig. 3.14. Photon energy into and out of a thin shell in the stellar interior.

the thickness of the shell. If each photon carries energy E_i into the shell and then loses that energy via interactions within the shell, the energy deposited within Δr is

$$\Delta E_\gamma = N_{\text{abs}} E_i = \rho\kappa\Delta r (N_{\text{inc}} E_i) \\ = \rho\kappa\Delta r E_\gamma,$$

where $N_{\text{inc}} E_i = E_{\text{inc,tot}} \equiv E_\gamma$ is the total energy carried by the net outward flux of photons as they cross the inside surface of the shell. Since each photon carries momentum $p_{\gamma,i} = E_i/c$, the total momentum transferred to the shell will be

$$\Delta p_\gamma = \rho\kappa\Delta r E_\gamma / c,$$

which over a time Δt means an average force on the shell due to the photons of

$$F_\gamma = \frac{\Delta p_\gamma}{\Delta t} = \frac{\rho\kappa\Delta r}{c}\left(\frac{E_\gamma}{\Delta t}\right).$$

Since $E_\gamma/\Delta t$ is nothing more than the luminosity at r, we have

$$F_\gamma = \frac{\rho\kappa\Delta r}{c} L_r.$$

We now turn our attention to what is happening on each surface of the shell (not in between). We see that there will be a photon pressure ($P = \frac{1}{3}aT^4$, see Sec. 1.3) on each surface due to T_1 and T_2 such that

$$P_{2,\gamma} = \tfrac{1}{3}aT_2^4 \qquad \text{and} \qquad P_{1,\gamma} = \tfrac{1}{3}aT_1^4.$$

The pressure difference across the shell will be

$$\Delta P_\gamma = \tfrac{1}{3}a(T_2^4 - T_1^4).$$

Then if, $T_2 \neq T_1$, the net force on the shell has a magnitude

$$F_\gamma = (\Delta P_\gamma)A = \tfrac{1}{3}a(T_2^4 - T_1^4)4\pi r^2.$$

If the shell is thin, then $T_2 = T_1 + \Delta T$ and we can write

$$T_2^4 - T_1^4 = (T_1 + \Delta T)^4 - T_1^4 = (T_1^4 + 4T_1^3\Delta T + \cdots) - T_1^4 \\ \approx 4T_1^3\Delta T.$$

Then we have

$$F_\gamma \approx \tfrac{1}{3}a(4T_1^3\Delta T)4\pi r^2 = \tfrac{16}{3}\pi a r^2 T^3 \Delta T.$$

The expression $F_\gamma = (\rho\kappa\Delta r/c)L_r$, which we derived by considering what was going on *inside* the shell, should be equal to the one above which takes into account what is going on at the surface of the shell, the latter in response to the internal absorption of photons. The result is a rather long-winded equation which stated in full reads

$$\frac{16\pi}{3} ar^2T^3\Delta T = \frac{\rho\kappa\Delta r}{c} L_r, \tag{3.14}$$

and is called the *radiative transport equation*.

Getting down to essentials we see that because there is absorption of photons, κ, particles to do the absorbing, ρ, and photons available for absorption, L_r, we have a temperature gradient across the shell,

$$\frac{\Delta T}{\Delta r} \propto \frac{\rho\kappa L_r}{r^2T^3}\ . \tag{3.15}$$

Indeed the above expression tells us that all three, ρ, κ, L_r, must be nonzero if we are to have a temperature gradient. If any one is zero, there is no gradient; $\Delta P/\Delta r$ is thus zero and we have free fall. Even if one is large, say L_r, which means a large photon flux, the other two may be so small ($\rho\kappa \ll 1$) that at best there will be a small gradient and thus collapse. On the other hand, if the product of the three is large, $\Delta T/\Delta r$ can be so high that the star expands or may even eject matter to form perhaps a ring nebula or supernova.

Let us expand a bit on the last two statements. Considering both particles and photons, the total internal pressure at radius r is

$$P_r = \frac{\rho k T_r}{\mu m_{\rm H}} + \tfrac{1}{3}aT_r^4.$$

The change in pressure across the shell will then be (recall $\Delta T^4 \approx 4T^3\Delta T$)

$$\Delta P = \frac{\rho k}{\mu m_{\rm H}}\Delta T + \tfrac{4}{3}aT_r^3\Delta T,$$

which give us a pressure gradient,

$$\frac{\Delta P}{\Delta r} = \left(\frac{\rho k}{\mu m_{\rm H}} + \tfrac{4}{3}aT_r^3\right)\frac{\Delta T}{\Delta r}\ .$$

Substituting $\Delta T/\Delta r \propto \rho\kappa L_r/r^2T_r^3$, the last equation becomes

$$\frac{\Delta P}{\Delta r} \propto \left(\frac{\rho k}{\mu m_{\rm H} T_r^3} + \tfrac{4}{3}a\right)\frac{\rho\kappa L_r}{r^2}\ .$$

For QHE we must have

$$\frac{\Delta P}{\Delta r} = -\rho\frac{GM_r}{r^2}\ .$$

Therefore, with reference to Fig. 3.9 and Eq. (3.4), radiative transfer implies that if

(i) $\rho\kappa L_r$ is too large, then $|\Delta P/\Delta r| > \rho(GM_r/r^2)$ and the star expands
(ii) $\rho\kappa L_r$ is too small, then $|\Delta P/\Delta r| < \rho(GM_r/r^2)$ and the star contracts
(iii) $\rho\kappa L_r$ is just right, then $|\Delta P/\Delta r| = \rho(GM_r/r^2)$ and the star is in QHE.

In principle we may use the radiative transport equation to determine the luminosity anywhere in the interior of the Sun. In practice however, we must know the detailed structure of the Sun if we are to get accurate results. Nonetheless we can make a rough estimate of L_r if we assume a constant-density Sun and a linear temperature gradient. For example, at $r = R_\odot/2$, the temperature will be $T_{R/2} = \frac{3}{4}T_c \sim \frac{3}{4}10^7$ °K (see Sec. 3.2) with the temperature gradient given by $\Delta T/\Delta r \cong T_c/R_\odot$. Plugging this into the radiative transport Eq. (3.14) and solving for L_r we get

$$L_r = 16\frac{\pi}{3}\left(\frac{R_\odot}{2}\right)^2 \frac{ac}{\rho\kappa}\left(\tfrac{3}{4}T_c\right)^3 \frac{T_c}{R_\odot}$$

$$\approx 16\left(\frac{7\times 10^8}{2}\right)^2 \frac{8\times 10^{-16}\times 3\times 10^8}{10^4}\left(\tfrac{3}{4}10^7\right)^3 \frac{10^7}{7\times 10^8}$$

$$\approx 4\times 10^{25}\ \text{J/sec},$$

where we have used $\rho\kappa \sim 10^4\ \text{m}^{-1}$. This time we were lucky; in spite of our crude estimates we get a result that is close to the luminosity of the Sun, $L_\odot \approx 4\times 10^{26}$ J/sec. Clearly the radiative transport equation is a marvelous tool for analyzing the behavior and structure of a star or cloud.

3.3.3. Self-Adjustment and Equilibrium: The Mass-Luminosity Relationship

There are other sources of luminosity besides the production of photons via the loss of gravitational PE. Fusion is one; the recombination of $H^+ + e^- \rightarrow H + \sum\gamma s$ is another. Each process can be represented by an appropriate L_r in the gradient equation $\Delta T/\Delta r \propto \rho\kappa L_r/r^2T^3$. The reason is the generation of γs and hence the luminosity L_r is usually a function of the temperature of the gas, i.e., $L_r \propto T^n$. The value of n in turn depends upon the type of process, whether fusion, recombination, etc., and can range from four for black body radiation to $n \sim 120$ (!) for some fusion reactions. The value of n should not concern us here; only that L_r does depend upon T. This is important because the average temperature of the star, as we saw earlier from the virial theorem, depends on $1/R$,

$$\langle T\rangle = \tfrac{1}{5}\,\frac{\mu m_{\mathrm{H}}}{k}\,\frac{GM}{R}\,,$$

where R is the radius of the star or cloud. The fact that $L \propto T^n$ and $T \propto 1/R$ means that L_r and the size of the star are linked. Thus if L_r is so *high* as to *expand* the gas, R goes *up*, causing T to go down, which in turn *lowers* L_r! If the expansion lowers L_r too much, $\Delta T/\Delta r$ will no longer be large enough to support the gas and *contraction* will again occur. This *reduces* R and raises $\langle T\rangle$, *raising* L_r. This increases $\Delta T/\Delta r$ and the gas may once again *expand*. In this fashion the star may oscillate about some equilibrium point (as possibly a Cepheid-variable star) until damping produces QHE. Thus, stars have built into them a self-regulating feedback mechanism through which they try to maintain the status quo. It is like a thermostat in a house that turns the heat on and off to maintain a constant temperature. The coupling between L, T, R, and $\Delta T/\Delta r$ does just this. It does not always work, though. Sometimes the initial expansion is so rapid, equilibrium is impossible to achieve and we have an explosion. But these extreme cases, although very important, are rare—so we expect QHE to be the norm—and it is.

As another application of the radiative transport equation, we recall from Sec. 2.6.2 that over much of the main sequence, the stars obey the mass luminosity relation, $L \propto M^3$. We can understand this in terms of the above relation $\langle T\rangle \propto M/R$ and the transport equation, $L \propto (R^2T^3/\rho)(\Delta T/\Delta R) \propto (RT)^4/M$, where to obtain the latter we have taken $\rho \propto M/R^3$, $\Delta T/\Delta R \propto T/R$ and assumed κ is about the same for such stars. Combining these two relations gives

$$L \propto \frac{(RT)^4}{M} \propto \frac{M^4}{M} = M^3.$$

Although this analysis is very crude, the results agree quite well with our experimental observations and illustrate the internal consistency between the virial theorem from which we derived $\langle T\rangle \propto M/R$ and the radiative transport equation.

When we last left the Sun it had just reached QHE, but its surface temperature was still ~ 50 °K. Inside of course it is very hot, the center being over 10^4 °K, but because $\rho\kappa$ is so large, these photons will not be able to carry that information to the surface for many hundreds of millions of years. Does this mean the surface will not brighten into a visible star for millions of years? The answer is no; it brightens up in a matter of years, maybe even days because of another method of transferring energy.

3.4. Convective Transfer of Energy

You have all seen convection at work as bubbles of gas rise to the surface of a boiling pot of water, or as warm air rises and cold air falls in a room with baseboard heating. For a cloud or star, we shall see that convection is a spectacular method of transferring energy and accounts for the sudden increase in luminosity of a cloud that has just arrived at QHE.

Consider again a thin spherical shell of matter inside the cloud. Within the shell we take a small element of volume δV as shown in Fig. 3.15. If this element is in equilibrium, we must have ($\delta m = \rho \delta V$)

$$F_{\text{net}} = F_2 - F_1 - \frac{GM_r \delta m}{r^2} = 0$$

$$F_2 - F_1 = \frac{GM_r \delta m}{r^2} .$$

Let us say, for some reason, another element, $\delta V'$ ($= \delta V$), in the same ring has a different density ρ', rather than ρ. We assume this is the only place where the densities are different. Since F_1 and F_2 are due to atoms hitting the outer and inner surface of each element, the values of these forces will be the same whether they are acting on element δV or element $\delta V'$. But F_g' is now different from F_g since $\rho' \neq \rho$ and therefore $\delta m' \neq \delta m$. Thus, element $\delta V'$ is not in equilibrium even though element δV is. We can see

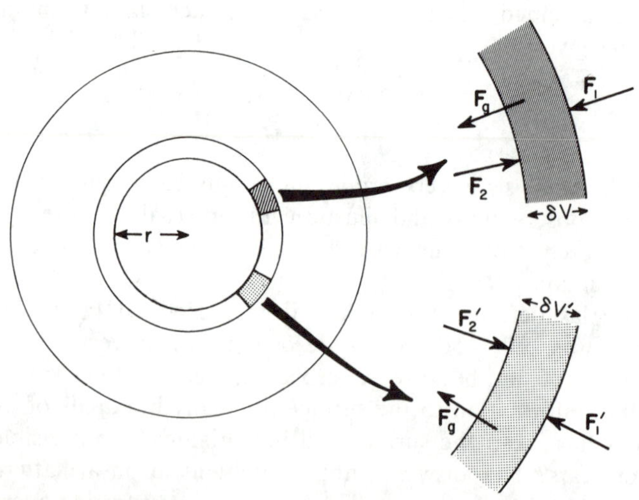

Fig. 3.15. Forces acting on the elements of a spherical shell in the interior of a stellar cloud. The indicated volume elements δV and $\delta V'$ have different densities.

this as follows:

(a) $F'_{\text{net}} = \delta m'\langle a\rangle$ as applied to element $\delta V'$ is

$$\delta m'\langle a\rangle = F_2 - F_1 - \frac{GM_r\delta m'}{r^2}.$$

(b) But as we saw above, $F_2 - F_1$ can be determined from the equilibrium state of element δV,

$$F_2 - F_1 = \frac{GM_r\delta m}{r^2} = \frac{GM_r\rho\delta V}{r^2}.$$

(c) Plugging this into the previous equation, we find that

$$\rho'\delta V\langle a\rangle = \frac{GM_r\rho\delta V}{r^2} - \frac{GM_r\rho'\delta V}{r^2},$$

where we now use $\delta V = \delta V'$.

(d) So then we find

$$\langle a\rangle = \frac{GM_r}{r^2}\left(\frac{\rho - \rho'}{\rho'}\right). \tag{3.16}$$

Thus if we have element $\delta V'$ at density ρ' surrounded by a gas of density ρ, and if $\rho' > \rho$, then $\langle a\rangle$ is negative and element $\delta V'$ will sink. If, on the other hand, $\rho' < \rho$, $\langle a\rangle$ is positive and the element will rise.

Assuming $\rho' < \rho$, so that the element rises, let us follow it for a bit. As the bubble rises, the density surrounding it will change. For a typical star or cloud, ρ in fact decreases as r increases. The bubble itself will expand since the pressure on the bubble decreases as r increases. Thus the density of the element also decreases ($\rho' = \delta m'/\delta V$) as r increases. The important question now is: Which decreases faster, ρ or ρ'? If ρ' manages to remain less than ρ, then the bubble will keep on rising until it reaches the surface of the star. This is illustrated in case 1 of Fig. 3.16. It is possible, however, that ρ

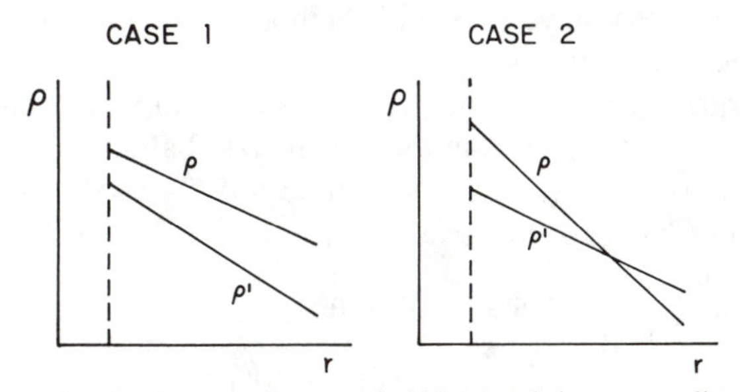

Fig. 3.16. Density distribution vs. r for a gas bubble ρ' and the surrounding gas ρ. In case 1 the bubble rises to the surface, in case 2 it oscillates about its equilibrium position.

decreases faster than ρ'. In this situation we will eventually reach a radius where $\rho = \rho'$ and thus $\langle a \rangle = 0$. The bubble, however, due to inertia, keeps on rising. But any further rise above the point where $\rho = \rho'$ places the bubble in a region where $\rho' > \rho$ and therefore where $\langle a \rangle$ is negative. The bubble must then slow up and possibly stop before it reaches the surface. In the latter case with $\langle a \rangle$ negative and $\mathbf{v} = 0$, the bubble heads back down towards the center of the star. It will now oscillate about the point $\rho' = \rho$ until frictional forces dissipate its translational kinetic energy and cause it to come to rest where $\rho' = \rho$.

To appreciate the energy transport by this convective process, we should ask about the temperature of the bubble vs. the temperature of the surrounding gas. Since F_1 and F_2 are the same on the bubble as on an equivalent amount of surrounding gas, the gas pressure on the bubble P' is equal to the pressure on the surrounding gas. Since both are ideal gases, this means that

$$\frac{\rho' k T'}{\mu m_{\mathrm{H}}} = \frac{\rho k T}{\mu m_{\mathrm{H}}} .$$

Therefore, we have

$$\frac{\rho}{\rho'} = \frac{T'}{T} .$$

Going back to Eq. (3.16), we find

$$\langle a \rangle = \frac{GM_r}{r^2}\left(\frac{\rho'(T'/T) - \rho'}{\rho'}\right) = \frac{GM_r}{r^2}\left(\frac{T' - T}{T}\right). \qquad (3.17)$$

Now we can focus our attention on the temperature. As long as $T' > T$, $\langle a \rangle$ is positive and the bubble rises. If T' should fall below T, $\langle a \rangle$ is negative and the bubble will oscillate. Thus a rising bubble must contain hotter atoms than the gas surrounding it. Hot gases are thus carried upward by the rising bubble, bringing energy to the surface from deep within the cloud. In this way we have a second method of transferring energy from one part of the star to another.

What is surprising and amazing is the speed with which the convective gases rise. Let us start a bubble in the Sun at $R_\odot/2$ and ask how long it will take for it to move $\frac{1}{10} R_\odot$, assuming a modest $\Delta T = T' - T = 1$ °K. Then $M_r \sim 10^{30}$ kg and $T \sim \frac{3}{4} 10^7$ °K give

$$\langle a \rangle = \frac{GM_r}{r^2}\left(\frac{T' - T}{T}\right) \approx \frac{6.7 \times 10^{11}\,(10^{30})(1)}{(7 \times 10^8/2)^2 (\frac{3}{4} \times 10^7)}\ \mathrm{m/sec^2} \sim \tfrac{2}{3} \times 10^{-4}\ \mathrm{m/sec^2}.$$

Assuming the acceleration is constant over $r = \frac{1}{10} R_\odot$, we have

$$\Delta t = \sqrt{\frac{2\Delta r}{\langle a \rangle}} \approx \sqrt{\frac{2 \times (7 \times 10^8/10)}{\frac{2}{3} \times 10^{-4}}} \sim 1.4 \times 10^6 \text{ sec}$$

$$\sim 3 \text{ weeks!}$$

Thus in a matter of months, convective gases, having high temperatures, can travel from deep inside the Sun to the surface. This is to be compared to photon radiative transfer times of 10^6 years!

3.5. Evolution from the Onset of QHE to the Main Sequence

The journey to the main sequence will now be governed by the various methods of energy transfer. With a luminosity of 10^{-4} $L_\odot$, a surface temperature of 50 °K, and a radius of 40 $R_\odot$, we are clearly a long way from our main sequence point on the H–R diagram. The initial phase of the journey will be a spectacular one, after which we will settle down into a paradox.

3.5.1. A 200-Day Wonder—The Convective Transfer of Energy

When we reach the dissociation point we create, just as we do in a pot of boiling water, hot gases. It was the astronomer Hayashi who first pointed out that these dissociated particles will form a hotter gas than their surroundings, resulting in rising convective currents. As we saw above, for a star the size of the Sun it would take ~3 weeks for the rising gases to traverse a region 1/10th the size of the star. For a cloud of 40 $R_\odot$, but at much lower internal temperatures, it takes about the same time. Riding convective currents, therefore, it will take the hotter gases only about 30 weeks (~200 days) to reach the surface. Thus, in a matter of a few months, the surface temperature of the "cloud" will shoot up drastically from 50 °K to over 4000 °K (remember, the gases will cool as they rise). During this time the radius will remain constant (how much could it change over 200 days?) and the luminosity will therefore increase by a factor of 10^6, from 10^{-4} $L_\odot$ to 10^2 $L_\odot$. The path of the star on the H–R diagram during this brief period is shown in Fig. 3.17. Notice that it may take the initial cloud several hundred million years to reach the point of collapse where $E = 0$, then only a few thousand years to hit QHE, and finally only a few months to brighten into a visible star.

Because of the rising currents, it is expected that hot gases will be hurled out of the star during and long after the initial convective phase has

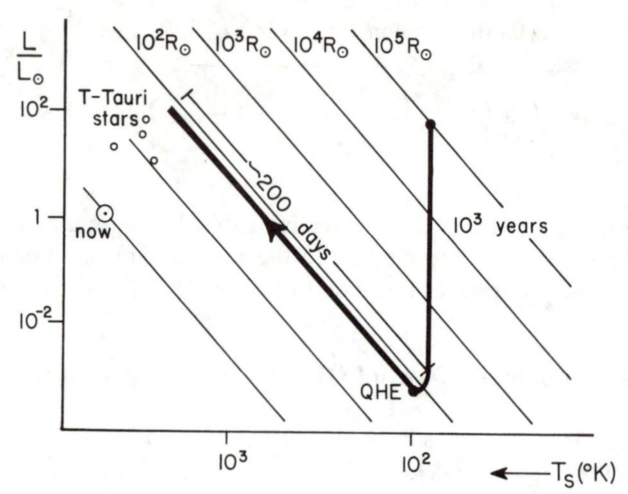

Fig. 3.17. H–R diagram showing the path of a stellar cloud from the onset of collapse, through QHE, until it brightens into a visible star.

brightened the Sun. It is satisfying to note, therefore, the existence of a group of stars called T-Tauri stars which appear to be ejecting hot gases, according to Doppler shift information, and which are located in the expected convective region of the H–R diagram (see black dots in Fig. 3.17).

At this point the star is entirely convective and the rate at which energy leaves the star (its luminosity) will depend upon the convective velocities. There is no radiative link between the center of the star and its surface because with $T_c \sim 10^5$ °K we are at the peak of the opacity curve. Photons would take over 10^8 years to "walk out" to the surface and are clearly, therefore, not competitive with the convective transfer of energy.

3.5.2. Becoming a Radiative Star—Heading for the Main Sequence

At point A in Fig. 3.18a, where the star has brightened into visibility, the Sun is fully convective. Gravity will continue to compress the stellar gas into a smaller and smaller volume. The contraction is slow enough so that we may use QHE to describe the gas configuration, i.e.,

$$T_c = \frac{1}{2} \frac{\mu m_H}{k} \frac{GM}{R} .$$

Thus as contraction proceeds, T_c goes up. In spite of the circulation of hotter gases, the surface temperature, interestingly enough, stays at 4000 °K. The reason is that the hotter temperatures below the surface where ρ is larger, produce an *increasing* value of κ. The increased absorp-

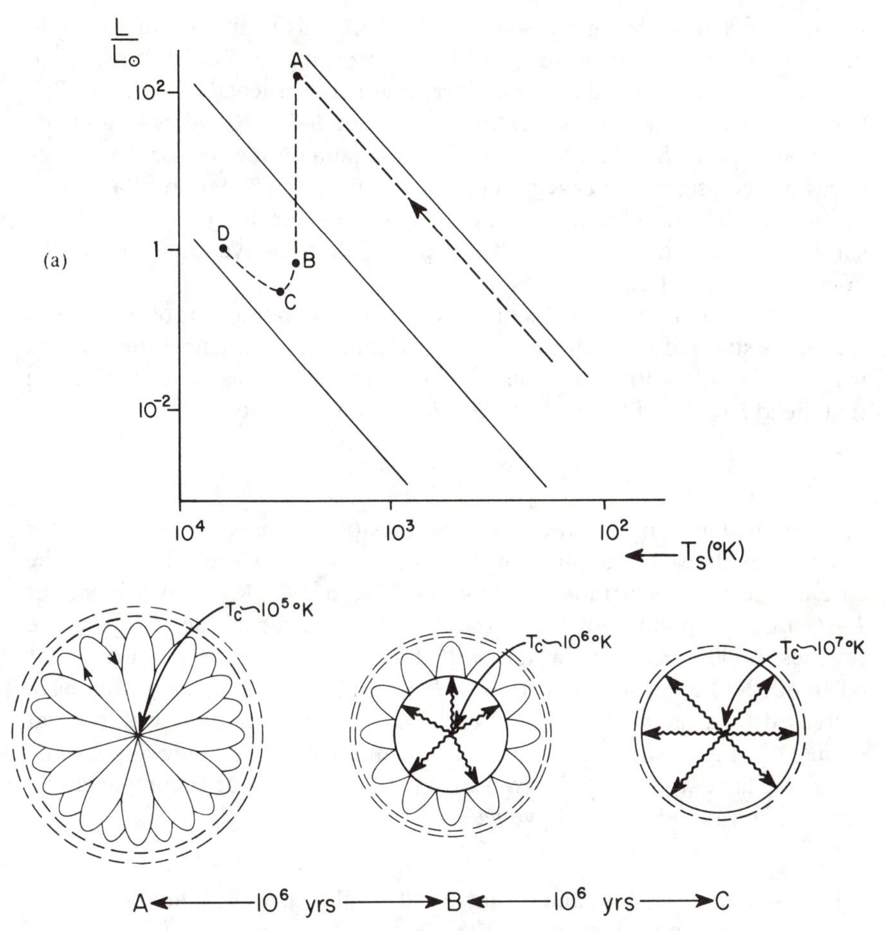

Fig. 3.18. (a) Path of the stellar cloud from the point of brightening into a visible star (A) until it reaches the vicinity of the main sequence (D). (b) Illustrating the interior of the star from its fully convective stage (A), through the onset of radiative transfer (B), until it is fully radiative (C).

tion blocks off the little energy that had previously reached the surface resulting in a surface radiation zone of constant temperature. Since the surface, luminosity is given by $L_s = 4\pi R^2 \sigma T^4$, the path of the convective star will be straight down the H–R diagram to point B.

How long will the convective transfer of energy last? After the star passes point A in Fig. 3.18, where the central temperature is $\sim 10^5$ °K, the absorption coefficient κ at the center of the star decreases, as shown in Fig. 3.13. Eventually we reach a value of κ where the walk time to the surface is a reasonably short 10^6 years. According to the κ vs. T curve of Fig. 3.13 this will occur when T_c has increased to $\sim 10^6$ °K. This is about a factor of 10 increase in central temperature and thus, according to $T \propto 1/R$, a tenfold decrease in the star's radius. Therefore, point B is reached when $R \sim 2R_\odot$. Computer calculations show that it takes about 2–3 × 10^6 years to go from point A to point B. The photons will now dominate the transfer of energy over an increasingly wider region of the star (see Fig. 3.18b) until a million years later the first photons, initiated by reactions in the center, reach the surface. At this time (point C in Fig. 3.18a), radiative transport is the dominant mode of energy transfer throughout the star.

Since the star's luminosity is now entirely due to photons (not hot gases), the entire star will respond to an increase in the central temperature, albeit with a delay of $\sim 10^6$ years. Because $T_c \propto 1/R$ this means, forgetting the time delay, $T_s \propto 1/R$, too. With $L \propto R^2 T^4$, we see that

$$L \propto R^2 \left(\frac{1}{R} \right)^4 = 1/R^2,$$

so that as the star contracts, its luminosity *increases*. This is another example of the self-regulating mechanism built into a star; the hotter the core the more the star radiates. At 4000 °K and $R \sim 2R_\odot$, L will be about $L_\odot/2$ (around point C in Fig. 3.18a). By the time the surface temperature reaches 6000 °K, the radius will have decreased by a factor of $6000/4000 \sim 1.5$, or to almost $\sim 1.3R_\odot$, while the luminosity will have increased by a factor of $(1.5)^2 \sim 2$ to $L \sim L_\odot$. Thus eventually we arrive at point D in Fig. 3.18a. Considering the accuracy of these crude approximations, we have apparently arrived at our destination, our present location on the main sequence—or have we?

3.5.3. A Paradox: How Can a Star Be 10 Million and 10 Billion Years Old at the Same Time?

Let us see if we can estimate how long it takes for the star to go from point C to point D in Fig. 3.18a. Since the luminosity of the star is due to the energy carried off by escaping photons, it also should be directly related to the change in the star's potential energy (remember, Δ(PE) is the original source of the photon energy). We can find the relationship between L and Δ(PE) as follows:

(i) From energy conservation we have $\text{KE}_{\text{tot}} + \text{PE}_{\text{tot}} = E$, where E is the total energy of the star which of course is negative. Since E decreases below zero as photons leave the star, we have

$$E = -E^{\text{lost}}_{\text{radiation}}.$$

(ii) From the virial theorem, $2\ \mathrm{KE}_{\mathrm{tot}} + \mathrm{PE}_{\mathrm{tot}} = 0$, so that $E = \frac{1}{2}\mathrm{PE}_{\mathrm{tot}}$ and thus

$$E^{\text{lost}}_{\text{radiation}} = -\tfrac{1}{2}\mathrm{PE}_{\mathrm{tot}}.$$

(iii) Now the energy radiated away over a time Δt is related to the average luminosity over that interval by

$$\langle L \rangle = \frac{\Delta E^{\text{lost}}_{\text{radiation}}}{\Delta t}.$$

(iv) Since $E^{\text{lost}}_{\text{radiation}} = -\frac{1}{2}\mathrm{PE}_{\mathrm{tot}}$,

$$\langle L \rangle = \frac{\Delta(-\frac{1}{2}\mathrm{PE})}{\Delta t} = -\frac{1}{2}\frac{\Delta(\mathrm{PE})}{\Delta t}. \tag{3.18}$$

Interestingly enough this last equation tells us that $\frac{1}{2}$ the lost PE is radiated away. The rest remains to produce the temperature gradient $\Delta T/\Delta r$. Returning to points C and D in Fig. 3.18a, we find:

(i) The average luminosity between t_{D} and t_{C} is

$$\langle L \rangle \cong \frac{L_\odot + \frac{1}{2}L_\odot}{2} = \tfrac{3}{4}L_\odot.$$

(ii) The change in PE, from $\mathrm{PE}_{\mathrm{C}} = -\frac{3}{5}GM^2/2R_\odot$ to $\mathrm{PE}_{\mathrm{D}} = -\frac{3}{5}GM^2/R_\odot$ is

$$\Delta(\mathrm{PE}) = \mathrm{PE}_{\mathrm{D}} - \mathrm{PE}_{\mathrm{C}} = -\frac{3}{5}\frac{GM^2}{R_\odot} - \left(-\frac{3}{5}\frac{GM^2}{2R_\odot}\right)$$

$$= -\frac{3}{5}\frac{GM^2}{2R_\odot}.$$

(iii) Solving for Δt from (3.18) above we have,

$$\Delta t = -\frac{1}{2}\frac{\Delta(\mathrm{PE})}{\langle L \rangle} = -\frac{1}{2}\frac{\left(-\frac{3}{5}(GM^2/2R_\odot)\right)}{\frac{3}{4}L_\odot}$$

$$\approx \frac{4}{3} \times \frac{3}{20}\,\frac{6.7\times10^{-11}(2\times10^{30})^2}{4\times10^{26}\times7\times10^{8}} \sim 2\times10^{14}\ \text{sec} \approx 10^7\ \text{yr}.$$

Very nice—we are 10 million years old. Right? Wrong! In the early part of the 20th century, physicists accepted the above calculation and believed that the age of the Sun was 10 million years. Then along came the geologists who said NO—the age of rocks "tell" us that we are much older, maybe 10 billion years old—a mere factor of 10^3 longer! Can't be, said the physicists. Must be, said the geologists. Of course we now know the geologists were right. But in a sense so were the physicists. Based on the laws of physics that they knew, only gravity could supply the luminosity of the Sun and therefore when R reached $R_\odot$ and L reached $L_\odot$ their calculations said we are 10 million years old. (Remember, the convective phase took only a few million years or so while the earlier stages only a few thousand.) What the turn-of-the-century physicists did not know and what we now know is that there is another source of energy—nuclear fusion—that can produce the Sun's luminosity and drastically increase its age. So we have a new lease on life—we have a right to be here, more than 10 million years old. How this comes about will be discussed in the next chapter.

3.6. The Virial Theorem and the Maximum Mass of Main Sequence Stars

Leaving aside the question of nuclear fusion, we can explain one property of main sequence stars already noted earlier—the maximum mass of observable main sequence stars, $M_{max} \sim 60\ M_\odot$. The minimum size of main sequence stars will be discussed in chapter 4. In either case the virial theorem plays a central role.

If one looks at the H–R diagram, one sees that the number of stars on the main sequence dwindles as M approaches 50 to 100 $M_\odot$. This indicates an upper limit to the mass of these stars. The intriguing question we might ask is, can we predict or estimate this upper limit? The one handle we have on the problem is that a cloud will not be bound if its total energy is greater than, or equal to, zero. The virial theorem of Sec. 3.2.4, however, tells us that if a cloud is composed of extremely relativistic particles, such as photons, then in fact the total energy is equal to zero. Now the number of particles in a cloud is proportional to M. The number of photons, however, will go as M^4 since, as we have seen for heavy main sequence stars, $L \propto M^4$. Thus a heavy star in QHE on the main sequence will be caught in a contradiction; dominated by photons, the virial theorem coupled to energy conservation says, "Your total energy must be zero. Thus, you must be unstable, and cannot be on the main sequence. You do not exist!!" And indeed this seems to be the case.

How do we go about estimating the grey region where the transition from a dominant particle cloud to a photon cloud takes place? Let us assume as we approach extreme relativistic conditions that the photon energy is at least on a par with the energy of the classical particles. We can establish the grey area quantitatively by estimating for a star on the main sequence,

$$\begin{aligned} KE_{p,tot} &\approx KE_{\gamma,tot} \\ \tfrac{3}{2}Nk\langle T\rangle &\approx a\langle T^4\rangle V. \end{aligned} \tag{3.19}$$

(For the Sun, where particles dominate the total kinetic energy, we have $KE_{p,tot} \gg KE_{\gamma,tot}$.) Since we are approaching the conditions where photons dominate, let us use the relativistic virial theorem, $KE_{tot} + PE_{tot} = 0$. With $KE_{tot} = KE_{p,tot} + KE_{\gamma,tot} = 2\ KE_{p,tot}$, the relativistic virial theorem becomes

$$2\ KE_{p,tot} + PE_{tot} = 0 \tag{3.20}$$

and with $N = M/m_H$, (3.20) reduces to

$$\begin{aligned} 2\frac{3M}{2m_H}k\langle T\rangle &= \tfrac{3}{5}\frac{GM^2}{R} \\ \frac{5k}{Gm_H}\langle T\rangle R &= M_{max}, \end{aligned} \tag{3.21}$$

where we have written M_{max} to indicate our first estimate of the maximum mass of a main sequence star. Unfortunately with $\langle T\rangle$ and R present there are too many variables for us to obtain a unique value of M_{max} from (3.21). Here is where (3.19) comes to the rescue. To evaluate (3.19), however, we need to know $\langle T\rangle$ and $\langle T^4\rangle$. Now that we are using the relativistic virial theorem we must alter the average value of T of Sec. 3.2.5 to $\langle T\rangle = \frac{4}{5}T_c$. But what, pray tell, is $\langle T^4\rangle$? Unfortunately answering this question is not so simple because, $\langle T^4\rangle \neq \langle T\rangle^4$.

We saw in Sec. 3.2.4 that at each point in a constant density star $KE_{p,r} \propto T_r \Delta V_r$ while $KE_{\gamma,r} \propto T_r^4 \Delta V_r$, so that summed over all points

$$KE_{p,tot} \propto \sum T_r \Delta V_r \qquad \text{and} \qquad KE_{\gamma,tot} \propto \sum T_r^4 \Delta V_r.$$

In terms of the averages defined in (3.19) we see that

$$\langle T\rangle = \frac{(\sum T_r \Delta V_r)}{V} \tag{3.22a}$$

and

$$\langle T^4\rangle = \frac{(\sum T_r^4 \Delta V_r)}{V}. \tag{3.22b}$$

Thus $\langle T^4\rangle \neq \langle T\rangle^4$ since $\langle T\rangle^4 = ((\sum T_r \Delta V_r)/V)^4$. We are now faced with the problem of finding $\langle T^4\rangle$. To do so we must evaluate the sum $\langle T^4\rangle = \sum T_r^4 \Delta V_r / V = 3\sum T_r^4 r^2 \Delta r / R^3$. This can be done by the methods of calculus, if we know how T varies as a function of R. A reasonable distribution is the exponential profile $T_r = T_c \exp(-\frac{5}{4}(r/R))$. When plugged into (3.22a) this distribution gives the correct value $\langle T\rangle = \frac{4}{5}T_c$.

Following this lead we plug the distribution into (3.22b) and obtain

$$\langle T^4 \rangle = 3T_c^4 \sum_0^\infty e^{-(5r/R)} \frac{r^2}{R^2} \Delta\left(\frac{r}{R}\right) = 3T_c^4 \sum_0^\infty e^{-(5x)} x^2 \Delta x$$
$$= \tfrac{6}{125} T_c^4.$$

Since $\langle T \rangle = \frac{4}{5} T_c$ we see that $\langle T^4 \rangle = \frac{6}{125} \frac{625}{256} \langle T \rangle^4 \approx \frac{1}{8.5} \langle T \rangle^4$. Assuming the validity of these conditions, and a constant-density star with the above temperature distribution, we are now ready to evaluate (3.19). We substitute $N = M/m_{\rm H}$ and $V = (4\pi R^3/3)$ and obtain

$$\langle T \rangle R = \left(\frac{9kM}{8\pi a m_{\rm H}} 8.5 \right)^{1/3}.$$

If we substitute the above value of $\langle T \rangle R$ into (3.21) we obtain an equation for $M_{\rm max}$ that contains only fundamental constants, the radius and temperature having dropped out,

$$M_{\rm max} = 8.5 \left(\frac{5^3 9}{G^3 8\pi a} \right)^{1/2} \left(\frac{k}{m_{\rm H}} \right)^2 \approx 70\, M_\odot.$$

Thus, as we move away from this mass region, we very rapidly ($\sim M^4$) move toward a photon dominated star where the virial theorem and energy conservation conspire to make $E = 0$. Within the limits of our assumptions we have found the maximum mass of a main sequence star that is consistent with observation.

There is, however, another more qualitative argument as to why heavy stars ($\sim 10^2\, M_\odot$) are not observed on the main sequence. Even if they did exist, then as we learned in Sec. 2.6.2, the upper main sequence mass-luminosity relation $L \propto M^4$ would imply that

$$L_{10^2 M_\odot} / L_{M_\odot} \sim (10^2 M_\odot / M_\odot)^4 \sim 10^8.$$

Such a relation requires that heavy *MS* stars burn up their nuclear fuel very quickly (see Sec. 4.1.1) and live only $\sim 10^6$ years. Thus such stars may live for too short a period to make a noticeable impression on the H–R main sequence line.

The above arguments should not be construed as indicating that stars of $M > 75\, M_\odot$ might not exist off the main sequence. Although they have not been detected, supermassive stars with masses in excess of 1000 $M_\odot$ have been theorized. The trick is to get these huge masses into a relatively small radius so that the internal gravitational forces are strong enough to prevent the radiation pressure from blowing the star apart and to take advantage of general relativity to reduce the photon pressure. For example, in regard to the stronger gravitational force, from our QHE conditions a star will feel a pressure gradient due to gravity of $\Delta P/\Delta r = -\rho G M_r / r^2 \propto M^2/R^5$. As we can see from Fig. 2.33, observed stars on the main sequence have $M/R = \text{constant} = M_\odot/R_\odot$ (we shall see why $M/R \approx M_\odot/R_\odot$ in the next chapter) so that as M increases from the lower to the upper end of the MS curve, M^2/R^5 decreases dramatically (from $10^3\ M_\odot^2/R_\odot^5$ for $M = 0.1\ M_\odot$ to $10^{-6}\ M_\odot^2/R_\odot^5$ for $M \approx 75\ M_\odot$). It is this weakening of gravity along with the increase in photon pressure ($L \propto M^4$) that limits main sequence stars. If, however, we can manage to get off the main sequence and stuff enough mass into a star so that, say, $M/R = 10^4 M_\odot/R_\odot$, then such a supermassive star with $M = 10^4\ M_\odot$ will have $M^2/R^5 = 10^8 M_\odot^2/R_\odot^5$. When gravity reaches these levels it is effective in stabilizing the star not only because the atoms are more tightly bound to each other, but because, under the conditions of general relativity, the photons lose energy to the gravitational field and thus have less momentum available to push on the atoms. Whether nature has produced such a beast remains to be seen.

Problems

1. We have suggested that stars are formed when disk material (gas and "dust") catches up with a density wave. Assuming that a newborn star continues about the galactic center with a circular speed appropriate to its distance from the center, how far will the star move from the spiral arm of its birth in 10^6 years?
2. We saw in the previous chapter that for a diffuse cloud, $KE_{tot}/PE_{tot} \sim 20/1$. If such a cloud were moving through a medium whose hydrogen density was 0.1/cc at a speed of 10 km/sec,
   a. at what rate would mass be added to the cloud?
   b. how long would it take, assuming the accretion rate is constant, for the cloud to become gravitationally bound?

   Take the cloud to be 15 LY across.
3. On the surface of a red giant, $T_s \sim 4000$ °K and $n_H \sim 10^5$/cc. How long would it take to form the seed of a graphite grain ($M_C = 12\ m_H$) which is 1/100 the final grain size?
4. Verify that the orbital angular momentum of Jupiter is 190×10^{41} kg-m/sec^2. Show that its spin angular momentum is negligible compared to its orbital angular momentum ($M_J \approx 2 \times 10^{27}$ kg, $R_J \approx 0.7 \times 10^8$ m, $T_J \approx 10$ hr, $r_{JS} \approx 8 \times 10^{11}$ m).
5. Show that a negative charge is also dragged in the spin direction of a corotating magnetic field.
6. We mentioned in the text that a globe-girdling wire which intercepts the Earth's corotating magnetic field has a major drawback as a power source. What is that drawback?
7. Assume the Sun reduced its angular momentum to its present value by shedding matter into a thin ring of radius r. If at the moment of collapse the radius of the cloud was $2 \times 10^5\ R_\odot$ and its equatorial rotational speed was 100 m/sec, and if it shed ~10% of its mass, what would be the value of r?
8. A globule has $M = 100\ M_\odot$.
   a. What would its radius have to be so that it is just capable of gravitational collapse? Assume a reasonable value of $\langle v \rangle$.
   b. What would its number density be?
   c. What would its luminosity be?
   d. What would its t_{ff} be?
9. Consider Fig. 3.8.
   a. Find the mass contained in a spherical shell of thickness 10 km if it is a distance $r = 10^6$ m from the star's center. Take $\rho_{star} = 2 \times 10^4$ kg/m^3.
   b. What is the gravitational force on the shell?

c. What is the magnitude of the pressure gradient across the shell, if the shell is in QHE?

10. The Sun is conjectured to be 75% hydrogen, 20% helium, and 5% heavy elements.
    a. Find the total number of particles in the Sun if none of the atoms are ionized.
    b. Find the total number of particles in the Sun if all of the atoms are ionized.
11. In the Sun the temperature at the center is $T_c \approx 1.5 \times 10^7$ °K and the central density is $\rho_c \approx 100$ gm/cc $= 10^5$ kg/m^3. Assuming the center to be 100% ionized *helium*, find the
    a. central gas pressure;
    b. central radiation pressure.
12. In the text we set up the pressure gradient equation for QHE. Using the ideal gas law (in modified form) and the equation for $\Delta P/\Delta r$, we can derive an equivalent temperature gradient equation for QHE.
    a. Show that QHE is also given by

$$\frac{\Delta T}{\Delta r} = -\frac{\mu m_{\mathrm{H}}}{k}\frac{GM_r}{r^2}.$$

    b. Returning to the data of problem 9, find the temperature gradient necessary to maintain QHE at $r = 10^6$ m if the material in the shell is 100% ionized hydrogen.
13. Assume that we have a star whose T vs. r curve is *linear* from the central temperature T_c out to the edge of the star at R_0 where $T = 0$.
    a. Show that $T_c = GM^2/NkR_0$ where M is the total mass of the star, R_0 its radius, and N the total number of particles.
    b. Show that the above result implies for a constant-density star that $\langle T \rangle = T_c/5$.
    c. Find that radius in the star where $T = \langle T \rangle$.
    d. Show that at this radius about $\frac{1}{2}$ the star's particles are inside and about $\frac{1}{2}$ outside this point.
14. Say we have a cloud with $M_{\text{cloud}} = 20\ M_\odot$ that is composed of 100% helium. Helium's ionization energies are: 24 eV for the 1st electron and 54 eV for the second.

    From the virial theorem and total energy conservation, estimate the radius of the cloud just at the onset of QHE.
15. Let us say we find a cloud whose mass is estimated at 10 $M_\odot$ and whose radius is measured to be $10^7\ R_\odot$. We believe it is on the verge of collapse.
    a. What would T have to be if we are correct in our assumption that it is about to collapse?

b. Find the luminosity of the cloud at the point of collapse.
c. What will be its early path on the H–R diagram? Explain.
d. Assuming the cloud is 60% H and 40% He, find the radius where QHE sets in.
e. What will be the cloud's central luminosity at this point?
f. What will be its central temperature at this point? How does this differ from the result for a star of 1 $M_\odot$? Explain.
g. Estimate how long it will take to go from 10^7 $R_\odot$ to the point calculated in (d).

16. This problem explores the Compton scattering process.
a. By using conservation of energy, show that a photon that is Compton-scattered off an electron must always lose KE.
b. If an incident photon has an energy of $\frac{1}{2}$ MeV and the scattered electron (initially at rest) has a KE of 0.2 MeV, what is the KE of the scattered photon? What is its frequency and wavelength before and after scattering?
c. The Compton scattering formula reads

$$\lambda_{\text{scatt}} - \lambda_{\text{inc}} = \frac{hc}{m_e}(1 - \cos\theta),$$

where λ_{scatt} is the wavelength of the scattered photon, λ_{inc} the wavelength of the incident photon, m_e the rest mass of the electron, and θ the angle the scattered photon makes with the incident photon.

At what angle does the photon of (b) scatter?
d. Why doesn't Compton scattering off protons change the energy of the photon as much as Compton scattering off electrons?

17. Let us see if we can show that the momentum and total energy cannot be conserved if a photon is absorbed by a completely free electron. Recall, the relativistic expression for the total energy of a particle is given by $E = \sqrt{p^2c^2 + m_0^2c^4}$ where m_0c^2 is the particle's rest mass energy.
a. Write down the equation for the conservation of momentum for $\gamma + e \rightarrow e$. Recall $v_{e,\text{initial}} = 0$.
b. Write down the equation for conservation of total energy for the above reaction.
c. Show that the two equations lead to the result

$$h\nu + m_0c^2 = \sqrt{h^2\nu^2 + m_0^2c^4}\ .$$

d. Show that the above equation leads to the absurd result that either the electron must have zero rest mass or the photon must have zero frequency if momentum *and* energy are to be conserved in the reaction.

18. In Sec. 3.3.1 we noted for inverse bremsstrahlung that the absorption of a photon by a 'free'' electron caused the nearby nucleus to gain very little KE. Show why a nucleus can absorb momentum without a correspondingly large increase in kinetic energy.
19. Use the radiative transport equation to estimate the luminosity at $r = R/2$ for the Sun just as QHE sets in. Compare your result with that derived in Sec. 3.2.3, using the surface temperature of the cloud.
20. What is the *average* photon pressure inside the Sun? In terms of pressure, which are the dominant "particles" inside the Sun, photons or charged particles?
21. A bubble starts out at point a_0 with zero speed. Will it make it all the way to the surface (R_s) of the star if the T vs. r curves for the surrounding gas and bubble are as shown on the left? Explain.

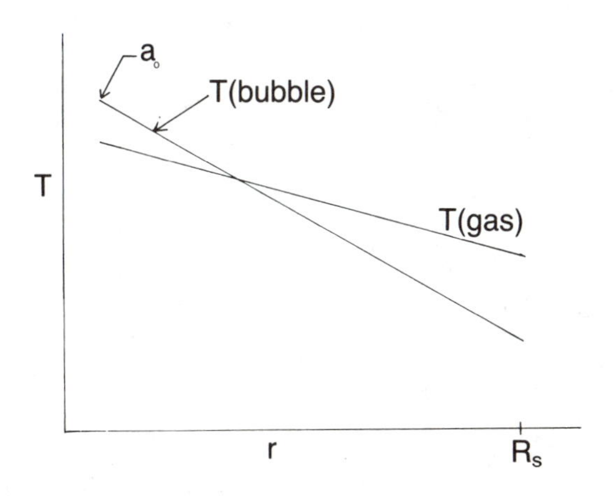

22. In Fig. 3.17 we noted the onset of QHE and the rapid brightening of a cloud of 1 $M_\odot$ into a star by the convective transfer of energy. Calculate the approximate time it will take a gas bubble to go from the center of the cloud to its surface, a distance of 40 $R_\odot$.
23. In problem 15 we calculated the conditions in a cloud of 10 $M_\odot$ at the onset of QHE. As the object continues towards the main sequence, find its luminosity, radius, and surface temperature at points A, C, and D of Fig. 3.18a.

    Estimate the magnitude of any dipole field the star will have at D. Clearly indicate what assumptions you made and what laws of physics are involved in your calculations.
24. We saw in the text that the luminosity of the Sun could be accounted

for by the continuous gravitational collapse of the star. Starting with $L = -\frac{1}{2}\Delta(\mathrm{PE})/\Delta t$, show that a collapsing star's luminosity is related to the instantaneous velocity of the collapsing surface, $v_R \approx \Delta R/\Delta t$, by

$$L = \frac{1}{2}\left(\frac{3}{5}\frac{GM^2}{R^2}\right)v_R.$$

In a recent fly-by of Jupiter it was found that Jupiter radiated more energy than it received. In fact the luminosity of Jupiter was 2×10^{18} J/sec. From this estimate find the collapse velocity v_R of Jupiter's surface.

(Note: $M_J = 2 \times 10^{27}$ kg, $R_J = 0.7 \times 10^8$ m)

25. Referring to Sec. 3.6, show that $\mathrm{KE}_{p,\mathrm{tot}} > \mathrm{KE}_{\gamma,\mathrm{tot}}$ for the Sun.
26. Show that the relativistic virial theorem leads to an average temperature $\langle T \rangle \approx (\frac{4}{5})T_c$.

CHAPTER 4

Stellar Evolution II: Away from the Main Sequence

The waves have a story to tell me,
As I lie on the lonely beach;
Chanting aloft in the pine-tops,
The wind has a lesson to teach;
But the stars sing an anthem of glory
I cannot put into speech.

Robert Service, *The Three Voices*

4.1. The Sun . . . on the Main Sequence

Like the ending in a scene from the *Perils of Pauline*, we left the Sun dangling in time at the conclusion of the last chapter. Will it survive? For how long? If it is to live a long time, what will come to its rescue?

4.1.1. A "New" Source of Energy

The answer to our mystery is found in the innocent "packing fraction" curve of Fig. 4.1. First we define the binding energy BE of a nucleus:

$$\mathrm{BE} = Zm_pc^2 + Nm_nc^2 - {}_ZM^Ac^2,$$

where $Z =$ number of protons, $N =$ number of neutrons, $A = Z + N$ $=$ number of nucleons, and ${}_ZM^A$ = mass of the nucleus.

The binding energy is that energy necessary to split the nucleus apart so that the nucleons are at rest and out of the range of each other's forces. Then the average binding energy is defined as $\langle \mathrm{BE} \rangle \equiv \mathrm{BE}/\mathrm{A}$ and is the total binding energy divided by the number of nucleons. A deuteron,

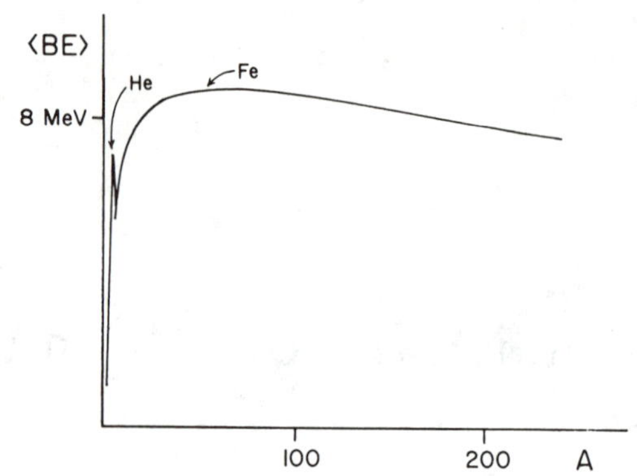

Fig. 4.1. Packing fraction curve of average binding-energy per nucleon vs. number of nucleons.

composed of one proton and one neutron, has an average binding energy of 2.22 MeV/2 = 1.11 MeV. In any nuclear reaction, if the average binding energy of the final state nuclei is greater than that of the initial nuclei, then a net amount of energy is released.

In *fission*, a single nucleus divides into roughly two equal pieces:

$$_{Z}M^{A} \rightarrow _{Z_1}M^{A_1} + _{Z_2}M^{A_2}.$$

If the original nucleus is well above A_{peak} (~60), then clearly the average binding energy of the fragments (where A_1 and $A_2 > A_{\text{peak}}$) is larger than the binding energy of the original nucleus and energy will be released.

In *fusion* two nuclei combine or fuse into a single nucleus:

$$_{Z_1}M^{A_1} + _{Z_2}M^{A_2} \rightarrow _{Z}M^{A}.$$

If the two original nuclei come from well below A_{peak}, then the final average binding energy (where $A > A_1$ and A_2 but $A < A_{\text{peak}}$) will be higher than the original average binding energy and again energy will be released.

Since the Sun has, at this stage, no heavy elements with $A > A_{\text{peak}}$, fission is not possible. Fusion of hydrogen, however, is possible although the protons must have enough kinetic energy to overcome their mutual Coulomb repulsion and get to within one nuclear diameter of one another ($d = 2 \times r_{\text{proton}} \approx 2 \times 10^{-15}$ m). We can estimate the minimum KE to fuse two protons using conservation of energy. Recall that when they are far apart, two protons have zero potential energy and when they just "touch," their kinetic energy is zero. Thus,

$$2\,\text{KE}_{0,\text{min}} + 0 = 0 + \text{PE}_{\text{touching}}$$

therefore,

$$\mathrm{KE}_{0,\min} = \frac{k_0 e^2}{2d} \approx \frac{9 \times 10^9 (1.6 \times 10^{-19})^2}{2 \times 2 \times 10^{-15}} \approx \frac{1}{2}\, 10^{-13}\ \mathrm{J},$$

where d is the distance between the centers of the nuclei when they touch. In electron-volt units this is

$$\mathrm{KE}_{0,\min} \approx \frac{\frac{1}{2} \times 10^{-13}}{1.6 \times 10^{-19}} \approx \tfrac{1}{3}\, 10^6\ \mathrm{eV} = \tfrac{1}{3}\ \mathrm{MeV}.$$

Thus we need KEs on the order of 1 MeV to overcome the Coulomb repulsion and produce fusion. In QHE the only source for this kind of KE would be thermal. This implies we need a temperature like: $\frac{1}{3}10^6$ eV $= \frac{3}{2}kT$ or $T \cong 3 \times 10^9$ °K. But remember we have that long exponential tail on the Maxwell energy distribution curve, Fig. 4.2, ($N_i \times \mathrm{KE}_i$ vs. KE_i). Thus the average KE does not have to be an MeV but can be much less—even a 100 times less—and still produce many fusion reactions. We found a similar situation in chapter 1 when we noted that the temperature (i.e., kT) at the surface of a star that produces appreciable ionization need only be $\sim 1/10$ of the atom's ionization energy. As we mentioned then, we will derive this reduction factor in chapter 7 when we discuss similar phenomena in cosmology. Applying a reduction factor of 100 in this case, we see that when the center of the star reaches a temperature of about 10^7 °K, fusion is ready to take place.

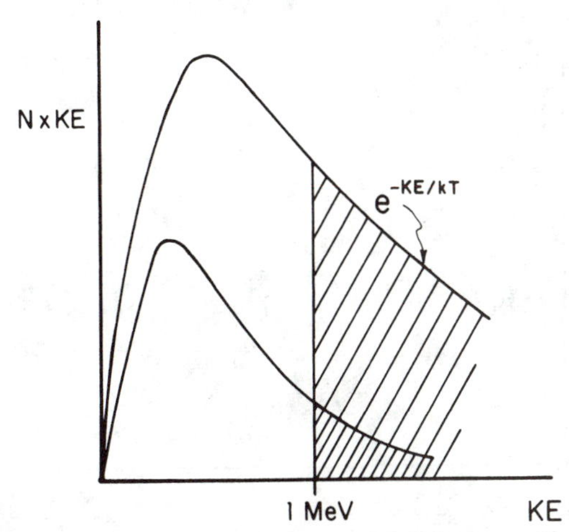

Fig. 4.2. Energy distribution curve vs. KE for two temperatures.

What then are the conditions in the center of the Sun 10 million years after its proton cloud began to collapse? You will recall that when we studied QHE in the last chapter we derived an expression for the temperature at the center of a cloud or star,

$$T_c = \tfrac{1}{2}\,\mu \frac{m_{\rm H}}{k} G \frac{M}{R}\,.$$

For the Sun with $\mu \sim 1$ this is

$$T_c \approx \tfrac{1}{2}\, \frac{1.7 \times 10^{-27} \times 6.7 \times 10^{-11} \times 2 \times 10^{30}}{1.4 \times 10^{-23} \times 7 \times 10^{8}} \approx 1.2 \times 10^{7}\ {}^{\circ}\mathrm{K}.$$

Thus the center of the Sun is at the right temperature for initiating nuclear reactions. The center of the Sun will have enough energy to commence burning some hydrogen just about the time the Sun is 10 millions years old and just when it arrives on the main sequence.

With this insight we should take another look at the significance of the main sequence. Since the Sun is on the main sequence and since the Sun's core is hot enough for fusion, very possibly the main sequence represents the point where all stars have commenced nuclear burning. If this is true, then all stars on the main sequence should have approximately the same core temperature ($\sim 10^7$ °K) and since

$$T_c \propto \frac{M}{R}\,,$$

they should have a constant ratio of mass to radius! The H–R diagram in Fig. 4.3 is a striking confirmation of this prediction: the stars on the main sequence (MS) do have close to a constant mass-to-radius ratio. Thus the

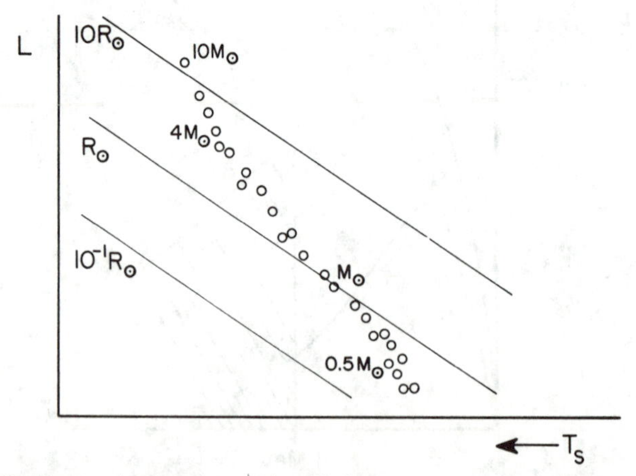

Fig. 4.3. H–R diagram showing the relationship between mass and radius for main sequence stars.

MS must be that line where the core temperature has risen to 10^7 °K and fusion is taking place. So with this new piece of information we can now go back to the drawing board and determine how long the Sun will live with a nuclear source of luminosity instead of the gravitational source that sustained it for its first 10 million years.

The sequence of events in fusion is not simple. It goes something like this:

One Part of the Star	Adjacent Part of the Star
$p + p \to d + e^+ + \nu + (\sum \text{KE})$	$p + p \to d + e^+ + \nu + (\sum \text{KE})$
$d + p \to \text{He}^3 + \gamma + (\sum \text{KE})$	$d + p \to \text{He}^3 + \gamma + (\sum \text{KE})$

where the KEs are shared by the particles in the final state and where the average neutrino energy, E_ν^{avg} is 0.26 MeV. Then the He^3s collide, giving us mainly

$$\text{He}^3 + \text{He}^3 \to \text{He}^4 + 2p + (\sum \text{KE}).$$

The He^4 is so stable that $\text{He}^4 + \text{He}^4$ will not fuse even at these temperatures —so the chain ends. In the case of the weakly interacting neutrinos, their KE is lost because νs escape the star without interacting. However the γ and charged particles, e^+, d, He^3, He^4 do interact with the star. The KE of the charged particles is further distributed via bremsstrahlung and so nuclear fusion becomes a new source of luminosity. You will notice that d and He^3 eventually disappear—they are merely catalysts in the chain. In effect we are taking six protons and making one He^4 and two protons, i.e.,

$$6 \text{ protons} \to \text{He}^4 + 2 \text{ protons} + 2 \text{ positrons} + (\sum \text{KE}).$$

In terms of conservation of energy:

$$6m_p c^2 \to m_{\text{He}} c^2 + 2m_p c^2 + 2m_e c^2 + (\sum \text{KE}).$$

Therefore, we find that

$$(\sum \text{KE}) = 4m_p c^2 - m_{\text{He}} c^2 - 2m_e c^2 = (3753.1 - 3727.5 - 1.0) \text{ MeV}$$
$$\approx 25 \text{ MeV} \approx 40 \times 10^{-13} \text{ J}.$$

So, we use up four protons and in the process release 25 MeV and create one He^4 nucleus. How long can this go on? Until we use up all the protons in the star. Since there are in a solar mass

$$N_{\text{proton}} = \frac{M}{m_{\text{H}}} \approx \frac{2 \times 10^{30} \text{ kg}}{1.7 \times 10^{-27} \text{ kg}} \cong 1.2 \times 10^{57},$$

and we use up four protons to produce one He, the number of reactions we can get out of the star is:

$$N_{\text{reactions}} \approx \frac{1.2 \times 10^{57}}{4} \approx 3 \times 10^{56}.$$

Moreover, 40×10^{-13} J is released for each reaction, so the star has the capability of releasing $3 \times 10^{56} \times 40 \times 10^{-13} \cong 1.2 \times 10^{45}$ J. At the present radiation rate, $L_\odot \approx 4 \times 10^{26}$ J/sec, the Sun has the capacity to shine for

$$\Delta t_\odot \approx \frac{1.2 \times 10^{45}\ \text{J}}{L_\odot} \approx \frac{1.2 \times 10^{45}}{4 \times 10^{26}} \approx 3 \times 10^{18}\ \text{sec}$$

$$\approx 10^{11}\ \text{yr!}$$

So not only does fusion supply the needed longevity, it overdoes it! Assuming we are $\frac{1}{2} \times 10^{10}$ years old, only 5% of our fuel has been used up! Unfortunately, long before we burn even a fraction of our hydrogen, other processes set in which cut short our life as a main sequence star. We shall discuss these processes later.

We can determine the maximum lifetime of stars other than the Sun from the general expression

$$\Delta t \sim \frac{Mc^2}{L}$$

and the mass-luminosity relations we gleaned from the H–R diagram in chapter 2. You will recall that for massive stars, $L = L_\odot(M/M_\odot)^4$, while for less massive ones, $L = L_\odot(M/M_\odot)^3$. In general then, the maximum lifetime of a star, assuming its luminosity has remained constant at its present value, will be

$$\Delta t \sim \frac{Mc^2}{L_\odot(M/M_\odot)^4} \sim \frac{M_\odot c^2}{L_\odot}\left(\frac{M_\odot}{M}\right)^3 = \Delta t_\odot\left(\frac{M_\odot}{M}\right)^3 \quad \text{for } M \gtrsim 3M_\odot$$

or

$$\Delta t \sim \frac{Mc^2}{L_\odot(M/M_\odot)^3} \sim \frac{M_\odot c^2}{L_\odot}\left(\frac{M_\odot}{M}\right)^2 = \Delta t_\odot\left(\frac{M_\odot}{M}\right)^2 \quad \text{for } M \lesssim 3M_\odot.$$

For example, a star of 10 $M_\odot$ will have a maximum lifetime of

$$\Delta t = \Delta t_\odot\left(\tfrac{1}{10}\right)^3 \approx 10^8\ \text{yr},$$

while one of 0.1 $M_\odot$ will have a maximum lifetime of $\Delta t = 100 \times (\Delta t_\odot) \approx 10^{13}$ yr! The upshot is, overweight stars do not live long while lightweight stars do (like people?).

4.1.2. Rate of Energy Generation vs. Temperature

In the previous section we saw that we could not initiate a fusion reaction until the particles coalesced. If we sit on nucleus 1 (charge Z_1e) and look at nucleus 2 (charge Z_2e) approaching with an initial relative velocity v_0, then classically fusion cannot take place unless we have

$$\mathrm{KE}_0 \geqslant k_0 Z_1 e Z_2 e / r_0,$$

where $\mathrm{KE}_0 = \frac{1}{2} m v_0^2$ is the initial KE of nucleus 2 (remember, we are sitting on nucleus 1 so to us its KE is always zero), m is the reduced mass ($m = m_1 m_2/(m_1 + m_2)$), and r_0 is the center to center separation between the particles when they finally touch. For the situation illustrated in Fig. 4.4, $\mathrm{KE}_0 < k_0 Z_1 Z_2 e^2 / r_0$ and classically fusion is not allowed.

In the early 1920s Louis de Broglie, working on his Ph.D. thesis in France, postulated that particles need not always behave in the manner predicted by classical physics. He associated with a particle a probability wave of wavelength λ where $\lambda = h/p$, p being the momentum of the particle. Under certain circumstances the interaction between particles would be determined not by the classical equation of motion but by their wave properties. In 1925 while scattering 54 eV electrons off Nickel crystals, C. J. Davisson and L. M. Germer of Bell Telephone Laboratories observed electrons in amounts in excess of the classical predictions coming off at 130° with respect to the incident beam. By applying the de Broglie wavelength hypothesis to these electrons, along with a wave-interference analysis, they were able to "predict" the angle of this anomalous scattering.

We also note that the quantization of angular momentum of the hydrogen atom which was postulated by Bohr (see Sec. 1.4.3) in 1916 can be derived from the de Broglie wave picture. The trick is to consider an atomic hydrogen energy level as allowed only if a whole number of de Broglie wavelengths can "fit" around the circumference of the orbit, i.e., $n\lambda = 2\pi r$. Since $\lambda = h/p$ and $L = pr$, we have $L = hr/\lambda = hr/(2\pi r/n) = nh/2\pi$, which is the Bohr quantization condition.

Returning to Fig. 4.4, we note that the region inside of r is not classically accessible since a particle in this region would have to have a negative KE. The de Broglie (quantum mechanical) picture on the other hand allows the particle to reach the attractive nuclear force at r_0 by "tunnelling" through the Coulomb barrier. The probability for doing this depends upon the ratio of the thickness of the barrier to the de Broglie wavelength of the particle:

$$\mathcal{P}_1 \propto e^{-2\pi(r - r_0)/\lambda},$$

where $r - r_0$ is the barrier thickness. Since we are only interested in the

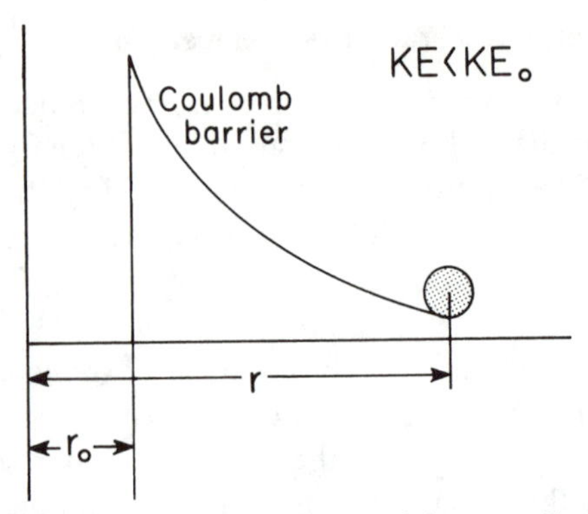

Fig. 4.4. Particle attempting to penetrate the potential energy Coulomb barrier as it approaches a nucleus of radius r_0.

qualitative features of barrier penetration we have simplified the problem by ignoring the fact that the particles slow down as they approach one another and have taken λ to be a constant where $\lambda = h/p = h/mv_0$. Since usually, $r \gg r_0$, we can write

$$\mathcal{P}_1 \propto e^{-2\pi r/\lambda} = e^{-2\pi r/(h/mv_0)}. \tag{4.1}$$

Now conservation of energy tells us exactly how close we can come to the nuclear radius r_0 for a given incident velocity v_0:

$$\tfrac{1}{2}mv_0^2 = k_0 Z_1 Z_2 e^2/r$$

so that

$$r = 2k_0 Z_1 Z_2 e^2/mv_0^2.$$

Plugging r into (4.1), we have

$$\mathcal{P}_1 \propto e^{-(4\pi k_0 Z_1 Z_2 e^2/mv_0^2)/(h/mv_0)}$$

$$\propto e^{-(4\pi k_0 Z_1 Z_2 e^2/hv_0)} \sim e^{-A/v_0}.$$

A plot of $\mathcal{P}_1$ is depicted by the rising curve in Fig. 4.5. Because of the minus sign in the exponential, the greater the velocity the greater the probability of a fusion reaction.

The other important factor is the number of particles with velocity v. This is given by the Maxwell–Boltzmann distribution. Calling this probability $\mathcal{P}_2$, we have

$$\mathcal{P}_2 \propto e^{-(\frac{1}{2}mv_0^2/kT)} \sim e^{-Bv_0^2}.$$

This descending curve is also shown in Figure 4.5.

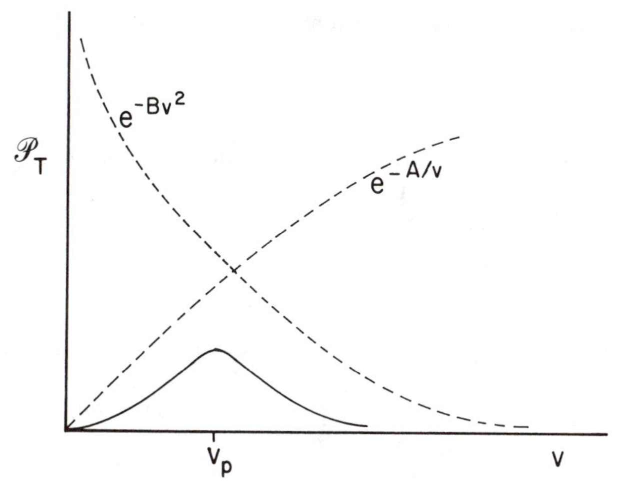

Fig. 4.5. A sketch of probability distributions $\mathcal{P}_1$ and $\mathcal{P}_2$ (dotted lines) and their product (solid line).

From the product of the last two expressions we get the probability that a reaction will go when the relative velocity between the particles is v_0,

$$\mathcal{P}_T \propto \mathcal{P}_1 \mathcal{P}_2 \propto e^{-(\frac{1}{2}mv_0^2/kT)} \times e^{-(4\pi k_0 Z_1 Z_2 e^2/hv_0)}.$$

The product curve is the solid line in Fig. 4.5. Amazingly enough it peaks at one particular point for a gas at a given temperature T. Thus in spite of v ranging from zero to infinity, the two probabilities, $\mathcal{P}_1$ and $\mathcal{P}_2$, conspire to produce a single most probable velocity for a temperature T. That velocity will occur approximately when $\mathcal{P}_1 \approx \mathcal{P}_2$. Solving in this manner for the velocity where the product curve peaks, v_p, we find (note: $k_0 = 9 \times 10^9$ Nt-m^2/Coul^2, $k \approx 1.4 \times 10^{-23}$ J/°K)

$$e^{-(\frac{1}{2}mv_p^2/kT)} \approx e^{-(4\pi k_0 Z_1 Z_2 e^2/hv_p)},$$

and therefore

$$\tfrac{1}{2}mv_p^2/kT = 4\pi k_0 Z_1 Z_2 e^2/hv_p.$$

Solving for v_p we find

$$\begin{aligned} v_p^3 &\approx (8\pi k k_0 Z_1 Z_2 e^2/hm)T \equiv CT, \\ v_p &\approx C^{1/3} T^{1/3}. \end{aligned} \tag{4.2}$$

We can now plug this velocity back into the probability expression to find the value of the probability $\mathcal{P}_T$ at temperature T:

$$\mathcal{P}_T \propto e^{-(\frac{1}{2}mC^{2/3}T^{2/3}/kT)} e^{-(4\pi k_0 Z_1 Z_2 e^2/hC^{1/3}T^{1/3})}$$

$$\propto e^{-(aT^{2/3}/T)} e^{-(b/T^{1/3})} \propto e^{-(a+b)/T^{1/3}}.$$

Now an exponential like this is a horribly complicated quantity to evaluate, but over a small enough temperature region it can be approximated by a power series in T, i.e.,

$$\mathcal{P}_T \propto e^{-(a+b)/T^{1/3}} \propto T^n$$

because both expressions are rising functions of the temperature. The power n will depend upon the constant C in (4.2), which in turn depends upon Z_1, Z_2, m and the temperature region in which we are interested. For the proton–proton chain ($Z_1 = 1$, $Z_2 = 1$, $m_p \approx 1.7 \times 10^{-27}$ kg) around 10^7 °K, it can be shown that

$$\mathcal{P}_T \propto T^4.$$

For the fusion of heavier particles which require higher temperatures, we will have much higher powers of n. For example, a reaction called the triple-alpha (3α) process, where He + He + He $\rightarrow C^{12}$, it can be shown that $\mathcal{P}_T \propto T^{40}$!

The quantity of interest to astrophysicists is not just the probability $\mathcal{P}_T$, but the rate at which energy is generated per kg of material. Usually given the symbol ϵ, we have for the p–p chain

$$\epsilon_{pp} \propto X_p X_p \rho \mathcal{P}_T \propto X_p X_p \rho T^4 \text{ J/kg-sec} \quad (\text{at } T \cong 10^7 \text{ °K}),$$

where X_p is the proton mass fraction, ρ the mass density of matter, and of course T is the temperature. Likewise for the 3α process one finds,

$$\epsilon_{3\alpha} \propto X_{\text{He}}^3 \rho^2 T^{40} \text{ J/kg-sec} \quad (\text{at } T \sim 10^8 \text{ °K}),$$

where X_{He} is the mass fraction of He nuclei.

Once the generation of nuclear energy begins, the bremsstrahlung process produces a nuclear luminosity, L_N, which carries energy away from the production site. When equilibrium is reached, the rate at which energy is generated ($\propto \epsilon_N$) is equal to the rate at which it is carried away ($\sim L_N$). Across a given shell we thus have

$$\Delta L = \epsilon_N \rho 4\pi r^2 \Delta r, \tag{4.3}$$

where ΔL is the increase in the luminosity between r_1 and r_2 due to the energy generated within δV, the latter depicted in Fig. 3.8. Thus unlike black body radiation, which always goes as T^4, the nuclear generation of luminosity can be a much more sensitive function of temperature.

There is another cycle, called the CN chain, that is very important in the evolution of massive stars ($M > 3M_\odot$). The CN reaction chain, considerably more complicated than the p–p cycle, is shown below:

$$p + C^{12} \rightarrow N^{13} + \gamma$$
$$N^{13} \rightarrow C^{13} + e^+ + \nu$$
$$p + C^{13} \rightarrow N^{14} + \gamma$$
$$p + N^{14} \rightarrow O^{15} + \gamma$$
$$O^{15} \rightarrow N^{15} + e^+ + \nu$$
$$p + N^{15} \rightarrow C^{12} + \text{He}^4.$$

In spite of the number of reactions, the CN chain still boils down to $4p \rightarrow \text{He}$ plus the release of about 25 MeV of energy, as was the case for the p–p chain. The temperature dependence of this cycle, however, is particularly strong with

$$\epsilon_{CN} \propto X_H X_C \rho T^{16} \text{ J/kg-sec} \quad (\text{at } T \cong 2 \times 10^7 \text{ °K}).$$

Clearly the star must contain carbon to begin with and as we shall see later in chapter 5, this suggests that the star was formed out of a supernova remnant. But even if the star contains carbon, the CN cycle will not be competitive with the p–p chain (due to the Coulomb repulsion between p and C) until the central temperature is around 2×10^7 °K. Since the central temperature of the Sun does not reach this value, the p–p chain dominates. But massive stars ($M > 3M_\odot$). do reach this temperature because the ratio M/R is not strictly constant but creeps up slowly as mass increases. This also helps explain why the luminosity of stars is so strongly mass dependent even though their central temperature changes by so little; the reaction rate of the CN chain is extremely sensitive to temperature. A slight increase in M/R causes a small increase in T_c, but this in turn causes a major increase in ϵ_{CN} and hence in the star's luminosity. In fact, it was while puzzling over how the relatively temperature insensitive p–p chain could lead to such an obviously strong mass-luminosity dependence that Hans Bethe came up with the CN cycle in 1938.

Equation (4.3) is the last equation we need for determining the structure of a star. How does one go about finding the pressure, temperature, luminosity, and mass distribution inside a star? First of all, one must either know or guess at the temperature dependence of ϵ and κ, the latter being the absorption coefficient we discussed in the last chapter. One must also assume values for the various mass fractions x, y, and z so that $1/\mu = 2x + \frac{3}{4}y + \frac{1}{2}z$ is determined. Then for a spherical volume of $4\pi r^2 \Delta r$ at radius r, the four gradient equations are

$$\frac{\Delta P}{\Delta r} = -\frac{GM(r)\rho(r)}{r^2}, \qquad \frac{\Delta M}{\Delta r} = 4\pi r^2 \rho(r),$$
$$\frac{\Delta T}{\Delta r} = -\frac{3L(r)\rho(r)\kappa}{16\pi acr^2T^3}, \qquad \frac{\Delta L}{\Delta r} = 4\pi r^2 \rho(r)\epsilon. \tag{4.4}$$

Starting, say, at the center of the star we take an educated guess of the values of L_c, M_c, T_c, and P_c, the latter two giving ρ_c via $P = \rho kT/\mu m_H$. The gradient equations generate the new values of L, M, T, and P a distance Δr from the center. The new value of T allows one to find ϵ and κ at this position. One can now plug this complete set of parameters back into the gradient equations to get the next value of L, M, T, and P along with ϵ and κ a distance Δr further out, now $2\Delta r$ from the center, and so on, until one maps out all values of these parameters from the center of the star to the surface. Here one can compare the known surface luminosity, surface temperature, mass, and size of the star with the calculated values. If these do not agree, one goes back to the center and picks new starting parameters and tries again. If it were not for computers this would be an impossible task. Even with computers it is not a trivial process to keep track of, and make sense out of all the parameters that are involved. Nonetheless many of the evolutionary tracks that we show you on the various H–R diagrams are the result of computer calculations associated with (4.4).

Using these techniques, the derived temperature and density profiles of

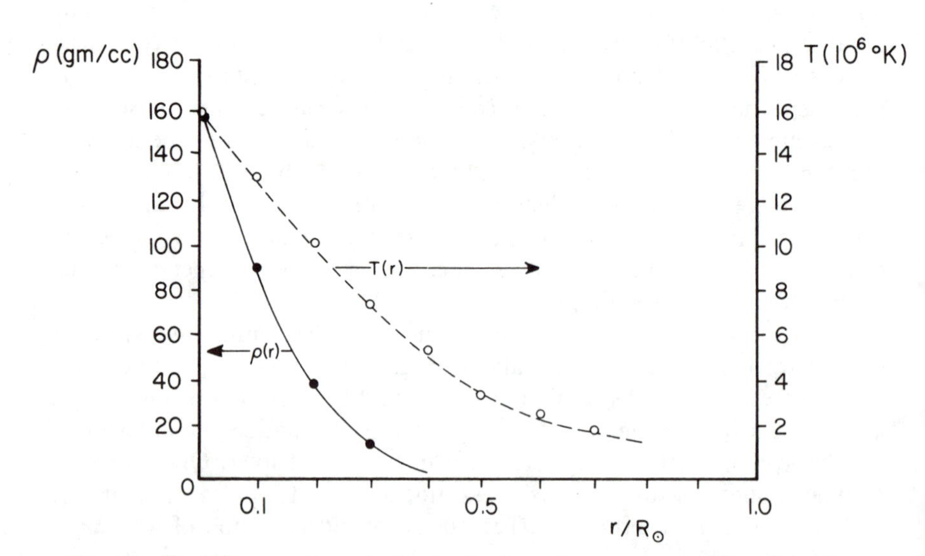

Fig. 4.6. Typical computer results for the density and temperature profiles of the Sun.

TABLE 4.1. SOLAR PARAMETERS

$M_\odot = 1.989 \times 10^{30}$ kg	$X = 72.6\%$
$L_\odot = 3.8 \times 10^{26}$ J/sec	$Y = 26.0\%$
$T_c = 1.53 \times 10^7$ °K	$Z = 1.4\%$
$T_S = 5800$ °K	$\rho_c = 160$ gm/cm^3
	$\kappa_c = 1$ cm^2/gm

the Sun are shown in Fig. 4.6. The present "accepted" values for some of the other parameters are tabulated in Table 4.1.

4.1.3. Where Are the Solar Neutrinos?

Just as soon as you think you really understand what is going on, that is when you usually get a big surprise. Within the last few years that surprise has come from the nuclear process itself. Recall that earlier we showed that photons take 10^6 years to leave the center of the Sun and reach the Earth. Thus, the luminosity we see is due to processes that were initiated in the center of the Sun 10^6 years ago. In the center of the Sun, however, we also have the reaction

$$p + p \to d + e^+ + \nu.$$

Now a neutrino rarely scatters off anything because it is a weakly interacting particle (lepton) involved only in radioactive processes. It has a very small reaction cross section so its path from the center of the Sun is direct —no random walk—which means the time for a neutrino to reach the surface of the Sun is just $\Delta t = R_\odot / c = 2$ sec! Thus, if we could detect neutrinos, they would allow us to monitor the fusion process in the center of the Sun *now*.

First let us estimate the expected neutrino flux ϕ_ν = number/m^2-sec reaching the Earth. The fundamental weak interaction producing νs is beta decay, $n \to p + e^- + \bar{\nu}$ or $p \to n + e^+ + \nu$ and this generates the basic fusion process $4p \to \alpha + 2e^+ + 2\nu$, releasing 25 MeV $= 40 \times 10^{-13}$ J of energy, where the α particle is a helium nucleus made up of $2p$ and $2n$. The number of fusion reactions driving the Sun's luminosity per second is then

$$N_{4p \to \alpha}/\text{sec} \approx \frac{L_\odot}{40 \times 10^{-13}\ \text{J}} \approx \frac{4 \times 10^{26}}{40 \times 10^{-13}} \approx 10^{38}\ \text{reactions/sec},$$

while the number of neutrinos leaving the Sun per sec. is twice this number, $N_\nu/\text{sec} \approx 2 \times 10^{38}$. The solar neutrino flux, the number of neutrinos reaching a 1-m^2 area on the surface of the Earth each second, is then

$$\phi_\nu = \frac{N_\nu/\text{sec}}{4\pi R_{ES}^2} \approx \frac{2\times 10^{38}}{4\pi\left(\frac{3}{2}\times 10^{11}\right)^2} \cong \tfrac{2}{3}\times 10^{15}\ \nu/\text{m}^2\text{-sec},$$

where R_{ES} is the distance between the Earth and the Sun. Note that ϕ_ν is an incredibly large number. Your body, which has an area around 1 m^2, is being inundated by $\sim 10^{15}$ neutrinos each second. It's a good thing neutrinos do not interact very strongly with matter!

But they do interact. One such interaction is

$$\nu + \text{Cl}^{37} \rightarrow \text{Ar}^{37} + e^-.$$

Since Ar is radioactive, we can detect the interaction by looking for the beta decay spectrum of Ar^{37}. This is what R. Davis and collaborators did about a decade ago. They placed a 100,000-gallon tank of Cl^{37} a mile underground in an abandoned salt mine (to make sure only νs reached the detector) and then periodically dumped the chlorine and scanned it for the Ar^{37} decay spectrum. Their only problem was that the rate of $\nu + Cl^{37} \rightarrow Ar^{37} + e^-$ is strongly dependent on the energy of the incoming neutrino. In the solar reaction $p + p \rightarrow d + e^+ + \nu$, the average energy of the neutrino is 0.26 MeV which is below the threshold of the $Ar + e^-$ reaction! Thus only a few neutrinos way up on the high-energy tail of the Boltzmann curve interact.

Along with $p + p \rightarrow d + e^+ + \nu$ is another reaction chain,

$$\text{He}^3 + \text{He}^4 \rightarrow \text{Be}^7 + \gamma + (\textstyle\sum \text{KE})$$

$$\text{Be}^7 + p \rightarrow \text{B}^8 + \gamma + (\textstyle\sum \text{KE})$$

$$\text{B}^8 \rightarrow \text{Be}^8 + \text{e}^+ + \nu \quad \text{with} \quad \text{E}_\nu^{\text{avg}} = 7.2\ \text{MeV}.$$

This neutrino has an average energy of 7 MeV and therefore has a much higher probability of being able to drive the chlorine interaction. Unfortunately the above chain is not the dominant reaction, though it should produce enough high energy νs to produce a detectable signal in Davis's apparatus. Enough, that is, based on the rate calculations from the luminosity produced from the p–p chain. Based on computer solutions of solar dynamics, we believe that the central temperature of the Sun is $T_c \sim 15 \times 10^6$ °K. This in turn predicts that only 0.01% of the neutrinos emitted, mostly from the B^8 decay, have high enough energy (HE) to convert Cl to Ar, with a corresponding reduced flux,

$$\phi_{\text{HE}\nu} = \phi_\nu \times 10^{-4} \sim 7\times 10^{10}\ \text{HE}\nu/\text{m}^2\text{-sec}. \tag{4.5}$$

These HEνs scatter off of Cl^{37} with a weak interaction cross section of order

$$\sigma(\nu + \text{Cl}^{37} \rightarrow e^- + \text{Ar}^{37}) \sim 10^{-46}\ \text{m}^2$$

and the counting rate per neutrino absorption per second per Cl atom is then

$$\text{rate} = \text{cross section} \times \text{flux} \sim 7 \times 10^{-36} \nu/\text{sec-Cl atom} = 7\ \text{SNU}, \quad (4.6)$$

where one solar neutrino unit (SNU) is $10^{-36}\nu/\text{sec-Cl}$ atom. In terms of Davis's tub of chlorine, this rate (4.6) gives

$$7 \times 10^{-36} \frac{\nu}{\text{sec-Cl atom}} \times \frac{1\ \text{Cl atom}}{37 \times 1.7 \times 10^{-27}\ \text{kg}} \times 0.8 \times 10^5 \frac{\text{sec}}{\text{day}}$$

$$\approx 10^{-5} \frac{\nu}{\text{day-kg of Cl}},$$

for about 10 high-energy neutrino absorptions in the 100,000-gal tank of Cl per day.

The rub is that the observed rates (between 1970 and 1978) are 1.7 ± 0.4 SNU instead of the 5 ± 2 SNU predicted from the standard solar model, the latter a refinement of (4.6).

What could be wrong?

(1) Davis's apparatus is not working properly.
Davis has checked and rechecked his equipment. He has invited others to do the same. Everything looks O.K.

(2) The inferred p–p rate is incorrect.
Since the rate of energy production depends on T_c, we could be using the wrong central temperature in the Sun. Recall that the νs we see now left the center of the Sun 10^6 years ago. So T_c may be different now. The ice ages come to mind.

(3) The physical principles we are using are wrong.
This is the most disturbing consequence of Davis's result. If something is wrong with the physical principles as applied to neutrino production, then this would imply a startling defect in our understanding of relativity and nuclear physics.

Recently, some evidence has come to light that the neutrino may have a mass. If so, the possibility of the new types of neutrino interactions could exist which would alter the solar neutrino flux here on Earth.

What can we do? One thing to do is to improve Davis's apparatus so that it is capable of detecting the low-energy neutrinos from the dominant p–p chain. This would allow us to look at the p–p chain directly. The other, besides inventing new physics (no easy task), or finding massive neutrinos, is to look for an explanation which would allow for temperature fluctuations in the center of the Sun of sufficient magnitude to account for a lower p–p rate. If it is lower now, then we are at least safe from "freezing" for about a million years, since it will take that long for the corresponding reduced *photon* flux to reach the surface of the Sun.

4.1.4. At Last . . . On the Main Sequence

When we last left the Sun, it was gravitationally contracting and its luminosity had reached the value $L \cong L_\odot$. Its radius and surface temperature were close to the current solar values as well: $T \cong T_\odot$ and $R \cong R_\odot$. Because $L \propto 1/R^2$ and $T_s \propto 1/R$, its luminosity and surface temperature would continue to increase. Indeed, with no other reactions besides gravity, our star would continue along the dotted line as shown in Fig. 4.7.

With the onset of nuclear fusion, we now have two energy sources: gravity and nuclear fusion. The balance between these two will determine the Sun's evolutionary track.

At first the production of nuclear energy is a small fraction of the production of gravitational energy. Still, the added energy is sufficient to cause the core temperature to rise even faster than it would if gravity were the only energy source. The added luminosity increases the temperature gradient, $\Delta T/\Delta r$, beyond that expected for gravity alone, which in turn slows down the *rate* of contraction, $v_R = \Delta R/\Delta t$, where v_R is the radial surface velocity of the Sun. The temperature gradient and the rate of contraction are tied together through the acceleration equation that we derived in Sec. 3.2.1. You will recall that the acceleration of any shell of material is given by $a = -(\Delta P(\Delta r)/\rho - GM_r/r^2$. Since $\Delta P/\Delta r \propto \Delta T/\Delta r$,

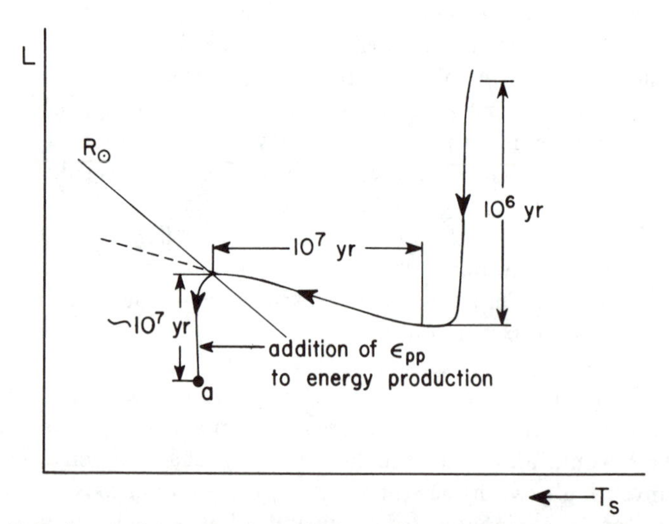

Fig. 4.7. H–R diagram for the Sun from the time it first visibly appears until the production of nuclear energy exceeds that of gravitational energy (point a).

it follows that as the magnitude of the temperature gradient increases, the acceleration is more negative, thus slowing up the positive outward radial velocity of the Sun.

The luminosity, still dominated by gravity, is related to the rate at which the Sun's potential energy changes. Then recalling (3.18) of Sec. 3.5.3, we have

$$L \approx -\tfrac{1}{2}\frac{\Delta(\mathrm{PE})}{\Delta t} = \tfrac{1}{2}\left(\tfrac{3}{5}GM^2\frac{\Delta}{\Delta t}\left(\frac{1}{R}\right)\right) = \tfrac{1}{2}\left(\tfrac{3}{5}\frac{GM^2}{R^2}v_R\right),$$

so that the surface luminosity, L_s, which had been practically constant, decreases as v_R slows down with time. The Sun continues to contract; it moves down the H–R diagram, falling below its present luminosity and radius. This break or turn in the H–R diagram is the first sign that the nuclear process has begun. This slowing down will continue for a few times 10^7 years until the rising core temperature results in the nuclear processes dominating gravity.

Once the nuclear production of energy, which is very temperature sensitive, takes over, the star remains in the vicinity of the main sequence for billions of years. Whereas the changes to the Sun up until now have taken "only" tens of millions of years, the interplay between fusion and gravity will produce a much more stable object. Let us see if we can outline why:

(a) At point *a* in Fig. 4.7, the production of nuclear energy exceeds that of gravitational energy. This will cause the core temperature to rise. This is an immediate and short-term effect. As we shall now see, if it were not for gravity this would lead to drastic short term changes.

(b) The rising core temperature increases ϵ_{pp} (remember, $\epsilon_{pp} \propto T^4$). The greater output of nuclear energy will in turn increase T_c, which will increase ϵ_{pp} even more, which will raise T_c, which will increase ϵ_{pp}, and so on.

(c) The output of nuclear energy produces a new source of luminosity L_N, where $L_N \propto \epsilon_{pp}$. Since $L \propto \Delta T/\Delta r$, the temperature gradient throughout the star will increase as ϵ_{pp} increases. As we showed in Sec. 3.3.2, a rising temperature gradient means a rising pressure gradient (recall, $P = \rho kT/\mu m$ so that $\Delta P/\Delta r \propto \Delta T/\Delta r$). From our discussion of QHE, if $|\Delta P/\Delta r| > |\rho GM_r/r^2|$, the material in the star will experience a positive acceleration. Thus the situation described in (b) above must lead to an expanding star if ϵ_{pp} were the only source of energy.

(d) And here is where gravity comes to the rescue. Any expansion will cause a decrease in the gravitational PE of the star. Since we are in

QHE and the virial theorem holds, we have $T_c \propto 1/R$. Thus, eventually, the core temperature must begin to go down.

(e) But this lowers ϵ_{pp}, which in turn lowers T_c, which lowers ϵ_{pp}, etc. Once again we are on the energy-temperature roller coaster, only now going downhill. The magnitude of the temperature gradients in the star become smaller until the star stops expanding.

(f) The core temperature is now so low that gravity takes over again as the main source of energy. Contraction will occur and T will again go up until we are back at step (a), ready for another cycle.

You will notice that neither gravity nor nuclear fusion can win this game. Too much of one leads to correcting effects due to the other. The overall result is to maintain the star at a constant radius and central temperature. In this situation, the strong temperature sensitivity of the nuclear reaction, $\epsilon_{pp} \propto T^4$, means that fusion turns on and off quickly as T changes. This, coupled with gravity, gives us a sensitive feedback mechanism that tries to maintain the star at a constant radius and temperature. Thus the star stabilizes with any future changes happening in much more subtle ways over much longer time intervals. Indeed, although it has taken the Sun only a few times 10^7 years to make the break and reach point (a), slightly below its present main sequence position, it will take over 4×10^9 years to move back to where $L = L_\odot$ and $R = R_\odot$.

4.2. The Red Giant Phase

Although the Sun will hover around the main sequence for a long time, it never comes to rest at any one point on the H–R diagram. Approximately 10^8 years after crossing the main sequence on the way down in luminosity, nuclear fusion will take over as the Sun's main source of energy production. From this moment on, the Sun slowly begins to move back toward the main sequence and then onward to its red giant phase.

4.2.1. For The Sun . . . A Move Upward

In spite of the trend toward stability, there is a process taking place in the bowels of the Sun that will start it on its way toward becoming a red giant. This long-term effect is the irreversible conversion of hydrogen to helium. Hydrogen will be burning to helium in the center of the star in a core region where the temperature is about 10^7 °K. Four protons disappear for each He nucleus that is formed. Since the pressure that a gas exerts is proportional to the number of particles it contains ($P \propto NkT$), the core pressure becomes inadequate to support the surrounding inert envelope (see

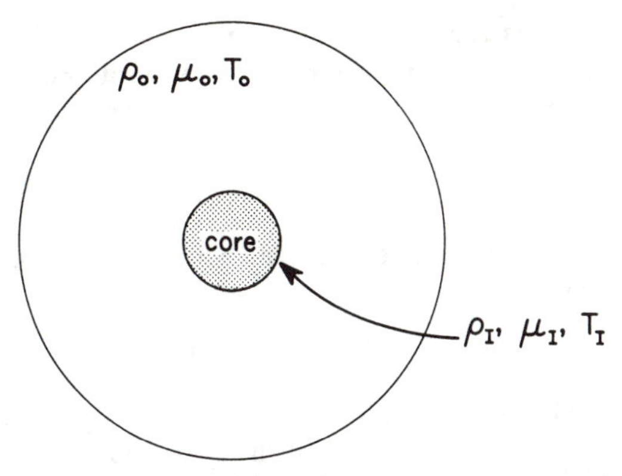

Fig. 4.8. Density, mean molecular weight and temperature parameters in the envelope and core of a star.

Fig. 4.8). The *core* is thus continually squeezed and contracts while $H \rightarrow He$. We can see this more formally as follows: Just outside the boundary between the rest of the star and the core, the ideal gas law gives us

$$P_0 = \rho_0 \frac{kT_0}{\mu_0 m_H},$$

while just inside the boundary we have

$$P_I = \rho_I \frac{kT_I}{\mu_I m_H}.$$

At the boundary we must have $P_0 = P_I$. If not, we would have $\Delta P/\Delta r \rightarrow \infty$ since across a "boundary," by definition, $\Delta r = 0$. Of course $\Delta P/\Delta r$ cannot be infinite since that implies an infinite acceleration, which is incompatible with QHE. Likewise at the boundary, $T_0 = T_I$, otherwise we would have $\Delta T/\Delta r \rightarrow \infty$. But $L \propto \Delta T/\Delta r$, and of course we do not have infinite luminosities. Thus across the boundary we must have

$$\frac{P_0}{T_0} = \frac{P_I}{T_I},$$

and taking P/T from the ideal gas law equations above, we finally have

$$\frac{\rho_0}{\mu_0} = \frac{\rho_I}{\mu_I}. \qquad (4.7)$$

Now for a gravitationally bound star with no nuclear burning, we would have $\mu_0 = \mu_I$ since the composition of the star is uniform. Thus $\rho_0 = \rho_I$. Since the stars we are talking about, however, have reached a phase where the core is burning a noticeable amount of H to He, the amount of hydrogen in the core is decreasing while the amount of helium is increasing. Outside the core these percentages remain constant. For example, in the non-nuclear burning part of the star, using the data from Table 4.1, we have

$$1/\mu_0 = 2 \times 0.73 + \tfrac{3}{4} \times 0.27 = 1.66,$$

while for the nuclear core, where hydrogen is being converted to helium, we *might* have at time t_1

$$1/\mu_I = 2 \times 0.70 + \tfrac{3}{4} \times 0.30 = 1.63,$$

while at a later time t_2, we might find $1/\mu_I = 2 \times 0.60 + \frac{3}{4} \times 0.40 = 1.50$. Thus μ_I, the mean molecular weight of the core, increases as a function of time while ρ_0 and μ_0 remain constant. From Eq. (4.7) we can write

$$\rho_I = \mu_I \left(\frac{\rho_0}{\mu_0} \right).$$

Thus, as more and more H burns to He, the density of the core must *increase* as μ_I increases. The mass of the core, however, remains constant. True, we are burning four protons to get one He nucleus and He is slightly less massive than four protons, but this is of no consequence (indeed, the mass used up is only 0.006 M_{core}) so we can take M_{core} to be constant vs. time. With the density of the core increasing and the mass of the core constant, we are left with only one conclusion: the core must be contracting while H burns to He. But a contracting core will release gravitational PE which means that an extra luminosity ($L_G \propto \Delta\text{PE}/\Delta t$) will be added to the nuclear luminosity ($L_N \propto \epsilon_{pp}$) so that the total luminosity outside the core will go up. This increases the temperature gradient $\Delta T/\Delta r$ throughout the rest of the star and since the acceleration of any layer is given by $a \sim (k/\mu m_{\text{H}})(\Delta T/\Delta r) - GM_r/r^2$, the regions outside the core will expand. That's right, we have a *contracting* core and an *expanding* envelope.

Because the envelope expands, the star does not heat up. The released gravitational energy is used up in external work but not in internal energy. Thus we can treat the average temperature of the star as a constant and write for the radius of the star

$$R = \tfrac{1}{5} \frac{GM_{\odot} m_{\text{H}}}{\langle T \rangle} \mu = \text{const} \times \mu,$$

the last expression focusing on the increasing mean molecular weight as the quantitative reason for the star's expansion. But the tale is only half told. As we shall see shortly, the temperature may not be constant in the core due to the nuclear reactions there. But first, what of the surface temperature and luminosity of the star?

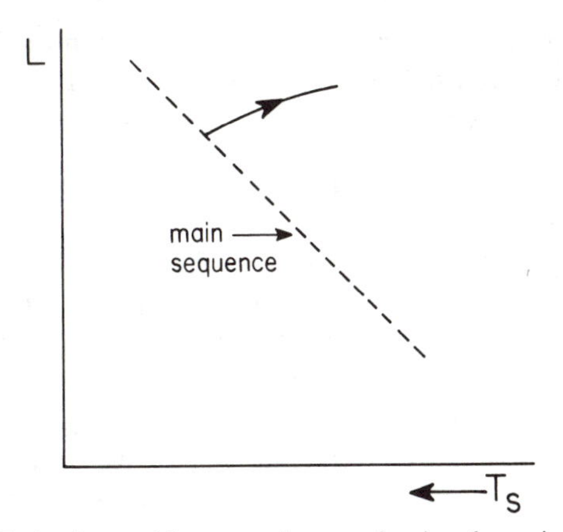

Fig. 4.9. Path of star with contracting core leaving the main sequence.

At the surface of the star, we have a different situation; the temperature behavior is not constant as over the rest of the star. Since no energy reaches the surface, it must expand in such a way that the total energy (E = KE + PE) in any region remains constant. But as the surface particles move further apart, the mutual gravitational potential energy ($-Gm_1m_2/r^2$) will become *less negative*. If PE increases while E stays constant, then KE must decrease. Since $\langle \mathrm{KE} \rangle = \frac{3}{2}kT$, the surface temperature goes down. With the star *expanding* and with T_s decreasing, the surface luminosity will be at the mercy of these two variables, each tugging at the luminosity in the opposite direction. This is reflected in the H–R diagram shown in Fig. 4.9, where there is a moderate increase in luminosity as the surface temperature falls.

As we mentioned previously, however, there is another factor that we must consider before we fully understand how the star's radius changes, and that is the generation of nuclear energy in the core. As H$\rightarrow$He, the mass fraction of hydrogen, X_p, goes down. Although this contributes to the increasing mean molecular weight, it also means there is less hydrogen to burn. Since $\epsilon_{pp} \propto X_p^2 T^4$, the generation of nuclear energy will decrease. Back on the downward-plunging roller coaster again, the core temperature and the nuclear luminosity will go down. If ϵ_{pp} were the only source of energy, the entire star would have to *contract* under the reduced temperature gradient. This of course raises the core and average temperature of the star vis-a-vis $T \propto 1/R$ until equilibrium is restored. Thus instead of an expanding envelope and constant average temperature when ϵ_{pp} is constant, we might have a contracting envelope and an increasing average temperature as ϵ_{pp} fluctuates.

Clearly we have two competing processes and the balance between them will determine the radius of the star; the radius tries to contract as X_p

decreases and ϵ_{pp} struggles back toward its equilibrium value, but also tries to expand under the released gravitational energy of the contracting core as μ increases. Taken together we obtain the radius

$$R \propto \frac{\mu}{T_c},$$

where μ and T_c are both going up as hydrogen burns to helium. Whether μ or T_c wins the race will depend upon the temperature sensitivity of the fusion process. In this regard we divide the stars into two categories: massive, when $M > 3\ M_\odot$ and low mass where $M < 3\ M_\odot$.

In the case of $M > 3\ M_\odot$, where the nuclear energy is derived from the carbon cycle, ϵ_{CN} goes as T^{16}. With this high sensitivity, the decrease in X can be made up with almost no core temperature increase. Thus the behavior of massive stars centers around μ and μ alone and will move off the main sequence as shown on Fig. 4.10.

For $M < 3\ M_\odot$ the nuclear energy comes from the p–p chain, where $\epsilon_{pp} \propto T^4$. The lower temperature dependence means that T_c will have to increase appreciably in order to make up for the depletion of protons, with the result that the average temperature of the star will have to go up. Thus μ and T are truely in competition for the star's radius. Computer calculations show that the increasing average temperature balances the increasing value of μ so that $R \sim$constant. The higher average temperature, however, means higher luminosities. Thus in its hydrogen-burning phase, the star will move almost parallel to the main sequence along a constant radius line as

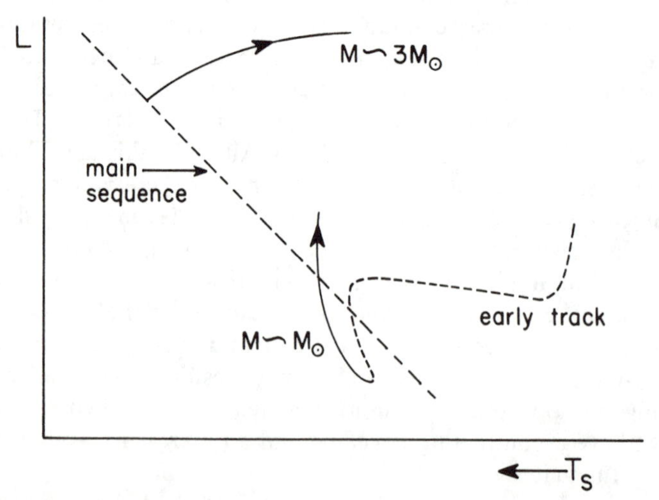

Fig. 4.10. Mass dependence of path of star with contracting core and nuclear generation of energy leaving the main sequence.

shown in Fig. 4.10. The Sun will thus begin to retrace its early evolutionary track, albeit at 1/100th the former rate, as it moves up the H–R diagram.

4.2.2. A Move To the Right

How long will a star of one solar mass keep moving almost parallel to the main sequence toward higher luminosities? As long as we continue to burn H to He, the track on the H–R diagram should be steady. It is only when all the hydrogen in the core burns to He that we can expect the next phase of the Sun's life to begin. When will that come about? Present estimates put $R_c \sim \frac{1}{10} R_\odot$ and $M_c \sim \frac{1}{10} M_\odot$. How much of the core have we used up in the past 5×10^9 years? That and the mass of the core will tell us how much longer we have to go before the hydrogen is depleted. Since $\Delta t_{\text{max}} \propto M_\odot$ and $\Delta t_{\text{now}} \propto \Delta M_{\text{used}}$, we have

$$\frac{\Delta M_{\text{used}}}{M_\odot} = \frac{\Delta t_{\text{now}}}{\Delta t_{\text{max}}} = \frac{5 \times 10^9}{10^{11}} = 0.05. \tag{4.8}$$

Therefore $\Delta M_{\text{used}} = 0.05\ M_\odot$ in the core, and

$$M_{\text{remaining}} = M_{\text{core}} - \Delta M_{\text{used}} = 0.1\ \text{M}_\odot - 0.05\ \text{M}_\odot = 0.05\ \text{M}_\odot .$$

Thus at the present rate of energy generation, it will be another 5×10^9 yr before all the hydrogen in the *core* will have been converted to He. Certainly the middle life of the star is a long and uneventful one. But the star's old age is a different story; it will be short (by comparison), hectic, and filled with adventure, as we shall see during the course of this chapter and the next.

Old age begins, therefore, with the formation of a solid He core. And immediately we have problems. First of all, the temperature to burn He (He + He + He → C) is too high ($\sim 10^8$ °K) for this process to start immediately in the hydrogenless core. Thus the core, which was an energy source when $p + p \rightarrow$ He, is now, except for gravitational energy, dormant. Since only 10% of the star's mass is in the core and since μ is no longer changing, the core will not be a gravitational source either. The consequences are profound. Because the core no longer generates energy, the luminosity inside the core will be zero; that is, $L_r = 0$ for $r < r_c$. But recall that $L_r \propto \Delta T/\Delta r$ and therefore inside the core $\Delta T/\Delta r = 0$; *the core becomes isothermal*, as shown in Fig. 4.11, and herein lies the second problem:

(1) The star would like to remain in QHE.
(2) In order to do this, it must maintain for all r,

$$\Delta P/\Delta r = -\frac{GM_r}{r^2}\rho.$$

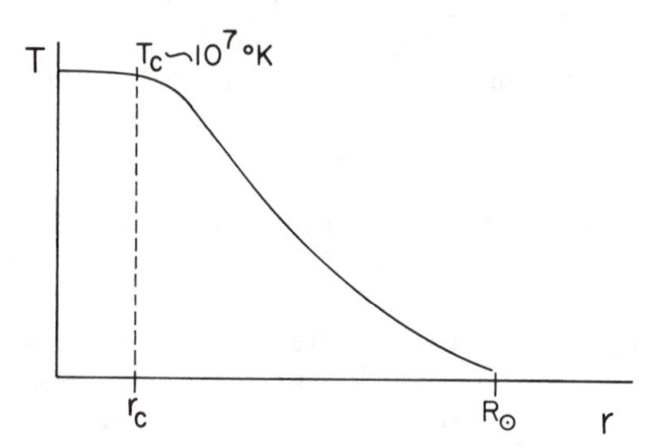

Fig. 4.11. Temperature profile in a star with an isothermal core.

(3) In the core it cannot do this because with $\Delta T/\Delta r = 0$, then $\Delta P/\Delta r$ is also zero.

However, there is a way to save the day. Since $P = \rho kT/\mu m_{\mathrm{H}}$ and since T_c is uniform, we can maintain a nonzero pressure gradient if there is a *density* gradient; i.e., if there is a $\Delta\rho/\Delta r$ of sufficient magnitude such that

$$\frac{\Delta P}{\Delta r} = \left(\frac{\Delta\rho}{\Delta r}\right)\frac{kT}{\mu m_{\mathrm{H}}} = -\rho\frac{GM_r}{r^2}\,. \tag{4.9}$$

Thus the star must develop a density gradient and a steep one at that if it is not to collapse. The next question is, how can it accomplish this?

The answer goes back again to the interplay among gravity, nuclear energy, and luminosity. As soon as the core becomes isothermal, it can no longer support the overlaying envelope. The star contracts and gravitational energy is released which initially, in fact, is the sole source of energy. Eventually the temperature just outside the core becomes high enough to ignite a thin *shell* of burning hydrogen. The star once again has a source of nuclear luminosity.

Before we study the consequences of this hydrogen-burning shell, we would like to follow the path of a star on the H–R diagram between hydrogen exhaustion in the core and the ignition of hydrogen in the shell. Here again we must consider massive and less massive stars separately.

For $M > 3\ M_\odot$, we saw that during the core hydrogen-burning stage, μ controlled the star's expansion. Because $\epsilon_{\mathrm{CN}} \propto T^{16}$, T_c was constant at about 10^7 °K. Since the shell will ignite at 2×10^7 °K, the star has a good deal of collapsing to do before the regions outside the core become hot enough for ignition to take place. The result is a turn toward smaller radii, as shown for path A in Fig. 4.12.

For $M < 3\ M_\odot$, the core temperature, due to the weak $\epsilon_{pp} \propto T^4$ dependence, continuously rises as μ increases. Then when exhaustion occurs, the

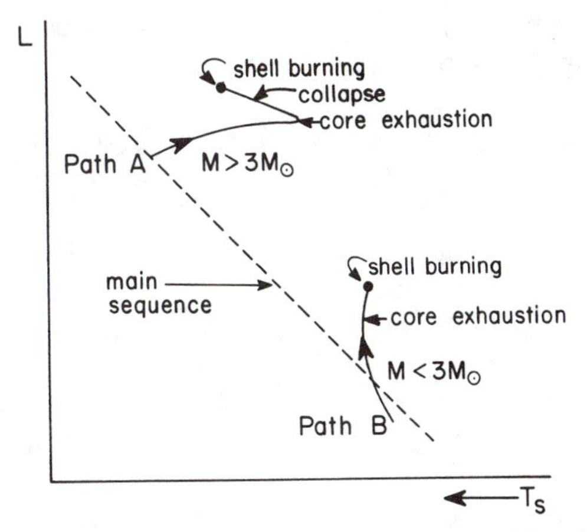

Fig. 4.12. Mass dependence of the transition to the shell-burning stage.

regions outside the core are close to the ignition temperature. There is thus a smooth transition to shell burning without the necessity of the whole star collapsing. The result is path B on the H–R diagram in Fig. 4.12.

What then are the consequences of the hydrogen-burning shell? First, for a less-massive star, the shell will be the sole source of the star's luminosity. At the shell radius r_s, the luminosity jumps from zero to approximately a constant value out to the stellar surface. This is depicted in Fig. 4.13. Second, the shell will be the cause of the density gradient that the star so

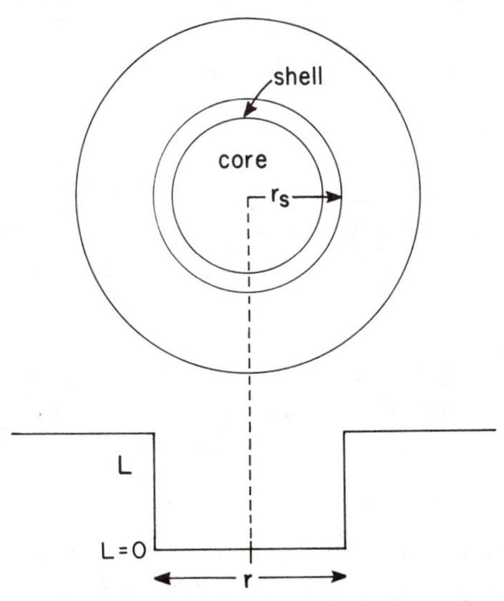

Fig. 4.13. Rapid increase in luminosity at the hydrogen-burning shell.

desperately needs in (4.9) to support the envelope outside the core. Third, the shell will supply the push that sends the star on its way toward its red giant destiny.

In order to understand the last two ideas, we must explore these ancillary points:

(i) If the radius of the shell should contract, the shell will move toward regions of higher density and temperature.
(ii) Since $\epsilon_{\text{shell},pp} \propto \rho T^4$, the inward motion of the shell will increase the rate of nuclear energy production.
(iii) The luminosity of the shell will increase until the rate at which energy is removed from the region ($\sim L$) equals the rate at which it is being produced ($\sim \epsilon_{pp}$).
(iv) At this elevated equilibrium temperature, however, the pressure on the shell ($P = \rho kT/\mu m_{\text{H}}$) is higher, pushing the shell back out. Thus the initial inward perturbation is counterbalanced by an outward force.
(v) If the shell should overshoot its original radius, it moves into a region of lower density and temperature, which results in a lower pressure; the shell falls back inward.

Thus through this feedback mechanism the shell is always pushed back toward its original position. Since there is also some friction, these oscillations will be damped out and the shell will tend to remain at a constant radius. It now follows that the temperature of the shell is constant. The virial theorem tells us that the average temperature inside radius r, which contains mass M_r, is proportional to

$$T \propto \frac{M_r}{r}\,.$$

Applying this expression to the core plus shell region, we see that, since the mass of the core and the mass of the shell are constant (recall the conversion of H to He changes the mass by less than 0.1%) and r_s is constant, the average temperature inside or at r_s must be constant too. Hence the shell temperature is also constant.

What do these facts buy us? Consider a core plus shell as shown in Fig. 4.13. Since r_s is constant, the enclosed volume, $(4\pi/3)r_s^3$, is also constant. As we mentioned above, the amount of mass in the region is also fixed. The average density, then for $r < r_s$, is fixed to be

$$\langle \rho' \rangle = M_{c+s}/(4\pi r_s^3/3),$$

where M_{c+s} is the mass of the core plus shell. The shell's temperature, however, is also constant. Therefore, a continuous stream of radiation is being emitted both inward and outward from the shell. The inward moving photons cause the core to be squeezed into a smaller and smaller radius. This raises the density at the center of the star, ρ_c. Because r_s and M_{c+s} do not change, the average density between the shell and center remains constant. Assuming that

$$\langle \rho' \rangle \approx \tfrac{1}{2}(\rho_c + \rho_s),$$

we see that the density at the shell must go down as the central density goes up. The result is a rising *density gradient*, the very gradient needed in (4.9) to support the star's envelope!

But there is more. The average density outside the shell ($r > r_s$) can be taken to be $\rho'' \approx \frac{1}{2}(\rho_s + \rho_{\text{surface}})$; it is clearly *decreasing* since ρ_s is going down while the density at the surface of the star is essentially zero. The outer envelope, however, encompasses a fixed amount of mass $M_{\text{env}} = M_0 - M_{c+s}$. With a fixed amount of mass and a decreasing average density, the envelope of the star expands; that is

$$R \sim \left(\frac{3}{4\pi} \frac{M_{\text{env}}}{\langle \rho'' \rangle} \right)^{1/3}. \tag{4.10}$$

This expansion will continue as long as the stationary hydrogen-burning ring squeezes the core, increasing the central density and decreasing the shell density.

Of course, the cause of the expansion is the outward flow of energy from the shell. Indeed, the shell monitors both the inward and outward behavior of the star; it squeezes the core, giving the necessary density gradient which causes the envelope to expand via (4.10).

The surface temperature, on the other hand, will go down for the same reasons we described in the last section; with no energy reaching the surface and with energy conserved, the average kinetic energy of the separating surface particles must go down. With the surface temperature going down, with R going up and with $L \propto T^4 R^2$, the surface luminosity will tend toward a constant. For the Sun, all of this means a sharp turn toward the right on the H–R diagram as shown in Fig. 4.14.

4.2.3. Becoming a Red Giant . . . A Move Upward

Thus it would appear that it is our fate to move horizontally on the H–R diagram. And this would be the case except for some interesting things happening in the hydrogen envelope. At and substantially below the stellar surface the temperatures are too low to ionize hydrogen. But as we go deeper into the star, the black body photons increase in average energy until we reach layers where $\langle E_\gamma \rangle \sim 3kT = 13.6$ eV. Generally these layers, at about 30,000 °K, will mark the boundary between ionized and un-

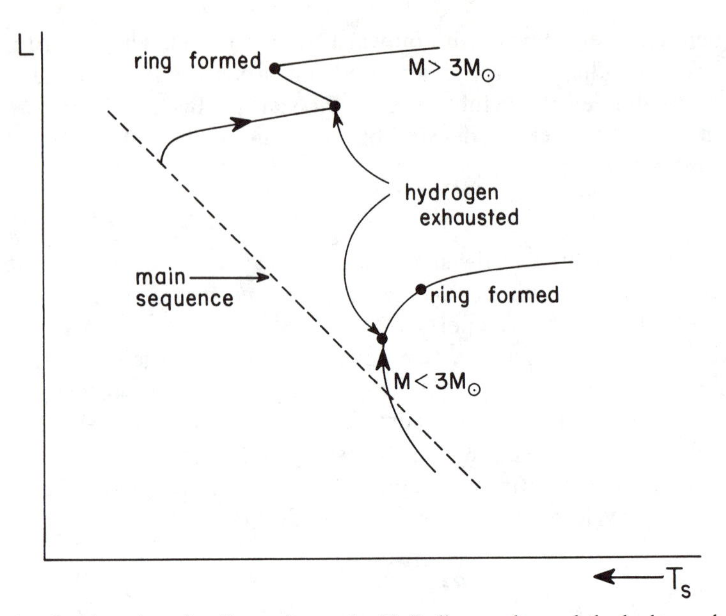

Fig. 4.14. Continuation of stellar paths on the H–R diagram beyond the hydrogen-burning shell.

ionized hydrogen. As the star expands, as shown by curves T_1 and T_2 in Fig. 4.15, the surface temperature decreases while the temperature at the ionization boundary stays fixed at 30,000 °K. Eventually we reach a radius where the curve T_2 falls below the adiabatic curve T'. Any bubbles that form will find their temperature (curve T') greater than that of their surroundings (curve T_2) and will shoot to the surface. Lo and behold we

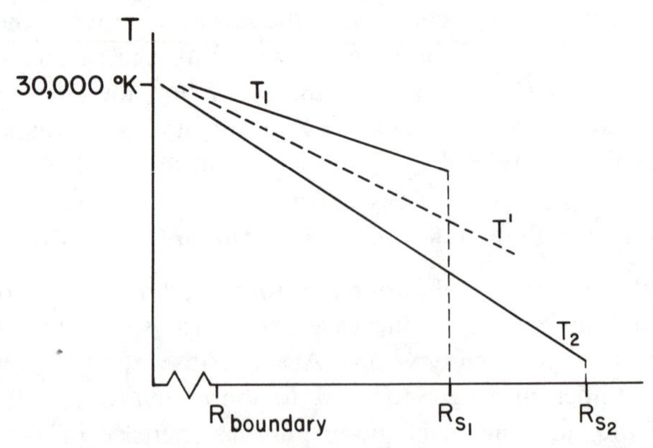

Fig. 4.15. Rough sketch of the temperature gradient as the star expands beyond the boundary between ionized and un-ionized hydrogen.

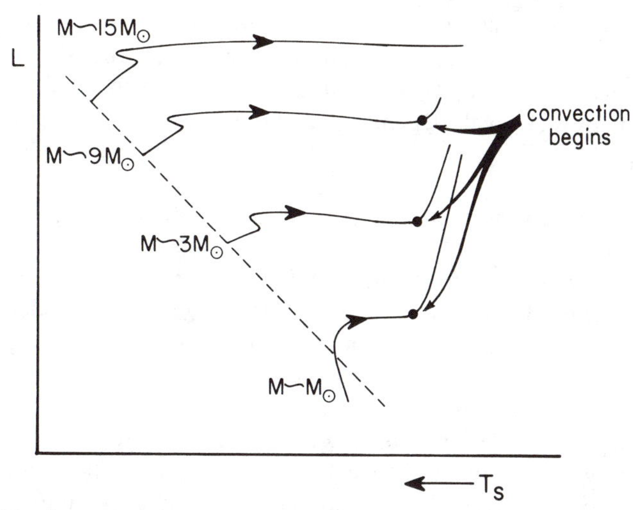

Fig. 4.16. Turning toward the red giant stage through the convective transfer of energy for all but the most massive stars.

have convection between the boundary layer and the surface. Hot gases will begin to rise rapidly due to the surface and the surface luminosity will no longer be controlled by black body radiation (except in the very thin, transparent outermost region). The rate of energy production will depend upon the speed with which hot gases rise to the surface. We saw in chapter 3 that convective flow is much faster than the flow of photons (remember, random walk). Thus the luminosity jumps way up, increasing by a factor of 10^3 in less than 10^9 years. With energy now reaching the surface, the temperature of the surface, which was decreasing, remains almost constant. The H–R diagram thus turns sharply upward for all but the most massive stars ($M \sim 15\ M_{\odot}$), as shown in Fig. 4.16. At this point, the temperature of the surface is about 4000 °K, yielding a $\lambda \sim 3 \times 10^7/4000 \approx 7000$ Å. The star is now red and very very large. It has become a red giant.

It is worth mentioning that it is in this environment of weak gravity ($F \propto 1/r^2$), that the seeds of "dust" grains can form and eventually be blown out by the radiation pressure of the underlying photons. As we shall see later, it is also because of the weak surface gravity that planetary nebulae are able to be ejected from red giants.

In principle it would seem that the upward branch on the H–R diagram would have no bound and that we ought to find stars going way up in size. But a look at the $M3$ globular cluster H–R diagram shows a dramatic cutoff near 100 $R_{\odot}$. Since all the stars in a globular cluster are about the same age ($\sim 10^{10}$ yr old), many should have developed well beyond

100 $R_\odot$. They have not, and to understand why we must leave classical physics and enter the realm of quantum mechanics and the concept of degenerate matter. In fact the subsequent evolution of the star into a white dwarf or neutron star will depend upon the behavior of degenerate matter.

4.3. Quantum Mechanics and Degenerate States of Matter

As the helium core of a star continues to contract, the core density of the gas increases rapidly. When is the gas in the core no longer "ideal"? What happens to matter under very high pressures? What equations govern the behavior of gases under such cramped conditions?

It turns out that such a gas no longer obeys the laws of classical physics, but instead is governed by the laws of quantum mechanics. While an understanding of the dynamical equations of quantum mechanics is beyond the intended level of this book, we may roughly summarize their content by adding to classical physics four radically new concepts:

(i) Uncertainty principle.
(ii) Ground state and discrete energy levels for bound particles.
(iii) Ground state and tunnelling for free particles.
(iv) Pauli exclusion principle.

Strictly speaking, these concepts are linked with a "wave-particle duality" of matter. We shall limit the discussion of the first three concepts primarily to the particle nature of matter, giving examples related to material already covered in this book. Armed with this brief insight, we shall then study in detail (iv), the exclusion principle, as it pertains to degenerate states of matter.

4.3.1. Uncertainty Principle

Take a series of measurements on identical objects, all with supposedly the same position x and magnitude of momentum $|p_x|$. Then the interaction between the measuring instrument and the objects will result in a spread of *simultaneously* recorded values of x and $|p_x|$ as shown in Fig. 4.17. Both quantum mechanics and experiments require the approximate uncertainty inequality

$$\Delta p_x \Delta x \gtrsim h/2, \tag{4.11}$$

where $h = 6.6 \times 10^{-34}$ J-sec is Planck's constant having the dimensions of angular momentum. This relationship is not a statement of poor experimen-

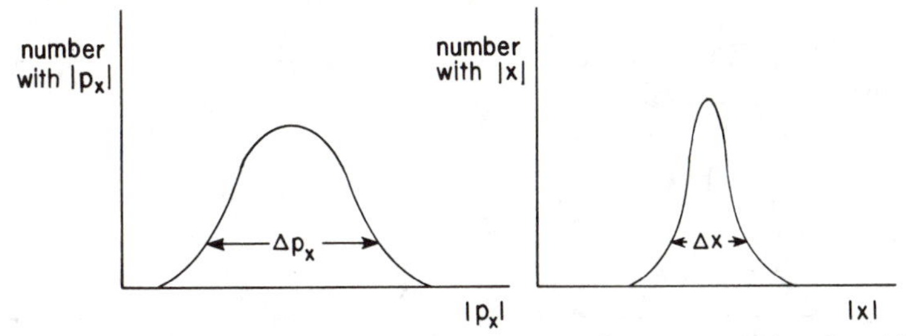

Fig. 4.17. Distributions in simultaneous measurements of momentum (a) and position (b).

tal technique. It is instead intrinsic to the measuring process—related to the wave-particle properties of photons, electrons, etc. An actual experiment yields values of $\Delta p_x \Delta x$ greater than $h/2$; the uncertainty principle (approximate) limit $\Delta p_x \Delta x \approx h/2$ is about the best we can do.

The Planck constant h sets the scale for the quantum domain. As $h \to 0$, $\Delta p_x \Delta x \to 0$ and p_x and x can be independently measured to any accuracy; the classical lower limit is then recovered. Historically, the first role of h in physics was in the explanation of the high frequency part of the photon black body curve (recall Sec. 1.4.2), with intensity proportional to $\exp(-h\nu/kT)$. First setting $h = 0$ in this exponential leads to the classical "ultraviolet catastrophe" as $\nu \to \infty$. In terms of wave-particle duality, h links the energy and frequency of photons and the momentum and wavelength of massive particles as

$$E = h\nu \qquad p = h/\lambda.$$

This duality also leads to tunnelling and nuclear energy generation in stars (recall Sec. 4.1).

4.3.2. Quantum Ground States

In the past we have dealt with bound systems—like the hydrogen atom (in Sec. 1.4.3)—and found the ground state to be the lowest possible energy state for that system. In the case of the hydrogen atom, the ground-state energy was determined experimentally to be -13.6 eV. In the ground state we expect a particle to have its smallest possible kinetic energy as well. Now what do we mean by "smallest possible kinetic energy"? Consider an electron in its ground-state orbit of radius $r = \Delta x$. Then the uncertainty in the electron's momentum is given by $\Delta p_x \geqslant h/2\Delta x$. A repeated set of measurements of the electron's momentum will yield this spread about some mean or average momentum $|\bar{p}|$. As Fig. 4.18 illustrates, consistent with this spread in momentum are many possible values of the average

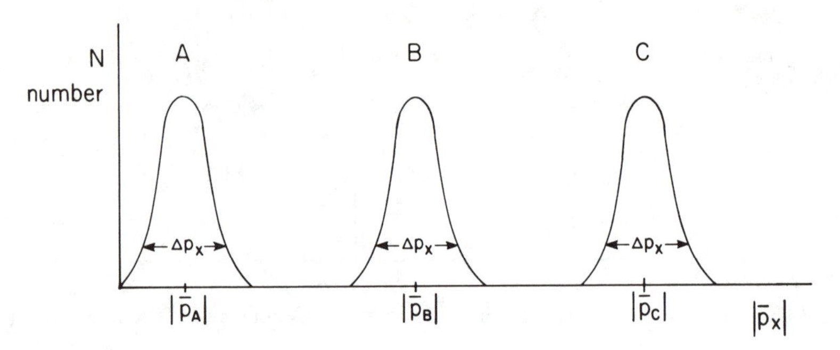

Fig. 4.18. Three possible values of $|\bar{p}|$ all with the same value of Δp_x.

momentum, but the *smallest* value of $|\bar{p}|$ is given by curve A where

$$|\bar{p}_{\min}| = \tfrac{1}{2}\Delta p.$$

To eliminate any confusion about sign we shall always consider the absolute value of the momentum vector (strictly speaking one should work with rms quantities as in Sec. 1.3). Thus the smallest average momentum allowed by the uncertainty principle and consistent with the size of the measured region is

$$|\bar{p}_A| = |\bar{p}_{\min}| \approx \tfrac{1}{2}\left(\frac{h}{2\Delta x}\right).$$

This, in turn, implies a smallest average $\overline{\mathrm{KE}}_{\min}$ of

$$\overline{\mathrm{KE}}_{\min} = \frac{|\bar{p}_{\min}|^2}{2\,m} = \frac{h^2}{32\,m(\Delta x)^2}\,.$$

The average momentum and kinetic energy can be bigger than those values, but never smaller. Now let us see if we can use this to find the size and ground-state energy of the hydrogen atom.

In the *ground state*, the average radial momentum is expected to be $|\bar{p}| = \tfrac{1}{2}\Delta p$, with $r = \Delta x$. The uncertainty principle, $|\bar{p}| = h/4r$, then sets up an outward "uncertainty principle pressure" which increases as the electron gets closer to the nucleus, ultimately balancing the electrostatic attraction between the electron and proton. Stated in terms of the virial theorem (3.9),

$$\mathrm{KE} = \tfrac{1}{2}|\mathrm{PE}|$$

$$\frac{h^2}{32 m_e r^2} = \frac{k_0 e^2}{2r}$$

or

$$r = \frac{h^2}{16 m_e k_0 e^2} = \frac{(6.6 \times 10^{-34})^2}{16 \times 9 \times 10^{-31} \times 9 \times 10^9} \approx 10^{-10}\ \mathrm{m}.$$

This is indeed the size of a hydrogen atom in its ground state! Likewise the virial theorem relates r to the total energy of the electron in orbit, now the ground state energy

$$E = -\tfrac{1}{2}\frac{k_0 e^2}{r} \sim -10 \text{ eV}.$$

The radius and energy equations are approximately equivalent to the more exact quantum analysis as given in Sec. 1.4.3. Here they follow from the uncertainty principle alone. More accurately, $E_1 = -E_I = -13.6$ eV where E_I is the ionization energy of hydrogen. As another remarkable application of the uncertainty principle, we now predict the minimum mass of a main sequence star.

Why would a star have a minimum mass at all? In principle, any mass can be bound if the kinetic energies of the particles composing the object are low enough. And here is the rub. Classically the average kinetic energy can be zero, in which case it would require literally no mass at all to bind the star. That is, $M_{\text{min}} = 0$ and the problem is of no interest. But the uncertainty principle says that we can never have zero average kinetic energy; there must be a minimum average kinetic energy and this implies a minimum mass is needed to bind the star. How do we go about finding M_{min} for a main sequence star (by main sequence here, we refer to an ideal gas of protons and electrons)? Again we use the virial theorem to establish the necessary equilibrium conditions. Unlike the case for the maximum size of such a star (Sec. 3.6), where the virial theorem was on the verge of becoming relativistic, for the minimum size, the particles are nonrelativistic and the usual virial theorem condition for QHE is valid: $\text{KE}_{\text{tot}} = \frac{1}{2}|\text{PE}_{\text{tot}}|$. For constant-density stars, we know that

$$|\text{PE}_{\text{tot}}| = \tfrac{3}{5}GM^2/R; \qquad \text{KE}_{\text{tot}} = \tfrac{3}{5}N(\overline{\text{KE}}_p + \overline{\text{KE}}_e); \qquad N = M/m_p.$$

But what we are interested in is the minimum value of KE and under these ground rules N and $\text{KE}_{p,\text{min}}$ are intimately related. How? You guessed it—by the uncertainty principle.

First consider $\text{KE}_{p,\text{min}}$. The uncertainty principle, $\Delta p \Delta x \geqslant h/2$, says the *minimum* momentum that the protons can have when they are a distance d apart is $|\bar{p}| = \frac{1}{2}\Delta p \approx h/4d$, giving a minimum KE_p of

$$\text{KE}_{p,\text{min}} = p^2/2m_p \approx h^2/32 m_p d^2.$$

On the other hand, if the protons are a distance d apart, they occupy an approximate volume d^3 such that the number of protons $N = M/m_p$ is

$$N = \frac{4\pi R^3}{3d^3}\,. \tag{4.12}$$

To maximize the effects of gravity for a star of a given mass, we want d to be as small as possible. This increases F_g via $1/r^2$ and allows us to bind the star with the smallest possible mass. For a main sequence star, we require that the protons get no closer than $d \approx 2 \times 10^{-15}$ m. If they get any closer than this, the short-range nuclear force will lock them together in a composite nucleus, violating our premise that the protons form an ideal gas. We now see that we can ignore KE_e since $\mathrm{KE}_p \gg \mathrm{KE}_e$. The reason is that the latter has the same uncertainty principle form with d replaced by $a_0 = 10^{-10}$ m, the dimensions of an atom, and $m_p d^2 \ll m_e a_0^2$.

Thus the virial theorem becomes

$$N(\mathrm{KE}_p) = \tfrac{1}{2} G \frac{M^2}{R}$$

and using (4.12) as an upper bound on N,

$$N \frac{h^2}{32 m_p d^2} \leqslant \frac{GM^2}{2R} .$$

Using (4.12) again we can further write

$$N = M/m_p \qquad \text{and} \qquad R = d(3M/4\pi m_p)^{1/3}.$$

Plugging all of this into the above equation and solving for M, we find

$$M^{2/3} \geqslant \frac{h^2 (3/4\pi)^{1/3}}{16 m_p^{2/3} d G}$$

and using $d \approx 2 \times 10^{-15}$ m this gives

$$M_{\text{min}} \approx \frac{h^3}{128} \left(m_p^7 G^3 d^3 \right)^{-1/2} \sim \frac{1}{10} M_\odot . \tag{4.13}$$

Thus with gravity pulling at a maximum and with the kinetic energy at a minimum, the smallest mass star consistent with the uncertainty principle and the virial theorem is about a tenth of a solar mass. From the H–R diagram in Sec. 2.5.3 we see that $\frac{1}{10} M_\odot$ is in fact near the lowest observed mass of main sequence stars.

A logical question to ask now is, what is the ground state, the lowest possible energy state, for a particle that is *not* bound, i.e., one that is free with PE = 0? For an ideal gas (no PE), the average energy of the system is $\langle E \rangle = \langle \mathrm{KE} \rangle = \frac{3}{2} kT$. Thus the lowest energy state would seem to be one where $T \to 0$. But classically

$$\langle \mathrm{KE} \rangle = \tfrac{1}{2} m \textstyle\sum v_i^2 / N$$

for an N particle system, so if $\mathrm{KE} \to 0$, the positive definite quantity $\sum v_i^2$ must also vanish. This requires each v_i to be zero, whlch *contradicts the uncertainty principle*. The latter says that there must be a momentum spread of at least $\Delta p_x \approx h/2\Delta x$. If $p_{\text{min}} = 0$ for some of the particles, others must have $|p_{\max}| = h/2\Delta x$. Therefore the $T = 0$ quantum ground state for free particles, the state with the lowest energy, must have a finite nonzero energy, exerting an outward nonzero pressure. There is no classical explanation for this—it is purely a quantum effect. In short, the uncertainty principle requires particles to be always "nervous," never settling into a state of zero kinetic energy, whether bound or free.

4.3.3. Exclusion Principle and Degenerate Matter at $T = 0$

The exclusion principle says simply that no two electrons can be in the same state at the same time. Pauli used it to explain the known electronic configurations of atoms. In the Bohr atom, such a state was defined by four quantum numbers n, l, m_l, m_s. If m_s is not defined, then two electrons can be in each state defined by the integers n, l, m_l such that $|l| < n = 1, 2, \ldots,$ and $m_l = -l, -l+1, \ldots, l$. The exclusion principle can be extended to any set of identical half-integer-spin particles, called fermions, including spin $\frac{1}{2}$ protons or spin $\frac{1}{2}$ neutrons, as well as spin $\frac{1}{2}$ electrons. The word "spin $\frac{1}{2}$," as we have seen in chapter 2, refers to a spinning particle of angular momentum $S = \frac{1}{2}(h/2\pi)$; spin 1 photons have angular momentum $S = 1\ (h/2\pi)$.

Now let us combine the exclusion principle for fermions with the notion of a $T = 0$ ground state as determined by the uncertainty principle. Consider first a one-dimensional degenerate gas of identical spin $\frac{1}{2}$ fermions in a box of length L at $T = 0$. Again, to avoid confusion about directions we shall only consider the absolute value of **p**. The first particle in the box occupies a *ground state* whose maximum momentum is $|p_{\max}| = \Delta p = h/2L$. A repeated set of measurements of x and $|p|$ will always find x between 0 and L and $|p|$ between 0 and $h/2L$. The second particle in the box can share the ground state, according to the exclusion principle, if its spin (angular momentum) is pointing in the opposite direction to that of the first particle—it cannot be pointing sideways relative to the first particle's spin because the rules of the game require only two possible spin $\frac{1}{2}$ states. This is depicted in Fig. 4.19.

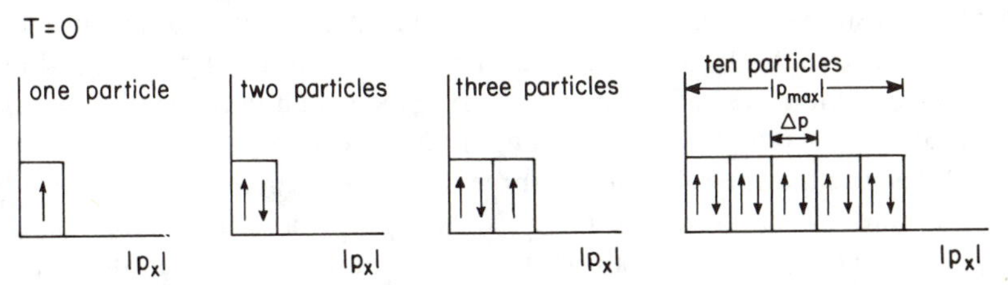

Fig. 4.19. Spin-orientation-dependent momentum distribution of spin $\frac{1}{2}$ particles at $T = 0$.

But what happens when a third particle is put in the box? The lowest level is filled, so it has to go into a higher momentum state in order to be consistent with the exclusion principle. The next highest state, with the lowest possible average momentum consistent with the uncertainty principle has $|p_{\min}| = \Delta p = h/2L$ and $|p_{\max}| = 2\Delta p = 2(h/2L)$. A fourth particle can go into this state too, but with spin pointing in the opposite direction. The pattern should now be clear; ten particles would have a maximum momentum of $5\Delta p = 5(h/2L)$, as shown in Fig. 4.19. In general, for an even number of particles N in this one dimensional box

$$|p_{\max}| = \frac{N}{2}\Delta p = \frac{N}{2}(h/2L)$$

or solving for N,

$$N = 2|p_{\max}|/\Delta p = 2|p_{\max}|(2L/h). \tag{4.14}$$

Thus, there is an intimate relation between the number of particles and $|p_{\max}|$ at $T = 0$. Even at this cold temperature, however, we can have a very large $p_{\max}$ and a correspondingly large average kinetic energy and pressure. In this case $\langle \text{KE} \rangle$ is not related to temperature but to the number of particles in the gas and the size of the box. This is what is meant by "degenerate matter"—it has no classical analog whatsoever! The ground state for N particles at $T = 0$ extends from $p = 0$ to $|p_{\max}|$ with no particles above $|p_{\max}|$.

In the physical world of three dimensions the rules of the game are set up in a similar fashion. For a box with dimensions L_x, L_y, L_z, the uncertainty principle requires a minimum uncertainty

$$\Delta p_x L_x \approx h/2$$
$$\Delta p_y L_y \approx h/2$$
$$\Delta p_z L_z \approx h/2,$$

the very lowest momentum state is still unique and is represented by the cube at the origin in Fig. 4.20. The first two particles will go into this state. Subsequent particles have a variety of states they can enter. For example, the third can keep $|p_x|$ and $|p_y|$ at their lowest value and move into the next highest $|p_z|$ state; or it can keep $|p_x|$ and $|p_z|$ at their lowest value and move into the next highest $|p_y|$ state, etc. Thus the lowest state contains two particles but the next highest contains six. Since each cube on the $|p_x p_y p_z|$ plot represents a possible "state" containing up to two particles in different spin states, the number of states in three dimensions increases dramatically. The volume of the smallest cube is $\Delta p_x \Delta p_y \Delta p_z$ and assuming $|p_{x,\max}| = |p_{y,\max}| = |p_{z,\max}| = |p_{\max}|$, the maximum "volume in momentum space" available up to radius $p_{\max} \equiv p_F$ (Fermi momentum) is $\frac{1}{8}(4\pi p_F^3/3)$, where

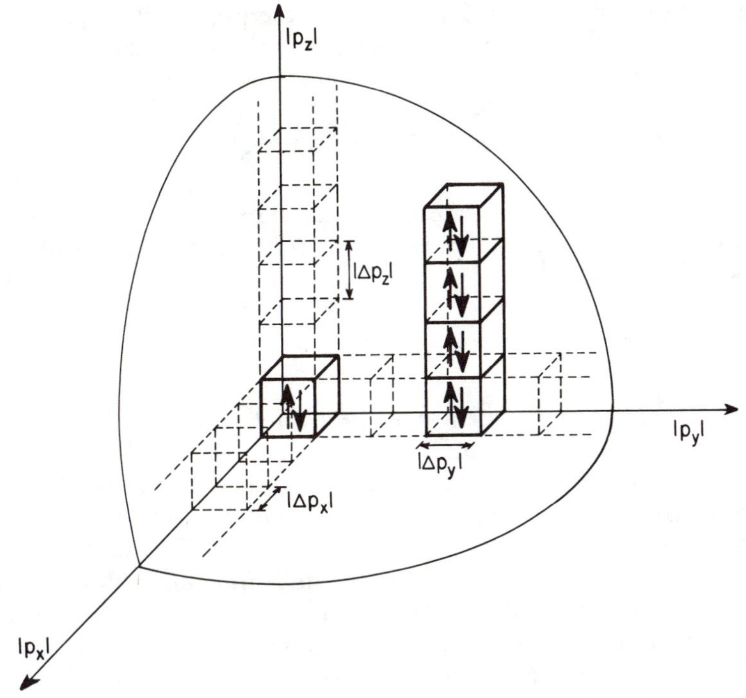

Fig. 4.20. Momentum distribution of fermions in three dimensions.

the factor $\frac{1}{8}$ selects out the positive quadrant in Fig. 4.20, as negative values of $|p_x|$, etc. have no meaning. The total number of particles that can fit into this "momentum space" of radius p_F at $T = 0$ is then

$$N = 2\left(\frac{4\pi p_F^3/3}{8}\right) \times \left(\frac{1}{\Delta p_x \Delta p_y \Delta p_z}\right) \tag{4.15}$$

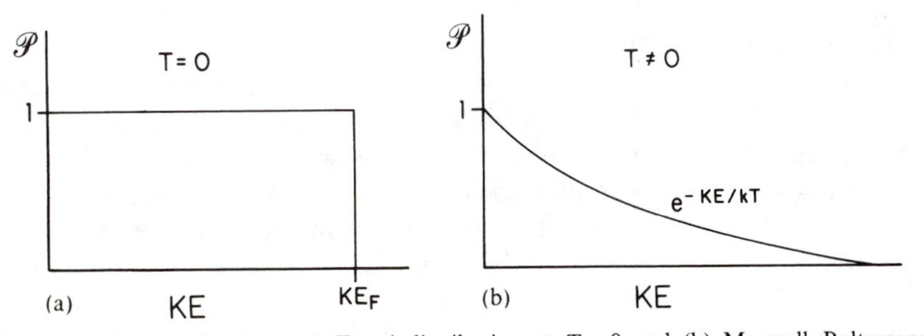

Fig. 4.21. Contrast between (a) Fermi distribution at $T = 0$ and (b) Maxwell–Boltzmann distribution at $T \neq 0$.

where the overall factor of 2 accounts for the two possible spin states in each momentum cell. Now the uncertainty principle implies a minimum cell volume of

$$\Delta p_x \Delta p_y \Delta p_z = \frac{h^3}{8L_x L_y L_z} = \frac{h^3}{8V}, \tag{4.16}$$

where V is the volume of the box. Defining $n = N/V$ as the number density of degenerate matter at $T = 0$, (4.15) and (4.16) lead to the important relation,

$$n = \frac{N}{V} = \frac{8\pi}{3} \times \frac{p_F^3}{h^3}. \tag{4.17}$$

In a more conventional derivation of (4.17), h^3 is the smallest volume in "phase space", while the "phase space volume" occupied by the particles is $(4\pi/3)p_F^3 V$. The number of states N is then given by $2(4\pi/3)p_F^3 V/h^3$, and when the latter expression is divided by V we obtain (4.17).

Notice that n now increases as the cube of $|p_{max}|$. Classically it was the temperature that controlled the momentum distribution with a long KE–Boltzmann tail out to infinity. For the degenerate gas at $T = 0$, it is the density of particles that controls the momentum and KE distributions with a sharp cutoff obtained from (4.17):

$$p_F = h\left(\frac{3}{8\pi}n\right)^{1/3}.$$

For convenience in problem solving, we convert this last expression to energy units by multiplication by the speed of light c:

$$p_F c = hc\left(\frac{3}{8\pi}n\right)^{1/3}, \tag{4.18}$$

where $hc \approx 12 \times 10^{-7}$ eV-m and $(3/8\pi)^{1/3} \approx \frac{1}{2}$. Then in eV units, $p_F c \approx 6 \times 10^{-7}\ n^{1/3}$. The number density is still expressed in atoms per cubic meter.

The concept of the "Fermi energy" is also quite useful. Since there exists a maximum momentum for a given number density n, there must also exist a maximum kinetic energy, called the Fermi energy, $\text{KE}_F \equiv \text{KE}_{max}$. A degenerate gas at $T = 0$ will have KE all the way from zero to KE_F, but then there will be no particles above KE_F. As shown in Fig. 4.21a, we express this fact in terms of the probability of finding a particle vs. its KE. For $\text{KE} < \text{KE}_F$ all states must be filled, so there is a 100% probability ($\mathcal{P} = 1$) of finding a particle below the Fermi energy. For $\text{KE} > \text{KE}_F$ the probability is zero.

Contrast this with the classical Maxwell–Boltzmann distribution for $T \neq 0$ depicted in Fig. 4.21b. In the latter case there are particles with KE from zero to infinity and probabilities ranging continuously from 0 to 1.

How does one obtain KE_F given $|p_{\max}| = p_F$? The general relationship between energy and momentum of a free particle is $E = [(pc)^2 + (mc^2)^2]^{1/2} = \mathrm{KE} + mc^2$, or solving for KE,

$$\mathrm{KE} = \sqrt{(pc)^2 + (mc^2)^2} - mc^2.$$

In the nonrelativistic limit $\mathrm{KE}_{\mathrm{NR}} = p^2/2m = (pc)^2/2mc^2$, while in the extreme relativistic limit $\mathrm{KE}_{\mathrm{rel}} = pc$. A rough rule of thumb for the dividing line between these two limits is $pc \cong mc^2$:

$$\begin{array}{ll} \underline{\text{Nonrelativistic}} & \underline{\text{Relativistic}} \\ p_F c < mc^2 & p_F c > mc^2 \\ \mathrm{KE}_F = \dfrac{(p_F c)^2}{2mc^2} = \dfrac{h^2c^2}{2mc^2}\left(\dfrac{3}{8\pi}n\right)^{2/3} & \mathrm{KE}_F = p_F c = hc\left(\dfrac{3}{8\pi}n\right)^{1/3} \\ \approx 2\times 10^{-13}\dfrac{n^{2/3}}{mc^2}\,(\mathrm{eV}) & \approx 6\times 10^{-7} n^{1/3}(\mathrm{eV}). \end{array} \tag{4.19}$$

For example, at $n = 10^{39}/m^3$ we have from (4.18) $p_F c \cong 6$ MeV. If the particles are electrons with $mc^2 \approx \frac{1}{2}$ MeV $\approx p_F c$, we must use the relativistic expression for KE_F: $\mathrm{KE}_F = p_F c \approx 6$ MeV. On the other hand, if the particles are protons with $mc^2 \approx 10^3$ MeV $\gg p_F c$, we must use the nonrelativistic expression, $\mathrm{KE}_F = (p_F c)^2/2mc^2 = 6^2/2\times 10^3 \approx 20$ keV. Note that p_F does not depend upon the type of particle or whether it is relativistic or not; it only depends upon the number density n in (4.18).

Also note the different dependence of KE on n; $\mathrm{KE}_F \propto n^{2/3}$ in the nonrelativistic limit and $\mathrm{KE}_F \propto n^{1/3}$ in the relativistic limit. Since all the states below KE_F are filled, it is possible to calculate a unique average kinetic energy $\langle \mathrm{KE} \rangle$ for a degenerate gas at $T = 0$. Due to the different n dependence of KE_F, $\langle \mathrm{KE} \rangle$ turns out to be different in the two extreme limits,

$$\begin{array}{ll} \underline{\text{Nonrelativistic}} & \underline{\text{Relativistic}} \\ \langle \mathrm{KE} \rangle = \frac{3}{5}\mathrm{KE}_F & \langle \mathrm{KE} \rangle = \frac{3}{4}\mathrm{KE}_F. \end{array} \tag{4.20}$$

Another point to note about a degenerate gas at $T = 0$ is that, since all the states below KE_F are occupied, the particles cannot lose energy. Why? Because to lose energy means a transition to a lower KE state—but those states are already occupied in the degenerate gas. Consequently at $T = 0$ the particles with $\mathrm{KE} < \mathrm{KE}_F$ move about energetically, accelerating but not radiating energy, because radiation means the creation of energetic photons with the subsequent loss of energy by the radiating particle. Since no lower energy states are open to the particle, there can be no radiation of photons. Thus at $T = 0$ the luminosity is zero—which makes sense classically too.

4.3.4. Degenerate Matter at $T \neq 0$

First of all the probability of finding a particle with $\mathrm{KE} > \mathrm{KE}_F$ will no longer be zero. Some of the states below KE_F will be vacant, as indicated by $\mathcal{P} < 1$ for $\mathrm{KE} < \mathrm{KE}_F$ in Fig. 4.22. Indeed, this must be so because for $T \neq 0$ we do get a luminosity, i.e., emitted photons. The bremsstrahlung-radiating charged particles must be losing energy and so must have lower states to fall into. Therefore there must be vacant states as pictured in Fig. 4.22 where $\mathcal{P}$ is less than one for $\mathrm{KE} < \mathrm{KE}_F$. As the object cools to $T = 0$, the states below KE_F will fill. Conversely, as we raise T, states below KE_F will become vacant and states above will become occupied.

Thus at $T \neq 0$ we have particles obeying two different laws of physics. Those below the Fermi level ($\mathrm{KE} < \mathrm{KE}_F$) will obey the laws of quantum mechanics while the rest, above KE_F, will obey the laws of classical physics. We can then imagine three situations:

(a) A *mostly degenerate* gas where few particles are above KE_F and therefore

$$\langle \mathrm{KE} \rangle \propto n^{2/3} \; (\text{or } \langle \mathrm{KE}_{\mathrm{rel}} \rangle \propto n^{1/3})$$

(see Fig. 4.23a).

(b) A *partially degenerate* gas where a substantial number of states below KE_F are vacant but where many are also filled. Thus we have two nonrelativistic situations to contend with, $\langle \mathrm{KE} \rangle \sim n^{2/3}$ and $\langle \mathrm{KE} \rangle = \frac{3}{2} kT$, which must be combined for the entire gas (see Fig. 4.23b).

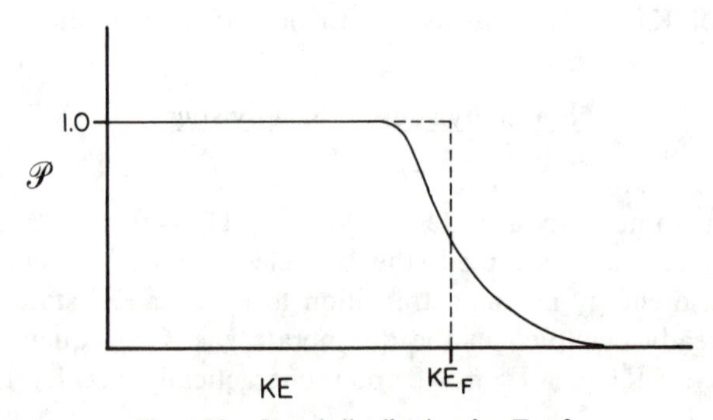

Fig. 4.22. Fermi distribution for $T \neq 0$.

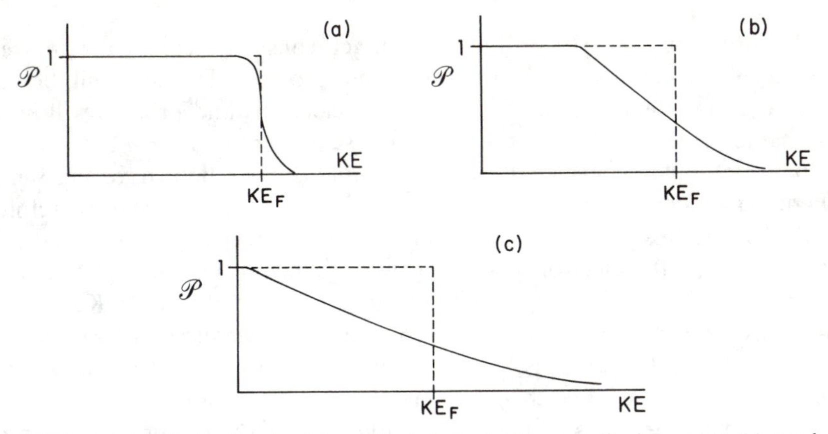

Fig. 4.23. Fermi distribution for (a) mostly degenerate, (b) partially degenerate, (c) nondegenerate gases.

(c) A *nondegenerate ideal* gas where few states below KE are occupied so that

$$\langle \mathrm{KE} \rangle = \tfrac{3}{2} kT$$

(see Fig. 4.23c).

How do we decide if a gas is degenerate, partially degenerate, or nondegenerate? The decision will depend upon T and n. The key is remembering that for an ideal gas $N \propto \exp(-\mathrm{KE}/kT)$ so that most of the particles above KE_F will be found in a band that is kT wide. For the degenerate gas the particles are all found in a band that is KE_F wide. These two curves (Fig. 4.21a and Fig. 4.21b) will exist in tandem. A measure of the number of particles in each state will be given by the widths, kT and KE_F, of the respective distributions. Thus,

(a) when $kT \ll \mathrm{KE}_F$, the gas is fully degenerate.
(b) when $kT \approx \mathrm{KE}_F$, the gas is partially degenerate.
(c) when $kT \gg \mathrm{KE}_F$, the gas is nondegenerate.

Thus the recipe for determining the degree of degeneracy of a gas is as follows:

(a) You have to be given T. Use this to find kT.
(b) You have to be given n. Use this to calculate p_F and then $\mathrm{KE}_F^{\mathrm{NR}} \approx 2 \times 10^{-13} n^{2/3} / mc^2$ (eV) or $\mathrm{KE}_F^{\mathrm{Rel}} \approx 6 \times 10^{-7} n^{1/3}$(eV).
(c) Compare kT and KE_F.

To show you how tricky all of this can get, consider a completely ionized hydrogen gas consisting of electrons and protons. The gas will be, of course, at one temperature. Thus both the electrons and protons will have the same value of kT. They even have the same density, $n_e = n_p$, since each atom contributes one electron and one proton. Thus both have the same Fermi momentum, $p_F c \approx 6 \times 10^{-7} n^{1/3}$ (eV). But $m_p > m_e$ and therefore $KE_{F,\text{proton}}$ is less than $KE_{F,\text{electron}}$. This means that for T such that $kT \ll KE_{F,\text{electron}}$, the electron gas is degenerate since it has the higher Fermi energy—while the proton gas could be nondegenerate with $kT \gg KE_{F,\text{proton}}$. It is also possible that, with n large enough and T small enough, both will be degenerate. The point is, that with two different types of particles, one may be degenerate while the other may be nondegenerate.

As an example let us look at the situation in the Sun (ignoring the dense, small core):

$$\langle \rho \rangle = M/V \approx \frac{2 \times 10^{30}\text{ kg}}{4(7 \times 10^8\text{ m})^3} \approx 1.4 \times 10^3\text{ kg/m}^3$$

$$n_e = n_p \approx 1.4 \times 10^3 / 1.7 \times 10^{-27} \approx 10^{30}/\text{m}^3.$$

Since $T \approx 10^7$ °K we have $kT \approx 0.8 \times 10^{-4} \times 10^7 = 800$ eV. For both particles we have $p_F c \approx 6 \times 10^{-7} n^{1/3} = 6 \times 10^{-7} (10^{30})^{1/3} = 6 \times 10^3$ eV, which is less than $m_e c^2$ and less than $m_p c^2$, so that e and p both can be treated nonrelativistically. Therefore, we have

$$KE_{F,\text{electron}} = \frac{(p_F c)^2}{2m_e c^2} \approx \frac{36 \times 10^6}{2(\frac{1}{2} 10^6)} = 36\text{ eV}$$

$$KE_{F,\text{proton}} = \frac{(p_F c)^2}{2m_p c^2} \approx \frac{36 \times 10^6}{2(10^9)} \approx 18 \times 10^{-3} \sim 2 \times 10^{-2}\text{ eV}.$$

Clearly both Fermi energies are less than $kT = 800$ eV and thus both particle gases are nondegenerate. But if we had somehow managed to increase n_H by a factor of 10^3 so that $n_e = n_p = 10^{33}/\text{m}^3$, then $KE_{F,\text{electron}} = 3600$ eV while $KE_{F,\text{proton}} = 2$ eV. If T is still 10^7 °K, then the electrons would be degenerate and the protons nondegenerate.

4.4. Between Red Giants and White Dwarfs

When we last left our red giant, it had a hydrogen-burning shell that was causing its nondegenerate hydrogen envelope to expand and cool. Its core of He was being compressed and slowly heated. The core was then composed of both He nuclei and electrons. Although it takes four protons to form a helium nucleus, there are not four free electrons per He nucleus. The reason is that the pair of initial $p + p \to d + e^+ + \nu$ reactions produce one positron for each reaction, or a total of two e^+s for every He nucleus that is produced. The two positrons annihilate two of the four electrons, leaving two e^-s left over—just enough to keep the net charge of He nucleus + electrons equal to zero with $2e^-$ for every α particle. Thus in the core we have the number density relation

$$n_e = 2n_{\text{He}}.$$

4.4.1. The Onset of Degeneracy and the Structure of the He Core

Clearly if anything is going to become degenerate, it will be the electrons in the compressed He core. We have already seen that in 8×10^9 yr 10% of the Sun's mass will have been burned to He so that $M_{\text{core}} = 0.1 M_\odot$ and $r_{\text{core}} \sim 0.1\ R_\odot$. As the Sun moves into the red giant phase, the mass of the core will remain approximately the same but the core will shrink. Although the size of the red giant core is not well known in the compressed state, it is believed to be about $\frac{1}{70} R_\odot \approx 10^7$ m with $M_{\text{core}} \sim 0.2\ M_\odot$ at the time convection begins (see Fig. 4.16), almost 10 billion years after the onset of hydrogen burning. At this time the temperature of the core will be about 2×10^7 °K.

From this information we can calculate the number density of He nuclei,

$$n_{\text{He}} = \frac{M_{\text{core}}}{m_{\text{He}}} \left(\frac{1}{V_{\text{core}}} \right)$$

$$\approx \frac{0.2 \times 2 \times 10^{30}}{4 \times 1.7 \times 10^{-27}} \left(\frac{3}{4\pi \times (10^7)^3} \right)$$

$$\approx 16 \times 10^{33} \text{ nuclei/m}^3.$$

The number density of electrons, therefore, is

$$n_e = 2n_{\text{He}} \approx 32 \times 10^{33}/\text{m}^3.$$

Now we find the Fermi momentum of the electrons,

$$p_F c = hc \left(\frac{3}{8\pi} n \right)^{1/3} \approx 6 \times 10^{-7} (32 \times 10^{33})^{1/3}$$

$$\sim 10^5 \text{ eV}.$$

Since $p_F c \ll m_e c^2$, we obtain the nonrelativistic Fermi energy as

$$\text{KE}_{F,\text{electron}} = \frac{(p_F c)^2}{2mc^2} \approx \frac{10^{10}}{2(\frac{1}{2} 10^6)} \approx 10^4 \text{ eV}.$$

Comparing this to $kT_{\text{core}} \approx 0.86 \times 10^{-4}(2 \times 10^7) \approx 2 \times 10^3$ eV, we see that the electrons in the core are partially degenerate. Detailed calculations show that 46% of the total pressure is supplied by the degenerate electrons. Note that with $m_{\text{He}} \approx 4 \times 10^9$ eV, the He nuclei are not degenerate.

With the onset of convection and further compression of the core, the degeneracy increases, eventually supplying close to 100% of the pressure in the star. What happens to the He nuclei and the structure of the core at these high densities?

We would like to approach this problem by first assuming that the electrons are absent! With this assumption, each He nucleus sees the full positive charge, $2e^+$, of each of his neighbors. The spacing between He nuclei is

$$l = \left(\frac{1}{n}\right)^{1/3} \approx \left(\frac{1}{16 \times 10^{33}}\right)^{1/3}$$

$$\approx \tfrac{1}{2} 10^{-11} \text{ m.}$$

With spacings this small, the doubly charged He nuclei will exert tremendous forces on one another with

$$F_{12} = k_0 (2e)^2 / l^2$$

and depending on the temperature they could freeze out into a lattice structure like the one shown in Fig. 4.24.

It is important to appreciate that a lattice structure is a very stable configuration. Consider nucleus #2 in Fig. 4.24. The *net* force acting on it will be zero when it is in the exact center of the lattice. If this nucleus should be perturbed to the right, however, then F_{23} will increase to the left while F_{12} will decrease to the right. The net force is no longer zero but points to the left and this results in a displacement back toward the

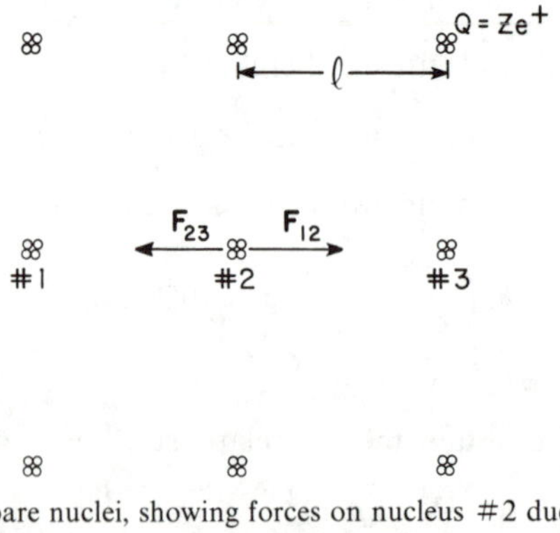

Fig. 4.24. An array of bare nuclei, showing forces on nucleus #2 due to nuclei #1 and #3.

equilibrium position. Any swing past equilibrium will always be greeted by a net force pointing in the opposite direction. Thus the middle nucleus is bound providing its initial KE is not too large. Of course if the temperature is high enough, then the average KE of the nuclei, which is proportional to kT, can cause the lattice to "melt" or "vaporize" by allowing the nuclei to move about freely.

The obvious trouble with this analysis is that we cannot ignore the electrons. In contrast to the above case, let us assume the electrons are tightly bound to the nuclei. Then the net charge each atom sees would be zero. There would be no repulsive forces, no lattice structure, and in fact we would have a gas whose average KE per particle would be $\sim kT$. Reality lies somewhere in between. The highly degenerate electrons, whose average KE exceed 10^4 eV, have almost no probability of being bound to a He nucleus whose atomic binding energy is only 23 eV. Indeed, we can imagine a uniform cloud of electrons surrounding each He nucleus so that the nuclei will see only a screened net charge, Q_{net}. It is estimated that the net charge is about $\frac{1}{8}$ the bare charge, $Q_{\text{net}} \sim \frac{1}{8} Q_{\text{bare}}$, giving an electrostatic potential energy $(\sim Q^2/r)$ between two lattice sights of $\text{PE} \propto Q_{\text{net}}^2 \propto \frac{1}{60} \text{PE}_{\text{bare}}$. If the vibrational kinetic energy of the nucleus about its lattice site, KE_{vib}, exceeds PE, then the lattice will melt. That is if $\text{KE}_{\text{vib}} \geqslant \frac{1}{60} \text{PE}_{\text{bare}}$, the lattice is a gas or liquid, whereas if $\text{KE}_{\text{vib}} \leqslant \frac{1}{60} \text{PE}_{\text{bare}}$, the lattice is a solid.

The variables then that determine the structure of the lattice will be the density n which determines l through $l \sim (1/n)^{1/3}$, the charge of the nuclei Z, and the temperature T. The first two determine $\text{PE}_{\text{bare}} = k_0 (Ze)^2/l$, the third $\text{KE}_{\text{vib}} \sim kT$. At a given temperature and density, it is possible to have an atom with low Z in the gas-liquid phase, while in another system with the same temperature and density, a high Z atom might end up as part of a solid. For our particular case we have $Z = 2$, $l = \frac{1}{2} \times 10^{-11}$ m, and $T \sim 2 \times 10^7$ °K. As you can easily verify, these numbers produce a gas-liquid core. But as the density increases, the core will pass over to the solid phase. We only need time.

The electrons, on the other hand, will always move about pretty much in a straight line since they have far too much degenerate KE (i.e., velocity) to be even minutely deflected by the force exerted between them and the nuclei ($\Delta v = F \Delta t/m$, $\Delta t \sim 1/v$ small, etc.). The electrons thus behave like a gas and can exert a pressure on the surrounding envelope. Now, however, the equation of state is not for an ideal gas but instead for a degenerate electron gas. For any nonrelativistic particle we saw in chapter 1 that

$$PV = \tfrac{2}{3} N \langle \text{KE} \rangle \qquad \text{or} \qquad P = \tfrac{2}{3} n \langle \text{KE} \rangle.$$

For a nonrelativistic degenerate gas, $\langle KE \rangle = \frac{3}{5}KE_F$ so that the degenerate pressure is

$$P_{deg} = \tfrac{2}{5}nKE_F \approx \tfrac{2}{5}n\left(\frac{2\times 10^{-13}}{mc^2}\right)n^{2/3}$$

$$\approx \frac{8\times 10^{-14}}{mc^2}n^{5/3}(\mathrm{eV/m^3}).$$

(Note: to get the pressure into units of N/m^2 $(= J/m^3)$ you multiply by 1.60×10^{-19} J/eV.) Notice that the pressure goes up as the $\frac{5}{3}$ power of the density. This is an outward pressure exerted on the base of the envelope. The inward force, as we have seen in chapter 3, is due to the gravitational attraction between successive layers of the envelope and the core and any external pressure on the surface of the star. The latter can be taken to be zero. Thus the pressure difference across the envelope is given by

$$\frac{\Delta P}{\Delta r} = \frac{0 - P_{deg}}{R - 0} \qquad (r_c \ll R_\odot).$$

This gradient may be able to support the star. Whether it can or not will depend upon the term $\rho GM_r/r^2$ summed over the envelope. The bigger the star, the larger will be the sum and therefore the harder it will be for the degenerate pressure to support the star. Detailed estimates of degenerate pressures are postponed until chapter 5. Here we shall start by considering two cases, $M_{star} < 0.8\ M_\odot$ and $M_{star} > 0.8\ M_\odot$.

4.4.2. A Planetary Nebula: $M < 0.8\ M_\odot$

Calculations show that with a star whose mass is less than 0.8 $M_\odot$, the degenerate pressure will support the envelope so that the core contracts very slowly and the core temperature rises very little. This very slow contraction of the core leads to a continuously expanding surface so that the star proceeds to climb up the H–R diagram to the region of super red giants with convection driving the luminosity as the star expands.

With continued expansion, interesting things occur. At each point below the surface, there is a temperature which is related to the average KE per particle at that point or layer by $\langle KE \rangle = \frac{3}{2}kT$. Now near the surface, KE is so low that most of the atoms are neutral since their collision energies are insufficient to produce ionization. As we go deeper into the star, T goes up and so does $\langle KE \rangle$. This means more and more atoms are ionized at lower layers since the higher $\langle KE \rangle$ implies more atoms have kinetic energies that exceed the ionization energy of the atom (remember, N vs. KE has that long exponential tail, $\exp(-KE/kT)$). Once we get beyond a point where $T{\sim}50{,}000$ to 100,000 °K, the atoms are mostly ionized—few are neutral.

Now, as our star expands, successive regions below the surface begin to cool. This means that the KE at these radii goes down and H^+ and e^- begin to recombine at a faster rate than their collisions would cause them to ionize. We find $H^+ + e^- \rightarrow H^* \rightarrow H + \gamma$s. The expanding and cooling envelope ironically becomes an energy source—a source of luminosity due to the emitted γs. Since $L \propto \Delta T/\Delta r$, we increase the local temperature gradient beyond the recombining layer, with the result that the envelope gets an extra push. Keeping the envelope from blowing away is gravity. But $F_g \propto 1/r^2$ and therefore as the star expands, the gravitational force holding the envelope weakens. Thus we have two effects tending to promote a more rapid expansion of the envelope outside of the recombining layers:

(a) an increasing $\Delta T/\Delta r$,
(b) a decreasing F_g.

The underlying layers will now cool faster, the rate of recombination will go up. The increasing luminosity from this local source will increase $\Delta T/\Delta r$ and decrease F_g even faster. Again we have a runaway situation, with greater recombination promoting a bigger L, larger $\Delta T/\Delta r$ and smaller F_g which together lower the temperature even more, thus promoting greater recombination etc. The result is that the envelope "blows" off.

Can we predict the size of the star when this happens? Consider a small volume near the surface where recombination is taking place. Then $T \sim$ 50,000 °K and the average kinetic energy per particle is

$$\langle \text{KE} \rangle = \tfrac{3}{2}kT \sim 7 \text{ eV}.$$

The average PE per particle is given by

$$\text{PE} = -GMm_{\text{H}}/R.$$

If the region is to be unstable and capable of being ejected, the total energy per particle ($E = \text{KE} + \text{PE}$) must be greater than or equal to zero. Using the minimum condition for ejection we have

$$0 = \text{KE} + \text{PE}$$
$$\text{KE} = -\text{PE}$$
$$\tfrac{3}{2}\text{kT} = +GMm_{\text{H}}/R.$$

Solving for R we have, for $M = 0.8\ M_\odot$,

$$R = GMm_{\text{H}}/\tfrac{3}{2}kT \approx \frac{6.7 \times 10^{-11}(2 \times 0.8 \times 10^{30})(1.7 \times 10^{-27})}{\tfrac{3}{2}(1.4 \times 10^{-23})(5 \times 10^{4})}$$

$$\approx 200\ R_\odot,$$

about the distance from the Earth to the Sun. We can also speculate about

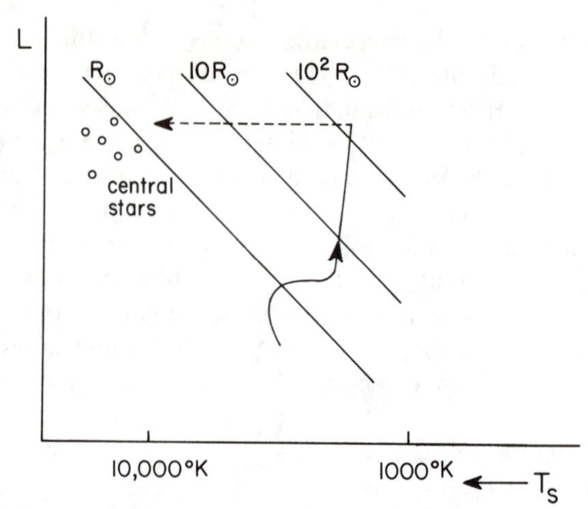

Fig. 4.25. Possible evolution between low mass red giant stars and the central stars in planetary nebulae.

the nature of the star left behind. Having had its surface blown away, the interior of the star is now exposed. Since the base of the envelope is about 50,000 °K, the remnant star would have that surface temperature and be very small. The overall luminosity is then roughly the same as before the explosion. The star is literally unmasked and flung across the H–R diagram to the region shown in Fig. 4.25. This small star will be degenerate.

What evidence is there then that this scenario actually takes place? Referring back to Fig. 1.18b we see the striking photograph of a planetary nebula. At the center of the nebular *shell* (it *is* a shell although it appears to be a ring because of the smaller optical thickness one encounters when looking at the center vs. along the edge) is a blue star. Even without a black body spectral analysis of the central star, we know it must have a surface temperature of over 25,000 °K. It takes at least 13.6 eV to ionize hydrogen, thus the central star must be rich in ultraviolet photons. How rich depends upon the star's luminosity, which in turn depends upon its surface temperature and its radius. We can get a handle on L by measuring the intensity of the visible light emitted by the nebula and use this to calculate the luminosity of the star. An analysis of T_s and L places the central star just where we theoretically predicted it should be, in the upper left part of the H–R diagram and directly in line with stars of $\sim 100\ R_\odot$. There are about 1000 detected planetary nebula and all have central stars in this region.

We can now turn our attention to the nebula. The Doppler-shifted emission spectrum of the nebula shows that it is expanding at the rate of about 30 km/sec. Since the shell is concentric with the central star, we can conclude that the shell is the remnant of the ejected envelope. We can even estimate the size of the star just before the explosion took place. Consider:

(a) To escape the star's surface, a molecule of mass m would have to have

$$E = \tfrac{1}{2}mv_0^2 - Gm\frac{M}{R} \cong 0.$$

(b) Doppler shift information for typical nebulae gives $v_{\text{expansion}} \sim 3 \times 10^4$ m/s. Assuming the shell has not slowed up too much since its ejection, we can use this as an estimate of the initial velocity v_0. Actually the expanding nebula, as it encounters the interstellar medium, will slow up so v_0 should be greater, but this is a good first order approximation.

(c) Solving for R we have

$$R = \frac{2GM}{v_0^2} \approx \frac{2 \times 6.7 \times 10^{-11}(0.8 \times 2 \times 10^{30})}{9 \times 10^8}$$

$$\approx 2 \times 10^{11}\,\text{m} \approx 300\ R_\odot,$$

which is consistent with our theoretical expectations.

Why then do we only see about 1000 planetary nebulae? The answer centers around the "lifetime" of the visible nebulae. Typically the radius of the shell is $\frac{1}{2}$LY$\sim 10^7\ R_\odot$. With an expansion velocity of 30 km/sec, such a nebula has been racing outward for $\Delta t = \frac{1}{2} \times 10^{16}$ m$/3 \times 10^4$ m/sec $\approx$ 5000 yr. After about 100,000 years, the nebula will have slowed up so that its velocity is indistinguishable from the random velocity of the interstellar medium; it will have disappeared. Clearly a "lifetime" of 100,000 years is not very long. In a sense it is surprising we see as many nebulae as we do. In fact with 1000 detected nebulae and these short lifetimes, they must be very common indeed. It would appear that most stars evolve through a planetary nebular phase.

A word of caution before we proceed. The evolution of stars between approximately 0.8 $M_\odot$ and 3 $M_\odot$ is fairly well understood. We shall pursue that topic next. At present, however, there is a great deal of uncertainty concerning the evolution of stars for $M > 3\ M_\odot$. The theories that apply here are interesting and varied, but at this point in the text they are beyond where we can discuss them intelligently. Therefore, to go through a detailed and complex analysis of the evolution of these massive stars would not be very productive. In light of this we shall only try and give a general, qualitative, brief, and hopefully noncontroversial picture of the development of massive stars.

4.4.3. 0.8 $M_\odot \lesssim M \lesssim 3\ M_\odot$: The Helium Flash—Planetary Nebulae with Carbon Cores

When M is greater than 0.8 $M_\odot$, the degenerate pressure of the electron gas is not high enough to support the envelope, so the core continues to contract and heat up. The heating, of course, only affects the metallic-like He core, and herein lies the problem. If the core were a gas, the buildup of heat would cause it to expand with a subsequent drop in core temperature (remember, from the virial theorem $R \propto 1/T$). We saw this earlier when we discussed the long-term self-regulating mechanism of the Sun. But as a solid, the He core does not expand very much when heated. Thus T_{core} continues to go up. When the temperature reaches 10^8 °K, the triple α fusion process begins:

$$\alpha + \alpha + \alpha \rightarrow \text{C}^{12} + 7.5 \text{ eV}.$$

And now because the core cannot expand, all hell breaks loose. The reaction produces energy at the rate

$$\epsilon_{3\alpha} \propto T^{40},$$

and is thus extraordinarily temperature sensitive. Normally this would not matter because the core would expand to lower T and stabilize the production of energy. But now the released 7.5 eV causes T_{core} to go up and this increases $\epsilon_{3\alpha}$ even more, releasing more energy, raising T_{core} even higher which releases energy even faster, etc. But the stable He lattice prevents expansion! There is no relief for the $T \rightarrow \epsilon_{3\alpha} \rightarrow T \rightarrow \epsilon_{3\alpha}$ cycle and the reaction runs away. The core explodes—according to the experts—within a few hours after the $3\alpha \rightarrow \text{C}$ starts! This is the helium "flash."

This explosive release of energy finally produces enough luminosity to expand the core and lower its temperature. This also lowers the density of the electrons so that $\text{KE}_{F,\text{el}}$ is now less than kT. The electrons are no longer degenerate, the core is no longer solid, and with the lowered temperature the 3α process shuts off. Even the hydrogen-burning rate in the outer shell goes down. Thus the internal luminosity drops way down and because $\Delta T/\Delta r \propto \kappa\rho L_r$, the temperature gradient throughout the star falls. Since $\Delta P/\Delta r \propto \Delta T/\Delta r$, the pressure gradient drops and the envelope can no longer be supported via $\Delta P/\Delta r \propto \rho G M_r/r^2$. The surface will contract as the core expands. This is opposite to the first red giant phase where the contracting core meant an expanding envelope. Interestingly enough with $L = 4\pi R^2 \sigma T_s^4$, the falling L and contracting R keep T_s approximately constant. This is shown in Fig. 4.26 between points A and B.

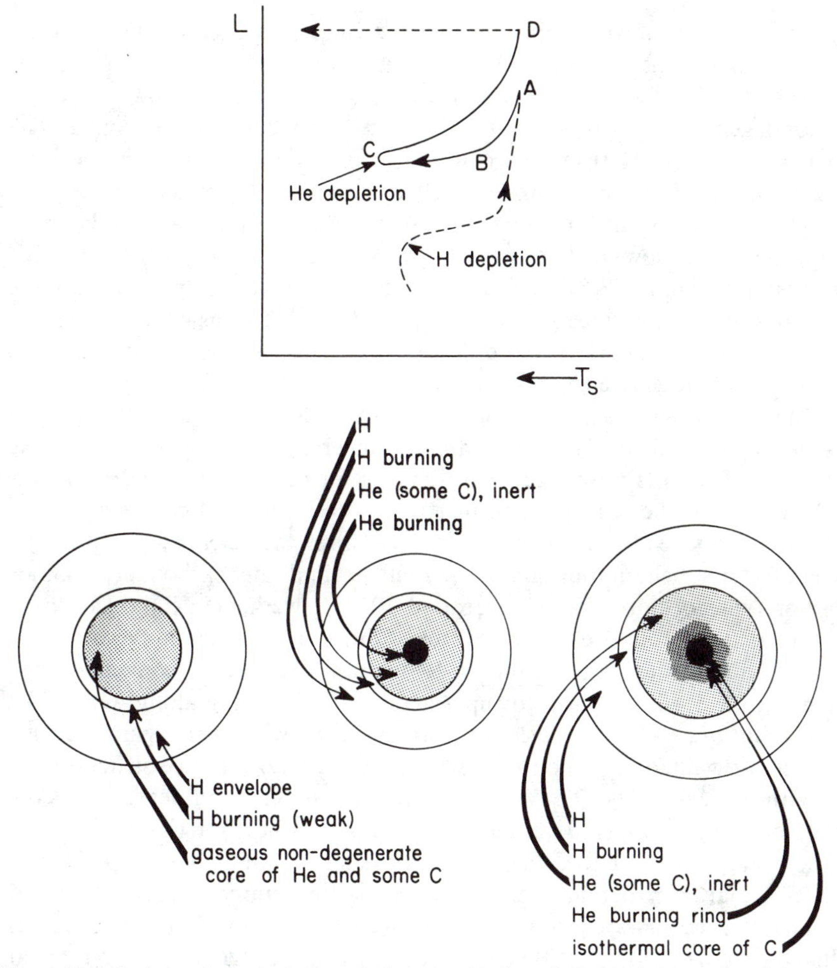

Fig. 4.26. Evolution from the helium flash (point A) through He depletion (point C) to the formation of a planetary nebula (point D). Also illustrated is the internal structure of the star between stages A and B, B and C, C and D, respectively.

The core expansion soon damps out—the pull of gravity within the core overcoming the momentum created by the explosion. Thus the explosion is confined and the star does not rupture or eject material. The expanded core is once again a gas composed mostly of He with some C left over from the He flash. The lowered temperature also lowers the rate of H burning so for all intent and purposes we have a star without a nuclear source—a star whose major energy source will be gravity. Between points A and B, as we have seen, the insufficient luminosity results in a contracting star. The virial theorem tells us, however, that the average temperature should begin to rise as the star collapses. Eventually the temperature at the center of the star rises to 10^8 °K. We again begin burning He ($3\alpha \rightarrow C^{12}$), but now the core is a nondegenerate gas so the rate of energy production is controlled. The increased luminosity from this new source of energy balances the decreasing luminosity from the hydrogen shell and we move to the left on the H–R diagram at a constant luminosity. We seem to be heading back to the main sequence—but now with a new major source of energy—the $3\alpha \rightarrow C^{12}$ process (see points B to C in Fig. 4.26). We might note here that the variation in the core mean-molecular-weight μ, which was so important for the evolution of the H$\rightarrow$He core (Sec. 4.2), is of minor importance here since for the heavier elements μ ($\cong 2$) is a constant.

Once again that magic moment comes when the core has used up its fuel —this time it will be He—and can no longer produce energy. The same sequence of events now takes place as took place when the hydrogen was exhausted and the core became inert He. The isothermal core means that the layers above it will begin to collapse until the released gravitational energy starts helium burning in a shell around the carbon core. Since carbon will not burn until T is around 10^9 °K, we have a long way to go before that nuclear fire will be started. The star once again expands and moves back to the right on the H–R diagram. Eventually convection will set in and we will shoot up to super-giant status. Moving along path C–D, shown in Fig. 4.26, the electrons in the core will once again become degenerate while the carbon core will stiffen up into a lattice structure.

With a mass in the 0.8 $M_\odot$–3 $M_\odot$ range, the electron-degenerate core will support the envelope enough so that T_{core} does not rise to the next fusion stage. Instead the star becomes a super giant and eventually produces a planetary nebula while at the same time unmasking a 50,000 °K surface that covers a carbon core. This will be the fate of our Sun. At this time (approximately 5×10^9 yr from now), its surface will expand out as far as the Earth (we will not be around to enjoy the spectacular sight of a massive red "balloon-like" surface engulfing us). Shortly afterward it will explode into a planetary nebula. Interestingly enough the Sun will end up in the same region of the H–R diagram as the $M{\sim}0.8\ M_\odot$ star, but with a carbon core.

4.4.4. 3 $M_\odot \lesssim M \lesssim 10\ M_\odot$: Carbon Flash—and Then?

If the initial mass of the star is in the above range, one is fairly certain of what happens next: the electron degeneracy will not support the envelope and the core will continue to contract and heat up. When the temperature of the core reaches ${\sim}10^9$ °K, the next set of nuclear reactions takes place.

Called carbon burning, it is characterized by these fusion processes,

$$C^{12} + C^{12} \rightarrow Mg^{24}; \qquad C^{12} + C^{12} \rightarrow Ne^{20} + \alpha; \qquad C^{12} + \alpha \rightarrow O^{16}.$$

Since by this time the core is a lattice of solid carbon, the reaction rate, which goes as T^{120}(!), is uncontrollable and the core explodes—a carbon flash. This course is similar to the path followed in reaching the He flash. But here theories begin to diverge.

According to one set of calculations, the energy released in the carbon flash is so great that the star is completely destroyed; no central star, nothing—poof, gone. We have a *full*-blown supernova explosion with only an expanding nebula to remind us of the event. Other theories suggest that at these high densities interactions like $p + e^- \rightarrow n + \nu$ followed by decay $n \rightarrow p + e^- + \bar{\nu}$ will moderate the explosion because of the energy carried away from the star by the *noninteracting* neutrinos. In this case the explosion will not be catastrophic and should leave behind a neutron star. Other theories also suggest the end result of the explosion will be a neutron star but for reasons other than neutrino loss. We shall take up neutron stars in detail in the next chapter and discuss in a little more detail the role supernova explosions play in neutron star formation. The point here is that no one is certain what happens after the carbon flash but since we have found neutron stars, degenerate carbon cores seem like a good place to produce them.

4.4.5. Beyond 10 $M_\odot$: Nondegenerate Stars, Nucleosynthesis, and Maybe Black Holes

Here things get murky indeed. One point that does seem clear is that when stars are this massive they do not develop degenerate cores during the course of their evolution. The reason is that their densities are too low. You will recall that on the main sequence $T \propto M/R$ is a constant. Since $\rho \propto M/R^3$ we see that $\rho \propto 1/M^2$. Thus as stars increase in mass, their densities correspondingly decrease. Beyond 10 solar masses it is estimated that the densities in the core will be too low for the electrons to become degenerate. This does not mean that these stars do not evolve through the various stages of nuclear burning. They do, but instead of going through violent internal explosion they heat up smoothly. At each stage gravitational contraction heats the core until a particular reaction commences. When the fuel has been used up, gravitational contraction will transfer burning to a ring about the core, whereupon further contraction of the core will lead to a higher temperature and the next stage of fusion. For stars between 10 and 20 solar masses, the core temperature will rise to a value high enough ($T \sim 2 \times 10^9$ °K) to burn oxygen and neon to form silicon, sulphur, and magnesium via

$$O^{16} + O^{16} \rightarrow Si^{28} + \alpha; \qquad O^{16} + O^{16} \rightarrow S^{32}; \qquad Ne^{20} + \alpha \rightarrow Mg^{24}.$$

For stars whose mass is greater than 20 solar masses, the core temperature can reach 3×10^9 °K so that Si can burn via

$$Si^{28} + Si^{28} \rightarrow Ni^{56}.$$

This is a particularly important reaction since nickel decays to cobalt via $Ni^{56} \rightarrow Co^{56} + e^+ + \nu$ and cobalt decays to iron via $Co^{56} \rightarrow Fe^{56} + e^+ + \nu$.

The decay chain ends here since Fe^{56} is stable. It is the most common isotope of iron found on Earth and certainly suggests our roots are to be found in the evolution of a massive star in this region of space. Something else happens here that is also very important. From the carbon-burning stage we have a large number of neutrons around from the reactions

$$C^{12} + p \rightarrow N^{13} + \gamma; \qquad N^{13} \rightarrow C^{13} + e^+ + \nu; \qquad C^{13} + \alpha \rightarrow O^{16} + n.$$

Neutrons combine to form heavy neutron-rich isotopes from $A = 20$ to $A = 209$, but this buildup is particularly effective for the iron group (i.e., $Fe^{56} + n \rightarrow Fe^{57} + n \rightarrow Fe^{58} + n$, etc.). As we know, neutron-rich nuclei are unstable and so will decay into stable nuclei with the same A as the original isotope. The reason we do not built up isotopes with $A > 209$ is that these isotopes are α-decay unstable which is a fast decay process. Thus they decay more quickly than they can be built up by neutron capture. The beta

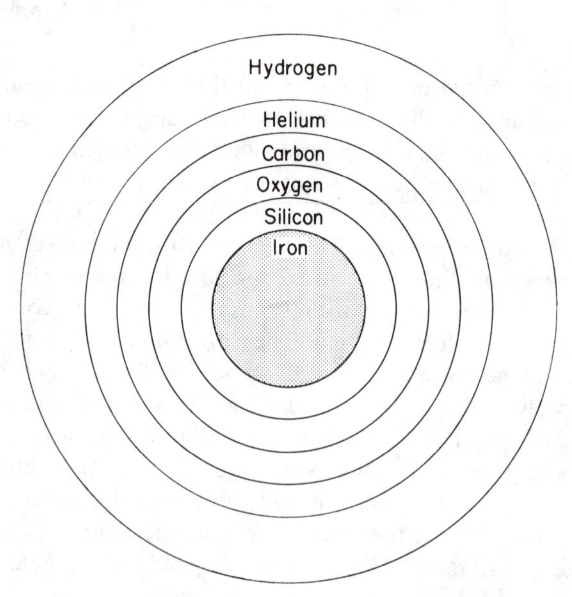

Fig. 4.27. Internal structure of a massive star.

decay process is much slower and thus allows for a buildup of elements with $A < 209$. When the core has exhausted its supply of Si, the Si + Si $\rightarrow$ Ni shuts off and the core of the star contracts until we get burning in a ring about the Fe core. The internal structure of these massive stars has been likened to the layers of an onion and is pictured in Fig. 4.27.

What happens as the temperature further increases in the iron core is subject to a great deal of controversy. Iron is at the top of the average binding energy curve we discussed in Sec. 4.1. Iron can only decompose into elements of lower average binding energy which means a net absorption of energy and the ultimate collapse of the core. But what does the collapse lead to? Some suggest a violent explosion, a supernova with a neutron star left behind. Others argue that no such explosion can take place and that a black hole must result when the iron dissociates. We shall look into each of these possibilities in the next chapter. Suffice it to say we should now turn our attention to the final configurations of matter.

"A BLACK HOLE — THAT'S THE ANSWER TO OUR RADIOACTIVE WASTE PROBLEM."

Problems

1. In Sec. 4.1 we made the statement that if the average binding energy of the final state nuclei is greater than that of the initial nuclei, a net amount of energy is released. Prove that this statement is equivalent to saying that the masses of the final nuclei must be less than that of the initial.

2. What is the binding energy per nucleon for $_{82}Pb^{204}$ ($M_{Pb} = 189949.5$ MeV)?
3. If the reaction $X^{120} \rightarrow Y^{119} + n$ were to go, what would the average binding energy of Y^{119} have to be if for X^{120} we have $\langle BE \rangle = 8.5$ MeV/nucleon?
4. When E. Rutherford's graduate students H. Geiger and E. Marsden scattered ~8 MeV alpha-particles (from the decay of Polonium) in 1913 off Gold ($_{79}Au^{197}$) they found particles that scattered backwards at 180°. Use this information to estimate the upper bound for the radius of the Au nucleus.
5. What temperature is necessary for two Helium nuclei to overcome their Coulomb repulsion and just "touch." (Note: $R_A = 1.2 \times 10^{-15} A^{1/3}$m).
6. Using the information in Fig. 2.33 and assuming hydrogen burning, calculate the lifetime of a star whose mass is 2 $M_\odot$ and one whose mass is 0.5 $M_\odot$.
7. What is the de Broglie wavelength of a proton whose KE$\approx$3 keV? How does the de Broglie wavelength compare with the distance of closest approach between this proton and a stationary proton? By how much would the incident velocity v_0 have to increase in order that the probability of penetrating the barrier increases by a factor of 2? At what temperature would the 3 keV proton be the most probable proton for producing a reaction? How does this compare to the central temperature of the Sun?
8. Find, by differentiating the equation $\mathcal{P}_T = e^{-A/v} \times e^{-Bv^2}$, the exact relationship between v and T.
9. Show that a curve $\propto e^{-(a+b)/T^{1/3}}$ and one $\propto T^n$ have similar shapes over small temperature regions.
10. Referring to Fig. 4.7, the luminosity of the Sun decreases by a factor of approximately 2 over ~10^7 years in going from the main sequence to point a.

    a. Why does the luminosity decrease instead of continuing along the dotted line?
    b. Using the above information, by what rate is the envelope of the Sun contracting?

11. At point (a) in Fig. 4.10, the evolutionary track of the Sun takes a turn toward higher luminosities.

    a. List the reasons why this turn takes place.
    b. Assuming the Sun at point (a) to be 100% hydrogen with

$R = 0.95\ R_\odot$ and $\langle T \rangle \approx 10^7$ °K, how big would the Sun be after burning 20% of its hydrogen to helium if T remains constant?

12. In Fig. 4.12 we compare the evolutionary track of two stars from the point of core hydrogen exhaustion to the point of hydrogen shell burning. Why are the tracks different?
13. Why is the radius of a hydrogen-burning shell constant?
    Why is the temperature of a hydrogen-burning shell constant?
    For a star of $\sim 1\ M_\odot$, the shell will be at $0.1\ R$ (where $R \approx 2\ R_\odot$ when shell burning commences) and $M_{\text{core+shell}} \sim 0.15\ M_\odot$.
    a. At what rate must the average density of the shell decrease in order to double the radius of the star in 10^9 years? (The latter number is based on computer calculations.)
    b. At what rate is the density in the core increasing?
14. Write down the equation that relates the number density n and the Fermi momentum p_F for electrons in a two-dimensional box.
15. Assume the core of a star composed of hydrogen had reached the density $\rho = 10^{10}$ kg/m^3.
    a. What would T_c have to be so that the electrons in the core would not be degenerate?
    b. What would be the average KE of these electrons?
    c. If we keep T_c fixed at the value found in (a), what would ρ have to be increased to in order that the core becomes an electron-degenerate gas?
    d. What would be the average KE of these electrons?
    e. Would the protons in the core described in (c) be degenerate? Explain.
16. Say we have a star whose core has reached a density of $\rho = 3 \times 10^5$ kg/m^3.
    a. What is the electron number density n_{elec}, if the core is hydrogen?
    b. What is the Fermi momentum of the electrons? of the protons?
    c. What is the Fermi energy of the electrons?
    d. What would the temperature of the core have to be in order that we are sure these electrons are degenerate?
    e. What would be their average kinetic energy?
    f. At this temperature are the hydrogen nuclei degenerate? What is their average KE?
17. As you know, copper (Cu) is a good conductor of electricity because each atom contributes one "free" electron to the wire.

a. What is the number density of free electrons in the wire, n_e? ($\rho_{Cu} = 9 \times 10^3$ kg/m^3 and $A_{Cu} = 64$)
b. What is the Fermi momentum and energy of the free electrons?
c. At room temperature, show that these electrons are degenerate. It may seem disconcerting that conduction electrons at room temperature are degenerate but this has important consequences. Consider the situation where no voltage ($V = 0$) is applied to the wire. As the Fermi energy indicates, the electrons are rushing about. Some will collide with the Cu atoms. Yet almost no heat is generated!! But apply a small voltage which increases the velocity of the electrons by only a small amount and heat is produced $\propto V^2/R$. We know this heat is produced due to collisions of the electrons with the atoms.
d. Why then, when $V = 0$, is no heat produced?

18. Verify that when $l = \frac{1}{2} \times 10^{-11}$ m and $T_c = 2 \times 10^7$ °K a helium core is in a gas-liquid state.
19. Estimate the number density of nuclei in a helium core if the core is a solid. Take $T_c \sim 10^7$ °K.
20. If carbon is contained in a core of $M_c \approx 0.4\ M_\odot$, $r_c \approx \frac{1}{10}\ R_\odot$ and $T_c \cong 4 \times 10^7$ °K, will the core be a gas-liquid or a solid?
21. Assuming that the velocity of a planetary nebula slows down linearly, how long will it take for a nebula to comingle with the 100 °K interstellar medium?

CHAPTER 5

Stellar Evolution III: Final Configurations of Matter

I've watched the big, husky sun wallow
    In crimson and gold, and grow dim,
Till the moon set the pearly peaks gleaming,
    And the stars tumbled out, neck and crop;
And I've thought that I surely was dreaming,
    With the peace o' the world piled on top.

Robert Service, *The Spell of the Yukon*

5.1. White Dwarfs

As we have seen, the remnant star of a planetary nebula *may* indeed be a core of degenerate material (from H to Fe). With luminosities similar to those of red giants, these stars lie in the upper left hand corner of the H–R diagram. Since these remnant stars do not lie on the main sequence, it is natural to guess and hope each represents an early stage of the development of a "white dwarf" which also lies off the main sequence curve. One can certainly imagine the evolutionary line shown in Fig. 5.1.

5.1.1. What Are They?

It is estimated that about 10% of the stars in the Galaxy lie in the white dwarf region of Fig. 5.1. They are approximately the size of the Earth with $R_{WD} \sim 10^{-2} R_\odot$ and have low luminosity with $L_{WD} \sim 10^{-3} L_\odot$. They have a bluish color because their surface temperature is about twice that of our sun, $T_{WD} \sim 10^4$ °K. Nevertheless they have become known as white dwarfs. That they are black body radiators can be seen as follows. We know such a radiator must satisfy the relationship $L = 4\pi\sigma R^2 T^4$. We know this holds for

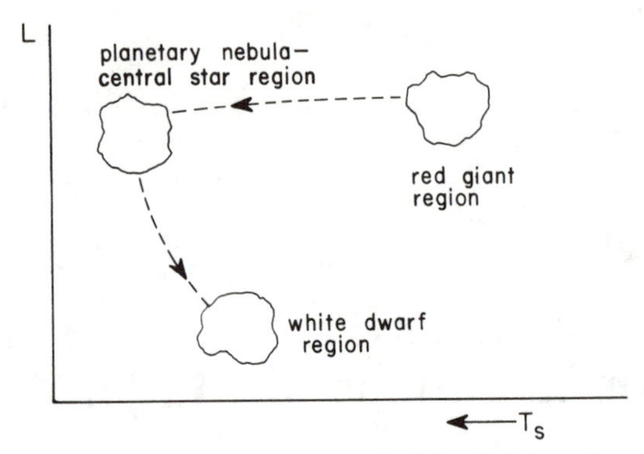

Fig. 5.1. H–R diagram for the likely evolution from the red giant to the white dwarf region.

the Sun, $L_\odot = 4\pi\sigma R_\odot^2 T_\odot^4$. So, assuming that the white dwarf is a black body, we find

$$\frac{L}{L_\odot} = \left(\frac{R}{R_\odot}\right)^2 \left(\frac{T}{T_\odot}\right)^4 \sim 10^{-4} \times 10 \sim 10^{-3},$$

which fits the experimental information to a tee. Thus $L = 4\pi\sigma R^2 T^4$ holds for a white dwarf too—it is a black body radiator.

One problem, however, is their mass. Observed white dwarfs in binary systems (recall the mass determination in Sec. 2.6) have between 1/10th and one solar mass, with an average of about $\frac{1}{2} M_\odot$. This means that some of the heavier stars which evolved into planetary nebulae had to eject quite a bit of matter in order to leave a core of less than one solar mass.

The low luminosity is another problem. How did L go from $10^3\ L_\odot$ (for a nebular remnant star) to 10^{-3} (for a white dwarf) while the surface retained such a high temperature?

We have implied earlier that the core of the remnant stars of planetary nebula are composed of degenerate H or He or C or Mg or Fe. Unfortunately one cannot look at a particular white dwarf and tell what it is made of. It could be C or it could be Fe—we just do not know. This is unfortunate, but a fact of life. A crucial point, then, in linking white dwarfs to planetary nebulae will be to try and deduce whether a white dwarf is composed of degenerate matter or not. If it is, then its relationship to planetary nebular central stars would appear to be on much firmer footing.

5.1.2. Degenerate or Not Degenerate . . . That is the Question

With a mass $M_{\mathrm{WD}} \sim \frac{1}{2} M_\odot$ and a radius $R_{\mathrm{WD}} \sim 10^{-2}\ R_\odot$, we have:

$$\rho_{\mathrm{WD}} = \frac{M}{V} \approx \frac{\frac{1}{2}(2 \times 10^{30}\ \mathrm{kg})}{(4\pi/3)(10^{-2} \times 7 \times 10^8\ \mathrm{m})^3} \approx 10^9\ \mathrm{kg/m^3}. \qquad (5.1)$$

This corresponds to an electron density of ($n_e = ZM/Am_{\rm H}V \cong \frac{1}{2}(\rho/m_{\rm H})$, where $Z/A \sim \frac{1}{2}$)

$$n_e \approx \frac{1}{2}\,\frac{10^9\ \mathrm{kg/m^3}}{1.7\times 10^{-27}\ \mathrm{kg}} \sim \frac{1}{3}\,10^{36}\ \mathrm{el/m^3},$$

which in turn leads to a Fermi momentum of

$$p_F c = hc\left(\frac{3}{8\pi}n_e\right)^{1/3} \approx 6\times 10^{-7} n_e^{1/3}(\mathrm{eV}) \sim \tfrac{1}{2}\ \mathrm{MeV}.$$

This makes the electrons relativistic (remember, $m_e c^2 \approx \frac{1}{2}$ MeV) so that their Fermi energy is

$$\mathrm{KE}_F = p_F c \approx \tfrac{1}{2}\ \mathrm{MeV}.$$

To be solidly degenerate requires $\mathrm{KE}_F \gtrsim 10\ kT$. What then is the temperature of a white dwarf? Both the planetary central star and the white dwarf are inert as far as nuclear energy generation is concerned. The Si + Si $\rightarrow$ Ni process goes at 3×10^9 °K. Thus we must be below that for a Ni core; C + C $\rightarrow$ Mg goes at 10^9 °K, so we must be below that for a C core; $3\alpha \rightarrow$ C goes at 10^8 °K, so we must be below that for an He core. Suffice it to say that our core temperature will not exceed 10^8 °K and will probably be lower (most experts think $T_{\rm core} \sim 10^7$ °K—like our sun's), so that an upper limit of kT is $kT \sim 10^4$ eV. Since KE_F is about 5×10^5 eV, we certainly satisfy the degeneracy condition we set for ourselves, that $\mathrm{KE}_F > 10\ kT$.

So to answer the question stated in the heading of this section . . . yes, a white dwarf would appear to be composed of a degenerate electron gas having H, He, C, . . . or even Fe in its core. White dwarf mass densities typically range from 10^8 kg/m^3 to 10^{11} kg/m^3, corresponding to nonrelativistic and extreme relativistic degenerate electron gases, respectively. It is likely that all remnant stars of planetary nebulae evolve into white dwarfs following the path along the H–R diagram indicated in Fig. 5.1.

5.1.3. Falling Luminosity and Internal Structure

Because $kT < \mathrm{KE}_F$, most of the electrons in a white dwarf are below the Fermi level and in fully occupied states. This is shown in Fig. 5.2. As we saw earlier, this means that none of these electrons can radiate! To do so means losing energy and falling into lower states, which is impossible because these are filled. Thus we have a hot core—10^7–10^8 °K emitting no radiation.

Now going back to our very luminous remnant star, two things will happen after it has ejected its envelope. One, it will continue to contract under gravity until possibly some point is reached where the increased degenerate electron pressure will balance the gravitational pull. Secondly, as the very luminous central star's electrons radiate energy, they will fall

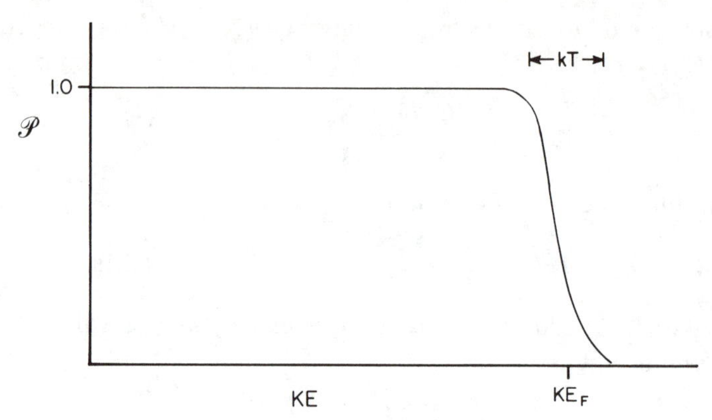

Fig. 5.2. Fermi distribution for $kT \ll \mathrm{KE}_F$.

into states below the slowly increasing Fermi level (recall, n_e increases as the star contracts) and no longer produce radiation. Thus we have the situation that the luminosity falls off faster than would be expected if only a smaller R were responsible. Indeed, classically a contracting star means an increasing surface temperature. On the contrary here, as R decreases, so does T_s. Only the electrons near the surface radiate since only here is the star nondegenerate (low n_e). But as these electrons fall into degenerate states, the star becomes dimmer and dimmer, fading from an initial white to yellow then red until it fades out—a "cold" lump of matter drifting through space.

As to the internal structure, we must first check to make sure that the nuclei are stripped of all their electrons and, indeed, that the electrons have almost no probability of being near a nucleus. We can take as the worst case, Fe with $Z = 26$. To free the last electrons requires (the electron being in an $n = 1$ state)

$$\mathrm{BE} = 13.6 Z^2/n^2 = 13.6\,(26)^2 \approx 10^4 \text{ eV}.$$

With $\langle \mathrm{KE} \rangle = (3/4)\mathrm{KE}_F = 3 \times 10^5$ eV for our typical white dwarf, we are clearly free of all atomic binding, $\langle \mathrm{KE} \rangle_{\mathrm{WD}} \gg \mathrm{BE}$. We thus have an electron gas with each nucleus, shielded by the uniform distribution of electrons, seeing some fractional charge of its neighbors (recall Sec. 4.4.1). Thus there is the tendency to "stiffen up" under the mutual net Coulomb repulsion between nuclei. As we saw earlier, a measure of this stiffness will be the mutual potential energy between nuclei, $\mathrm{PE_{bare}} = k_0(Ze)^2/l$. Trying to overcome this will be their average vibrational kinetic energy which is given by $\mathrm{KE_{vib}} \cong kT$. We recall that the dividing condition between a crystalline and noncrystalline phase is given by $\mathrm{KE_{vib}} \approx \frac{1}{60}\mathrm{PE_{bare}} = \frac{1}{60} k_0(Ze)^2/l$; if $\mathrm{KE_{vib}} \lesssim \frac{1}{60}\mathrm{PE_{bare}}$ it is a solid, if $\mathrm{KE_{vib}} \gtrsim \frac{1}{60}\mathrm{PE_{bare}}$ it is a liquid or gas. The variables involved in determining just what the structure will be are T, Z, and l. But l depends upon the density of the white dwarf since the average spacing will decrease as ρ increases:

$$l = \left(\frac{1}{n_{\text{nucl}}}\right)^{1/3} = \left(\frac{Am_{\text{H}}}{\rho}\right)^{1/3}.$$

Thus, with $A \sim 2Z$, we have

$$\text{PE}_{\text{bare}} = \frac{k_0 Z^2 e^2}{(Am_{\text{H}})^{1/3}}\rho^{1/3} \approx \frac{k_0 Z^2 e^2 \rho^{1/3}}{(2Z)^{1/3} m_{\text{H}}^{1/3}}$$

$$\text{PE}_{\text{bare}} \approx 2\times 10^{-19} Z^{5/3} p^{1/3}\ (\text{J}) \approx Z^{5/3} p^{1/3}\ (\text{eV}).$$

So the variables we are interested in are Z, T, and ρ.

Because the electrons are degenerate and do not either absorb or emit radiation, the star will be quite isothermal throughout its interior. (We have already suggested that this temperature will be around 10^7 °K.) Since T is constant, the burden of determining the structure will fall upon Z and ρ. For any given Z (say a carbon star), we might find a solid core—where ρ is high—which reaches halfway to the surface, and a liquid or gaseous envelope—where ρ is much lower—that extends out the rest of the way. But given the *same density profile*, an iron ($Z = 26$) white dwarf might be solid almost to its surface. For example, given an oxygen core at 10^7 °K, we can find the critical density between the solid and gas–liquid phases from

$$kT = \tfrac{1}{60}\cdot Z^{5/3}\rho_c^{1/3}\ [\text{eV}]$$

$$10^{-4}\times 10^7\ \text{eV} \approx \tfrac{1}{60}(8)^{5/3} p_c{}^{1/3}\text{eV or}$$

$$p_c \approx (\tfrac{60}{32}\times 10^3)^3 \sim 7\times 10^9\ \text{kg/m}^3.$$

All we need is the density profile of the star to determine the radius at which the phase transition takes place. Had the star been made of Fe, however, the value of Z in the above equation would have been 26 and ρ_c would have been 300 times smaller. Thus the radius at which the phase transition takes place would have been close to the surface of the star.

5.1.4. Maximum Mass of a Stable White Dwarf

Although we have electron degenerate pressures supporting a white dwarf, there still is gravity trying to compress it. As we add more mass, we need a larger and larger pressure gradient in order to maintain QHE. The maximum pressure will occur for relativistic electrons since they clearly have the highest KE. The pressure gradient equation is difficult to deal with, but because the virial theorem gives the same results, we can use the latter instead. As the mass of the white dwarf increases, gravity will finally overcome the exclusion principle outward kinetic pressure, even of a relativistic electron gas, and the star will become unstable and collapse. The situation corresponds to the onset of the relativistic virial theorem discussed in Sec. 3.2.4 with $E_{\text{tot}} = 0$ corresponding to the instability condition.

For the particular case of a white dwarf, we can again take the total PE as found for a star of constant mass density:

$$\text{PE}_{\text{tot}} = -\tfrac{3}{5}\,\frac{GM^2}{R}\,,$$

where the star's mass is dominated by *nucleons*. The total KE, however, of all the particles in the star, electrons, protons, α particles, etc., is

$$\text{KE}_{\text{tot}} = \text{KE}_e + \text{KE}_p + \text{KE}_\alpha + \cdots .$$

With the electron gas relativistic and degenerate, while the heavier particles are not, we have

$$\text{KE}_p + \text{KE}_\alpha + \cdots \ll \text{KE}_e.$$

Thus the electrons still dominate the KE of the star,

$$\text{KE}_{\text{tot}} \approx \langle\text{KE}\rangle_e \approx \tfrac{3}{4} N_e hc\left(\frac{3}{8\pi}\,n_e\right)^{1/3},$$

since $\text{KE} = pc$ for relativistic electrons. Then the onset of the relativistic virial theorem

$$E = \text{KE}_{\text{tot}} + \text{PE}_{\text{tot}} = 0,$$

balancing the electron-dominated KE_{tot} against the nucleon-dominated PE_{tot},

$$\tfrac{3}{4} N_e hc\left(\frac{3}{8\pi}\,n_e\right)^{1/3} \simeq \tfrac{3}{5}\,\frac{GM^2}{R}\,, \tag{5.2}$$

determines the criteria of an unstable white dwarf. The next job is to reduce this expression to the fewest possible fundamental variables and see what quantitative relationship exists amongst them. It turns out we can do this most easily by expressing N and n_e in terms of M and R: $N_e \approx M/\mu m_{\text{H}}$ and $n_e = N_e/V = M/\mu(\frac{4}{3}\pi R^3 m_{\text{H}})$. Substituting these into equation (5.2) the relativistic virial theorem implies

$$\tfrac{3}{4}\left[\frac{M}{\mu m_{\text{H}}}\right]\cdot hc\left[\frac{3}{8\pi}\,\frac{M}{\mu(4\pi/3)R^3 m_{\text{H}}}\right]^{1/3} \approx \tfrac{3}{5}\,\frac{GM^2}{R}\,. \tag{5.3}$$

In this rather complicated equation there are only (a) universal constants, and (b) the variables M, R, and μ. But notice on the left we have $[1/R^3]^{1/3} = 1/R$ which cancels the $1/R$ on the right. The instability condition is independent of R! Solving for the mass we have

$$M_{\mathrm{WD}}^{\max} \approx \left(\frac{5hc}{2G}\right)^{3/2} \cdot \frac{3}{16\pi}\left(\frac{1}{\mu m_{\mathrm{H}}}\right)^2. \tag{5.4}$$

For a white dwarf with a core heavier than hydrogen, $\mu = 2$, (5.4) becomes

$$M_{\mathrm{WD}}^{\max} \approx 1.7\, M_{\odot}.$$

A more accurate estimate gives 1.2–1.4 $M_{\odot}$, usually referred to as the Chandrasekhar limit for the maximum mass of white dwarf stars.

With regard to the formation of white dwarfs, if a planetary nebular central star has a mass $M < 1.4\, M_{\odot}$, it will stabilize as a white dwarf. Its temperature will slowly sink to absolute zero and the star then *dies*, becoming a cold hunk of matter being supported against gravitational collapse by the exclusion principle pressure of the electrons. We truly have reached a final state of matter. But what if $M > 1.4\, M_{\odot}$? Then gravity is the winner and the star keeps on condensing—but into what?

5.2. Neutron Stars

Until 1932 there was no reason to expect any other stable configuration of degenerate matter beyond a white dwarf. However with Chadwick's discovery of the neutron that year, another stable stellar state of matter with $M > 1.4\, M_{\odot}$ could be predicted.

5.2.1. The Making of a Neutron Star

Since the neutron was found to have spin $\frac{1}{2}$ (just like the electron, proton, and neutrino), it was possible to predict a reaction between the free electrons and the protons in the nuclei of the elements that made up the star:

$$e^- + p \rightarrow n + \nu.$$

This reaction satisfies charge and baryon number conservation. It needs a good deal of energy to go, however, because the rest mass of the neutron is greater than that of the electron plus proton. Recall: $m_n c^2 \approx 939.6$ MeV; $m_p c^2 \approx 938.3$ MeV; $m_e c^2 \approx 0.5$ MeV. Thus even if we produce a neutron at rest by slamming into a nuclear proton with a fast-moving electron, relativistic energy conservation demands the minimum kinetic energy of the electron to be $E_i = E_f$ or

$$\mathrm{KE}_{e,\min} + m_e c^2 + m_p c^2 \geqslant m_n c^2$$
$$\mathrm{KE}_{e,\min} \approx 939.6 - (938.3 + 0.5) \approx 0.8 \text{ MeV}.$$

Thus in a white dwarf star, the degenerate electrons will crash into nuclei and produce the reaction if their KE exceeds 0.8 MeV. And this will happen only if the maximum KE of these degenerate electrons (e.g., within the interior of a white dwarf), their Fermi energy, exceeds 0.8 MeV.

It is important to note that the reaction $e^- + n \to p + \nu$ does *not* go—because of charge conservation—and the reaction $e^- + n \to \bar{p} + \nu$ does *not* go—because of baryon number conservation. (For n and p, baryon number is $+1$; for $\bar{p}$ and $\bar{n}$, -1). Thus the interaction of the degenerate electrons with the nuclei is one sided; *e*s slamming into nuclei will "create" neutrons out of protons but not protons out of neutrons. If the reaction $e^- + p \to n + \nu$ does go, then our nuclei will become "neutron rich."

What we want then is an average kinetic energy for the electrons that is greater than 0.8 MeV. This means our degenerate electrons will have to attain a Fermi energy of $KE_F \geqslant \frac{4}{3}\langle KE \rangle \approx 1.1$ MeV. They can reach this relativistic energy ($KE > m_e c^2$) if n_e, the electron density becomes high enough, i.e., if the white dwarf keeps on contracting! Since $KE_F = hc[(3/8\pi)n_e]^{1/3} \approx 6 \times 10^{-7} n_e^{1/3}$ (eV), this means we will need a minimum number density of

$$n_{e,\min} \approx \left(\frac{KE_F(\text{eV})}{6 \times 10^{-7}}\right)^3 \approx \left(\frac{1.1 \times 10^6}{6 \times 10^{-7}}\right)^3 \sim 6 \times 10^{36}\ \text{el/m}^3.$$

In terms of the star's *mass* density, where $n_e = \frac{1}{2}(\rho/m_H)$, we have

$$\rho_{\min} = 2m_H n_e \approx 2(1.7 \times 10^{-27}\ \text{kg})(6 \times 10^{36}\ \text{el/m}^3)$$
$$\approx 2 \times 10^{10}\ \text{kg/m}^3.$$

This exceeds the average density of a white dwarf by about a factor of 20. If however, $M > 1.4\ M_\odot$, ρ will continue to increase until it reaches the above value; at this point the nuclei begin to undergo neutron enrichment via $e^- + p \to n + \nu$.

Now normally a neutron-rich nucleus is unstable. We saw this in the formation of the elements for $A > 200$. With too many neutrons ($N = A - Z$) a nucleus (X) will emit a series of electrons, via a process called beta decay, ${}^N_Z X^A \to {}^{N-1}_{Z+1} X^A + e^- + \bar{\nu}$, eventually producing a stable nucleus, but with fewer neutrons. A degenerate star is different, however, because the nuclei *cannot* beta decay. The "created" electrons that must be emitted in the beta decay process have no available states to go into—these states are totally filled by the degenerate electrons already there. True, the electrons with $KE > 0.8$ MeV are being absorbed, but the beta decay electrons have KEs less than $\frac{1}{2}$ MeV. In fact as electrons with $KE > 0.8$ MeV are absorbed, the escaping neutrinos carry energy away from the star. The absorbed electrons lower the electron density. This reduces the degenerate pressure, allowing for an increased rate of compression, thus increasing ρ even faster, accelerating in turn the reaction rate for $e^- + p \rightarrow n + \nu$. We shall return to this last point when we discuss the theories of neutron star formation from supernovae explosions.

We can therefore imagine a sea of degenerate electrons in which neutron-rich nuclei are imbedded. Now we must remember that both neutrons and protons have spin $\frac{1}{2}$ so the Pauli exclusion principle must apply to them as well as to the electrons. This will be true even when the *n*s and *p*s are all crowded into the nucleus and so confined to a space of linear dimension $\Delta x = 2A^{1/3} \times 10^{-15}$ m. Moreover there will be a ground state minimum momentum for the nucleons given by the uncertainty principle,

$$\langle p_{\min} \rangle c = \frac{\Delta p \cdot c}{2} \approx \frac{hc}{2\Delta x},$$

just as we saw in chapter 4 when we began our discussion of degenerate matter. For a carbon nucleus where $A = 12$, we have $\Delta x \sim 5 \times 10^{-15}$ m and therefore $\langle p_{\min} \rangle c \approx 10^8$ eV, which gives us a ground state neutron KE of $p^2c^2/2m_nc^2 \approx 10^{16}/2 \times 10^9 = 5$ MeV, where m_nc^2 is the mass of a neutron, 10^9 eV. Since the nuclear force is attractive, the total energy of the ground state (GS) will be

$$E_{\mathrm{GS}} = \mathrm{KE}_{\mathrm{GS}} - |\mathrm{PE}_{\mathrm{GS}}|.$$

To be more specific we consider a star with a carbon core. Because the carbon nucleus is stable, we must have the ground state energy E_{GS} less than zero, or $|\mathrm{PE}_{\mathrm{GS}}| > \mathrm{KE}_{\mathrm{GS}}$. But the carbon nucleus has six protons and six neutrons. They cannot all be crowded into the lowest energy level because of the Pauli principle. The lowest nuclear level can only accommodate two protons (↑ ↓) and two neutrons (↑ ↓). The third neutron and/or proton would have to go into the next nuclear energy level—as shown in Fig. 5.3a. For carbon the ground state would have to have three levels, each

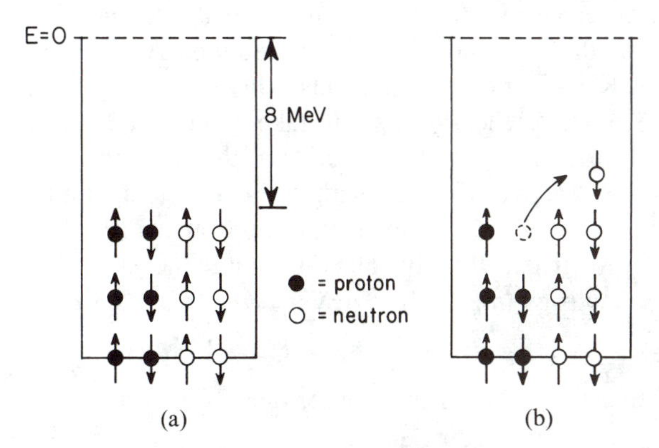

Fig. 5.3. Depicting the occupied nuclear energy levels for (a) the ground state of carbon and (b) a state where one proton has been converted into a neutron.

with two protons and two neutrons with the Pauli principle telling us that each higher level has a larger KE. This means that the ground state PE must be very negative. The question of interest is, how far is the third level from the continuum where $E = 0$? The answer comes from the average binding energy curve we studied earlier. Experimentally this curve shows that the average binding energy ($\langle \text{BE} \rangle = \text{BE}/A$) is about 8.5 MeV, which implies it takes about 8.5 MeV to remove the nucleon nearest the continuum from the carbon nucleus. Thus, after we have stuffed 6 *p*s and 6 *n*s in C^{12}, the last ones to go in are approximately 8.5 MeV below the $E = 0$ level.

Now as protons within the nucleus are turned into neutrons, the new neutrons have to go into higher energy levels—closer to the continuum (as Fig. 5.3b indicates) because the Pauli principle closes off the lower energy states. The higher states have more KE, and the neutrons occupying these states, seeing the same nuclear PE, will find themselves much less tightly bound. Some will begin to "evaporate" off the surface of the nucleus. This will be compensated by the capture of free neutrons back into the nucleus. Nonetheless, free neutrons are added to the "soup" inside the star so that the star now consists of

(a)	neutron-rich nuclei	(nondegenerate),
(b)	free neutrons	(nondegenerate),
(c)	electrons	(degenerate).

The rate of neutron evaporation will keep pace with the rate of condensation until the $\langle \text{KE} \rangle$ of all the free neutrons reaches ~8.5 MeV. At this point they can no longer be captured since $\langle \text{KE} \rangle > \text{BE}/A$. When this happens, the nuclei quickly disintegrate as $p \to n$ (inside the nucleus), and the *n*s evaporate. What determines the KE of the free neutrons and the point where their KE exceeds 8.5 MeV?

Like the free electrons, the free neutrons will become degenerate when the density of the star becomes high enough. As the density further increases, the KE of the free neutrons will increase until their $\langle \text{KE} \rangle$ is greater than 8.5 MeV. Clearly this will take a very high density because of the large mass of the neutron. Nonetheless as the star collapses, this state of matter should be reached. (Recall, with an internal temperature of $10^7 - 10^8$ °K, an average KE of 8.5 MeV means the neutrons will be degenerate.) We can mark the p when this happens because the nonrelativistic Fermi energy of the neutrons ($m_n c^2 \approx 10^3$ MeV KE~8.5 MeV)

$$\text{KE}_F = \tfrac{5}{3}\langle \text{KE} \rangle = \tfrac{5}{3} \times 8.5 \approx 14 \text{ MeV},$$

constrains a degenerate neutron number density n_n via

$$\text{KE}_F = \frac{p_F^2}{2m_n^2} = \frac{h^2c^2}{8m_n c^2}\left(\frac{3}{\pi} n_n\right)^{2/3}, \tag{5.5}$$

or with $hc \approx 1.2 \times 10^{-12}$ MeV-m,

$$n_n \approx \frac{1}{(hc)^3}\left(8m_n c^2 \mathrm{KE}_F\right)^{3/2}$$

$$\approx 2 \times 10^{43} \text{ neutrons/m}^3.$$

The corresponding neutron mass density is then

$$\rho_n = n_n m_n \approx 2 \times 10^{43} \times 1.7 \times 10^{-27} \approx 3 \times 10^{16} \text{ kg/m}^3. \tag{5.6}$$

When ρ_n exceeds this value we will soon have a degenerate neutron gas. The nuclei will have evaporated, the electrons will be almost gone and only a few protons will remain. The star has become in every sense a neutron star—supported against the inward pull of gravity by the outward degenerate pressure of the spin $\frac{1}{2}$ neutrons.

5.2.2. Mass and Size of Neutron Stars

A neutron star composed of nonrelativistic neutrons must be governed by the virial theorem $\mathrm{KE}_{\mathrm{tot}} = \frac{1}{2}|\mathrm{PE}_{\mathrm{tot}}|$. As the density of neutrons increases, however, the onset of the relativistic virial theorem, $\mathrm{KE}_{\mathrm{tot}} = |\mathrm{PE}_{\mathrm{tot}}|$, leads to an unstable star with $E = 0$. The situation is similar to the maximum size of a white dwarf, the difference being that $\mu = 1$ ($\rho = m_n n_n$) for a neutron star, whereas $\mu = 2$ ($\rho = 2m_{\mathrm{H}} n_e$) for white dwarfs with cores heavier than hydrogen. We may therefore borrow the result (5.4), modified by Chandrasekhar for white dwarfs,

$$M_{\mathrm{NS}}^{\max} \approx \left(\frac{5hc}{2G}\right)^{3/2} \cdot \frac{3}{16\pi}\left(\frac{1}{\mu m_n}\right)^2$$

$$\rightarrow 4 \times (1.4\, M_\odot) \approx 5.6\, M_\odot, \tag{5.7}$$

which is independent of the size of the star. Thus if the mass of the original star exceeds $1.4\, M_\odot$, it will pass beyond the white dwarf stage and become a neutron star. If the mass of the star exceeds $\sim 5.6\, M_\odot$, it cannot stabilize as a neutron star. Gravity will continue to crush it until . . . .

Actually we expect the maximum mass to be somewhat less than $5.6\, M_\odot$ because the attractive nuclear force between the nucleons is helping gravity to overcome the degenerate neutron pressure. The nuclear force was neglected in the derivation of $M_{\mathrm{NS}}^{\max} \sim 5.6\, M_\odot$; including it in a model-dependent manner reduces the maximum mass of a neutron star to $M_{\mathrm{NS}}^{\max} \sim 3\text{–}4\, M_\odot$. Unfortunately, general relativity complicates the issue even further. In fact it is possible to obtain $M_{\mathrm{NS}}^{\max} < M_{\mathrm{WD}}^{\max} \approx 1.4\, M_\odot$. Most experts, however, believe $3\, M_\odot \sim M_{\mathrm{NS}}^{\max} > M_{\mathrm{WD}}^{\max}$ and we shall adhere to this value.

The maximum density of a neutron star occurs when the nucleons are essentially touching one another, about 10^{-15} m apart. This spacing gives

$$n_n^{\max} \sim (1/10^{-15}\ \mathrm{m})^3 \sim 10^{45}\ \mathrm{neutrons/m^3},$$

corresponding to a mass density

$$\rho^{\max} = m_n n_n^{\max} \sim 2 \times 10^{-27} \times 10^{45} \sim 2 \times 10^{18}\ \mathrm{kg/m^3}.$$

As can be seen from (5.6), this mass density is about 100 times larger than the mass density needed for the onset of degeneracy for a neutron gas. In this case the neutrons are degenerate and *relativistic* because

$$p_F c = hc\left(\frac{3}{8\pi}\right)^{1/3} n_n^{1/3} \sim 6 \times 10^{-7}(10^{45})^{1/3}\mathrm{eV} \sim 600\ \mathrm{MeV},$$

which is near the rest mass energy of the neutron, 940 MeV. At this maximum relativistic density, neutron stars can achieve their maximum mass of $\sim 3\ M_\odot$. This mass and density are related by $M = 4\pi R^3\rho/3$, which suggests that a typical radius of a relativistic neutron star is

$$R_{\mathrm{NS}} = \left(\frac{3M_{\mathrm{NS}}}{4\pi\rho_{\mathrm{NS}}}\right)^{1/3} \sim \left(\frac{3 \cdot 2 \times 10^{30}}{4 \cdot 2 \times 10^{18}}\right)^{1/3} \sim 10^4\ \mathrm{m} = 10\ \mathrm{km!} \qquad (5.8)$$

This radius is $1/600\ R_E$ and $10^{-5}\ R_\odot$.

For nonrelativistic neutron stars with masses less than $3\ M_\odot$, the conditions of QHE lead to the virial theorem, $\mathrm{KE_{tot}} = \frac{1}{2}|\mathrm{PE_{tot}}|$. Since both $\mathrm{KE_{tot}}$ and $\mathrm{PE_{tot}}$ contain a factor of $\frac{3}{5}$, we can write the nonrelativistic stability condition as

$$N\frac{p_F^2}{2m_n} = \frac{1}{2}\frac{GM^2}{R}$$

and then apply $N = M/m_n$, $p_F = h[(3/8\pi)n_n]^{1/3}$, and $n_n = 3M/4\pi R^3 m_n$ to obtain

$$\left[\frac{M}{m_n}\right]\frac{1}{2m_n}\left[h^2\left(\frac{3}{8\pi}\,\frac{M}{(4\pi/3)R^3 m_n}\right)^{2/3}\right] = \frac{1}{2}\frac{GM^2}{R}.$$

Note that the factor of R does not cancel from the virial relation $[(1/R^3)^{2/3} = (1/R^2) \neq (1/R)]$, as it did in the case for $M_{\mathrm{WD}}^{\max}$ and $M_{\mathrm{NS}}^{\max}$. Solving, cancelling and rearranging we find instead

$$M^{1/3}R = \frac{h^2}{G}\left(\frac{9}{32\pi^2 \cdot m_n^4}\right)^{2/3} \sim 2 \times 10^{14}$$

in MKS units. (We can derive a similar relation for a stable nonrelativistic white dwarf whose mass is $\sim 1.4\ M_\odot$. We leave this as an exercise for the reader.) For $M_{\mathrm{NS}} \approx 2\ M_\odot \sim 4 \times 10^{30}$ kg, the above relation gives a radius for a nonrelativistic neutron star of

$$R_{NS} \sim 10^{14} \times 10^{-10} \sim 10^4 \text{ m} \sim 10 \text{ km}.$$

Since this radius is the same as for a maximum density, relativistic neutron star, we conclude that most neutron stars are about the same size, $R_{NS} \sim$ 10 km, with varying mass densities of 2×10^{16} kg/m^3 to 2×10^{18} kg/m^3 and with a mass between 1.4 $M_\odot$ and about 3 $M_\odot$. Neutron stars with $M < 1.4\ M_\odot$ cannot be ruled out, however, as they might be the remnant of a violent supernova explosion, to be considered shortly. An absolute lower bound of a neutron star is correlated with the remaining protons and electrons needed to block the β decay of the heavier and more numerous neutrons via the Pauli exclusion principle. Detailed calculations suggest $M_{NS} \gtrsim \frac{1}{5} M_\odot$.

5.2.3. The Core of Neutron Stars: Superfluids or Quarks?

The center of a neutron star will be a wonderfully fascinating place (especially to a high-energy particle physicist). At densities approaching 10^{18} kg/m^3, the Fermi momentum of the degenerate particles (either *e*s or *n*s) will rise toward 10^9 eV/c. With this momentum, the electron can initiate all kinds of reactions. For example, one can have

$$e^- + p \rightarrow \Lambda + \nu$$

$$e^- + n \rightarrow \Sigma^- + \nu,$$

where Λ and Σ^- are unstable particles (hyperons) whose masses are heavier than the proton mass: $M_\Lambda \approx 1116$ MeV, $M_\Sigma \approx 1197$ MeV, respectively. In order to produce a reaction where $e + p$ (or n) $\rightarrow X + \nu$ with X heavier than the proton, the KE of the electron must make up the mass difference (assuming the proton is at rest); that is,

$$\begin{aligned} \text{KE}_{\text{el,min}} &= M_X c^2 - m_p c^2 \\ &= 177 \text{ MeV for } \Lambda \text{ production} \\ &= 259 \text{ MeV for } \Sigma \text{ production.} \end{aligned}$$

With KEs approaching 600 MeV, the relativistic *electrons* at the center of a neutron star can produce most of the particles we can make in our electron accelerators (Stanford's two-mile-long linear accelerator for example). At this point the neutron star will be composed mainly of neutrons, but with *e*s, *p*s, Λs, Σs, etc., added to flavor the soup. Since $\text{KE}_F \gg kT$ (remember, $T \sim 10^7$ °K), the core will take on properties similar to those of He^3 when it is cooled to near absolute zero (where $\text{KE}_{F,\text{He}^3} \gg kT$), since the spin of He^3, having an odd number of nucleons, is $\frac{1}{2}$, the same as for a neutron. In the vernacular, this is called a superfluid and so the core of a relativistic neutron star is considered to be a dense superfluid.

It is now believed, however, that there may exist another substructure of matter, even more dense than the nucleons or hyperons, call quarks. During the past decade experimental evidence has pointed toward the existence of quarks as point-like constituents within nuclear matter. If quarks do exist then they represent a new and fascinating ingredient for the core soup of a neutron star and may mean that the maximum mass of a neutron star is greater than 4 $M_{\odot}$.

A striking confirmation of the quark model came about in a way very reminiscent of Rutherford's discovery of the nuclear structure of matter in 1913. In 1972, using a beam of protons with an energy of 200 billion eV (compared to Rutherford's alpha particle "beams" of 4 million eV) physicists at the CERN proton accelerator in Geneva Switzerland found a large number of wide-angle scattering events, many more than were expected for proton–proton direct interactions. The interpretation of these events was similar to the interpretation Rutherford gave to the anomalously large-angle scattering events he saw in 1913, i.e., that the proton, like the nucleus, must have a "point"-like structure.

Although free quarks have yet to be isolated in the laboratory, the quark model accounts for a remarkable number of properties of the many nucleons, hyperons, etc., in terms of just a few quarks, having spin $\frac{1}{2}$, called "up," "down," "strange," "charm" (and perhaps "bottom" and "top"). In particular the proton is thought to be made up of three quarks in the combination uud, the neutron of udd, the Σ^+ of uus, and the Σ^- of dds, etc.

Now if the neutron is composed of two down quarks and one up quark, and if the neutrons in the center of a neutron star are further squeezed, then these neutrons could decompose into quarks. This Fermi sea of spin $\frac{1}{2}$ quarks would be degenerate and could support the envelope of the neutron star. We shall leave it as a problem for you to show why degenerate quarks will allow a greater maximum mass for a stable neutron star composed only of neutrons.

So with general relativity indicating $M_{\max} < 1.4\ M_{\odot}$ and with quarks pointing towards $M_{\max} > 5\ M_{\odot}$, we shall still stick to our happy median of $M_{\max} \sim 3\text{–}4\ M_{\odot}$ where neutron-degenerate pressure supplies the repulsive force that supports the outer envelope.

5.2.4. Magnetic Field and Rotation of Neutron Stars

Because of the conservation of magnetic flux, we expect that the neutron star will possess a large magnetic dipole field. You will recall that as long as we have charged particles present, magnetic flux is conserved with

$$B_f = B_i\left(\frac{R_i}{R_f}\right)^2,$$

as depicted in Fig. 3.5. Since $R_f = 10$ km and R_i—say for our sun—is 10^9 m, we should find

$$B_{NS} \approx 10^{10} B_i. \tag{5.9}$$

Since the surface fields of most stars are between 1 and 100 gauss (the average field of the Sun is 1 gauss but in regions of sun-spot activity the field exceeds several thousand gauss), we expect the neutron star to possess a surface field of between 10^{10} and 10^{13} gauss!! The field will be generated by the highly mobile electrons in the interior of the neutron star. Since the electrons are degenerate, they normally will not lose energy when they collide with protons, since to do so would mean falling to an occupied lower-energy state. Only collisions near the Fermi level, where the sites are partially filled, will result in energy loss. Because these latter collisions are infrequent, the electron currents in the neutron star will circulate without dissipation, maintaining the magnetic field for times that have been estimated to exceed the age of the universe by a factor of 1000.

We also expect that the neutron star will be rotating, especially if it is the remnant of a supernova explosion. Assuming the star has stabilized, it must be rotating at such a rate that it is not shedding matter. As we have seen many times before, this means the equatorial particles of mass m must move in circular paths so that the net force equals mv^2/R. Since the net force is supplied by gravity, we have

$$mv^2/R = GMm/R^2.$$

With $v = R\omega$, the above becomes

$$R^2\omega^2/R = GM/R^2$$

and cancelling and solving for ω we have

$$\omega = \sqrt{GM/R^3}\,. \tag{5.10}$$

With a typical neutron star having a mass of around $2\,M_\odot$ and a radius of $\sim 10^4$ m, the maximum angular velocity it can have without falling apart is

$$\omega_{\max} \approx \sqrt{\frac{6.7\times10^{-11}\times2\times2\times10^{30}}{(10^4)^3}} \approx 1.6\times10^4\ \text{sec}^{-1},$$

corresponding to a *minimum* period of rotation of

$$\tau_{\min} = \frac{2\pi}{\omega_{\max}} \cong 4\times10^{-4}\ \text{sec} \sim \tfrac{1}{2}\ \text{msec}.$$

Minimum or not, this is a fast rotation for so large an object. We must also keep in mind that the axis of rotation and the direction of the dipole field may not line up so that this huge magnetic field may also be a co-rotating field resulting in large flux changes around the neutron star.

By the 1940s, one had all the physical tools for describing such a neutron star. One knew when it would be stable ($M \sim 3\ M_\odot$), its size ($R \sim 10$ km), the strength of its magnetic field ($B \sim 10^{12}$ gauss), and its period of rotation ($\tau \sim \frac{1}{2}$ msec). But does such an object exist and if so how would we "see" it? Unfortunately no one at that time had the insight how to find a neutron star and since no one had any strong motivation for looking for one, the neutron star remained a mere curiosity. Soon people forgot all about it . . . .

5.3. Supernovae

What people could see and did not forget about was a phenomena called a supernova. Visually a supernova is an outpouring of electromagnetic radiation from a star over a very short period of time. The star's luminosity will increase by a factor of 10^8 (!) and often become as bright as an entire galaxy, and then die out over a period of days. Because of their extraordinary light output, these "exploding" stars are easy to detect as one can see in Fig. 5.4.

One of the earliest recorded supernova explosions was that of the Crab which was observed by Chinese and Japanese astronomers on July 4, 1054. A picture of the remnant of the Crab, the Crab Nebula, is shown in Fig. 5.5a. Who else may have seen the original explosion has been a fascinating topic of investigation over the past few years. Indeed, it had long been a mystery as to why the Crab, which was initially brighter than Venus and which was visible at night for 6 months, was not recorded by European and Arabic astronomers. In 1978 in the scientific journal *Nature*, however, the Liebers and Brecher reported finding a reference to a "spectacular (athārī) star (kawkab)" in an ancient Arabic medical biography. Their careful analysis of this text pinpoints the date of the "athārī kawkab" to the summer of 1054! Closer to home, a petroglyph found in Navajo Canyon, Arizona shows the juxtaposition of a quarter moon and a bright star (see Fig. 5.5b). Tracing back in time, astronomers have shown that this drawing might represent the appearance of the moon and the Crab on the morning of July 4, 1054 as viewed from western North America. To top it off, archeologists have shown that the site of the petroglyph was occupied between 900 and 1300 A.D. It is intriguing to imagine that in their own way the Chinese, Japanese, Arabs, and Arizonans were referring to the same object, now thought to be the supernova explosion in the center of the Crab nebula.

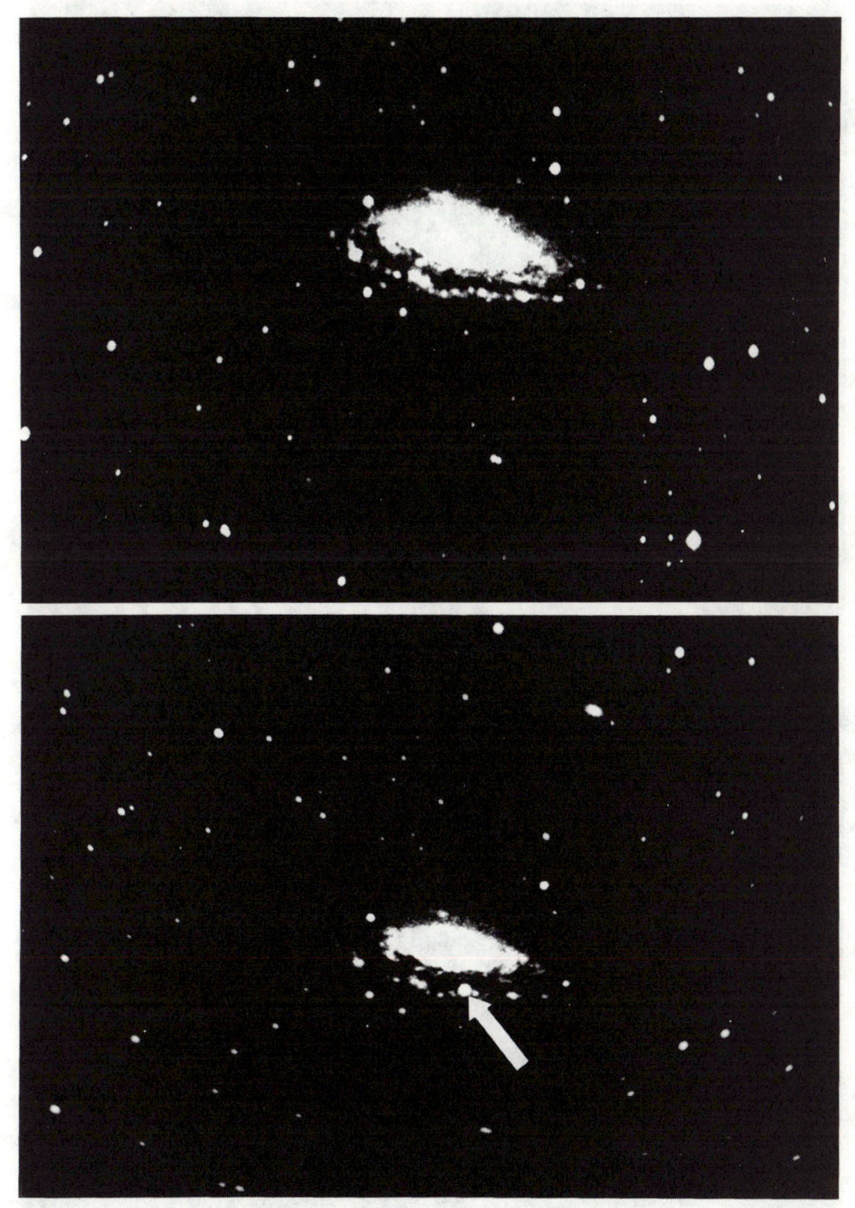

Fig. 5.4. Supernova explosion appearing in galaxy NGC 7331 in 1959 (Lick Observatory photograph).

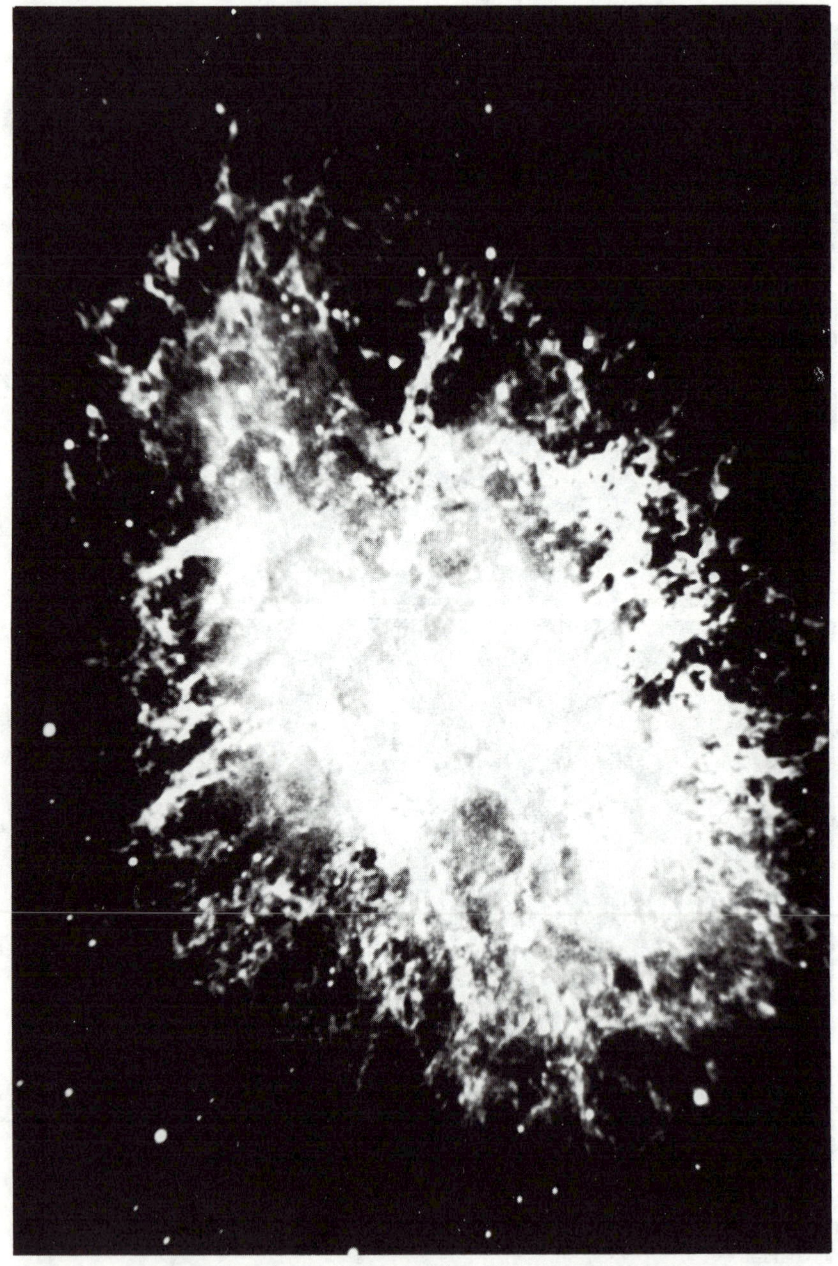

Fig. 5.5. (a) The Crab Nebula (Lick Observatory photograph).

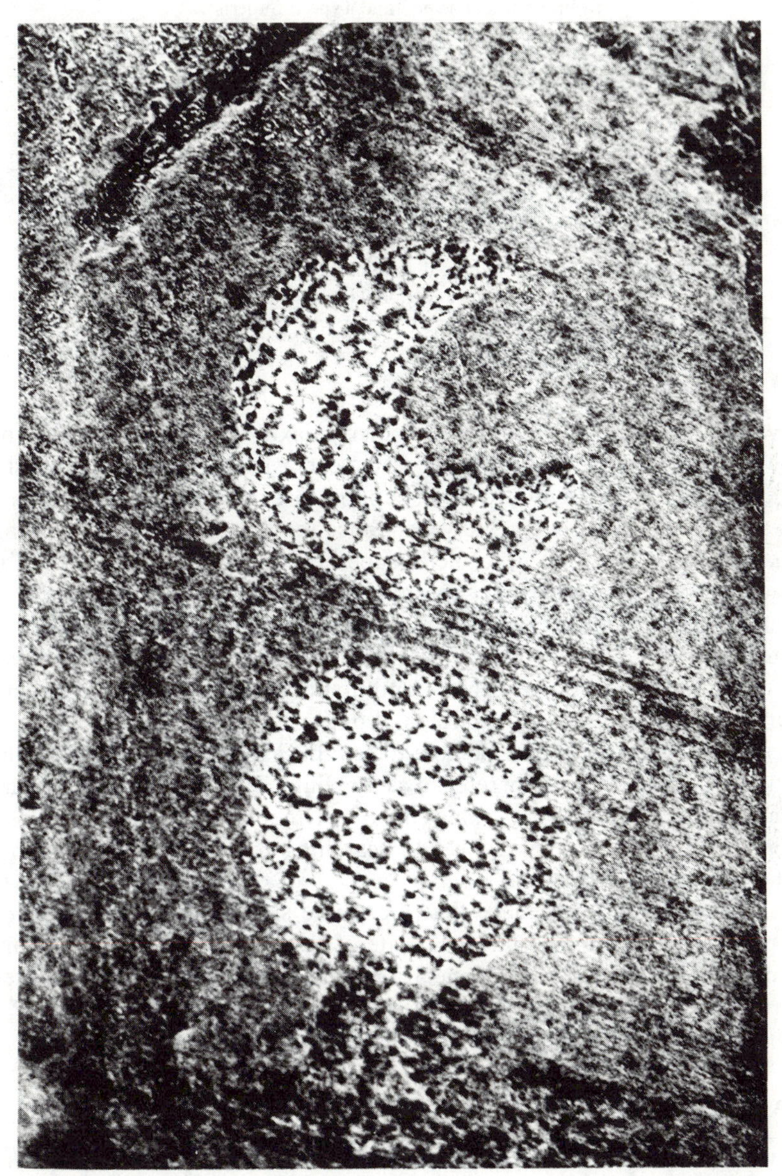

Fig. 5.5. (b) Navajo Canyon petroglyph believed to depict the explosion of the Crab Nebula in 1054 (Courtesy of Wm. C. Miller).

TABLE 5.1.

Properties of Type I and Type II Supernovae

	Type I	Type II
Total energy ejected	10^{42} J	10^{43} J
Average ejected mass	$\frac{1}{2} M_{\odot}$–$2 M_{\odot}$	2–5 $M_{\odot}$
Peak luminosity	$10^9 L_{\odot}$	$10^8 L_{\odot}$
Falloff of luminosity vs. time	Regular	Irregular
Spectrum of radiation	No hydrogen lines	Strong hydrogen lines
Location	With old and young stars and globular clusters	With young stars

5.3.1. Classification of Supernovae

Over 400 supernova explosions in other galaxies have been detected over the past century, about one every 50 years / galaxy. Yet in the last 20 centuries, we have seen only seven supernova explosions originating in our galaxy, a rate of one every 300 years. Since the most recent local supernova was seen in 1604 (called Kepler's supernova), statistically we are about ready to witness another one, you might say, any day now. From the data that has been gathered on these supernovae, it appears they fall into two categories—unimaginatively called Type I and Type II. Their properties are listed in Table 5.1. Along with the outburst of EM radiation, matter will be hurled into space by the explosion. Long after the explosion has visibly died away, one should still be able to see this expanding filament or cloud of matter which we usually refer to as a nebula. The most famous and most studied is the Crab Nebula, the remnant of the 1054 supernova explosion. A great deal of time has gone into searching for the nebular remains of other supernova explosions.

Since one does not believe that the whole star disintegrates, one would hope to find a remnant star at the site of the supernova explosion—much like the central star of a planetary nebula. One of the early mysteries surrounding supernovae was the fact that there did not seem to be *any central stars*!

5.3.2. The Nebula

If we have missed the explosion, then what we learn about the supernova will have to come from studying the expanding nebula. But to study the nebula, we will have to be able to see it. This can come about in two ways, the first with which we are familiar; but the second will require a bit of explanation.

First of all, the ejected matter quite literally sweeps up and heats the interstellar medium. As the expanding nebula crashes into the lower density interstellar medium, a shock wave occurs. About 10,000 years after the explosion, when enough matter has been swept up and heated by the shock wave, X-ray radiation will appear. Subsequently the expanding shock wave slows and the emitted radiation is produced at longer and longer wavelengths. After about 100,000 years most of the energy of the original ejected matter has been radiated away. At the back edge of the nebula where the ionized gases can cool and recombine, one can see the emission spectrum of hydrogen. This line spectrum is dominated by the red Balmer, H-α line. As shown in Fig. 1.18c, the Veil Nebula, sometimes called the Cygnus loop, is a beautiful example of this latter point.

But there is more to the radiation picture than the thermal spectrum we have just discussed. In both the Veil and Crab Nebulae (the latter shown in Fig. 5.5a), there is a component of radiation that does not fit a characteristic thermal curve. Besides the thermal distribution, there is a nonthermal component of the frequency spectrum which rises as one goes toward lower frequencies compared to the falling thermal spectrum. This is illustrated in Fig. 5.6. Moreover the nonthermal radiation is polarized. That is, the electric vector lies in a particular plane. Thermal radiation due to random motion, however, is never polarized. Let us see if we can understand the origins of the nonthermal radiation.

This radiation was first observed in Earth laboratories in electron synchrotron accelerators and is therefore also called synchrotron radiation. In a synchrotron, electrons are constrained to move in a horizontal circle by a vertical magnetic field. In light of this it is easy to see why synchrotron radiation is polarized. Since the electrons are confined to move in a plane,

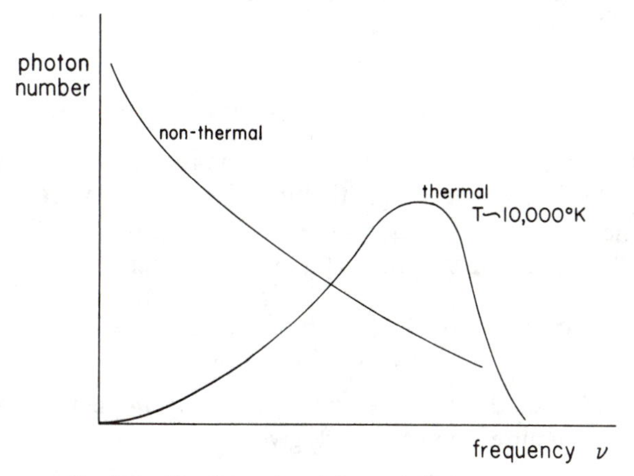

Fig. 5.6. Nonthermal and thermal photon spectra.

they will appear to an observer looking at that plane edge on as if they were moving back and forth, i.e., oscillating. Now any accelerating charge will produce electromagnetic radiation with the electric vector of the radiation proportional to the acceleration of the charge, i.e., $\mathbf{E}_{rad} \propto \mathbf{a}$. Since the acceleration vector of the electron is confined to the circular plane, the $\mathbf{E}_{rad}$ vector must also lie in that plane. Hence the polarization.

The frequency distribution of the radiation is a bit harder to understand. If we stick to the model of the circular motion being equivalent to a back-and-forth oscillation, then we expect the frequency of the emitted radiation to be equal to the frequency of the oscillations. But this frequency is just $1/T$, where T is the period of the circular motion of the electron in the magnetic field. It is both interesting and instructive to calculate this frequency classically. Looking down on the electron's orbit with $\mathbf{B}$ pointing up, the centrally directed magnetic force is

$$\mathbf{F} = e\mathbf{v} \times \mathbf{B}.$$

Since in a synchrotron $\mathbf{v}$ and $\mathbf{B}$ are at right angles to one another, $|\mathbf{v} \times \mathbf{B}| = vB \sin\theta = vB$. Moreover, since the electron moves in a circle, we know that $a = v^2/R$ so that $|\mathbf{F}_{net}| = m|\mathbf{a}|$ gives us

$$mv^2/R = evB.$$

Solving for v/R we get

$$v/R = eB/m,$$

and since the angular velocity is $\omega = v/R$, we have

$$\omega = 2\pi\nu = eB/m,$$

where in this case m is the mass of the electron.

Interestingly enough this classical derivation shows that ω is independent of the electron's energy so that all the particles should emit monochromatic photons of energy

$$E_\gamma = \hbar\omega = \hbar eB/m. \tag{5.11}$$

Let us see if we can estimate what E_γ might be for an electron (any energy now) moving in a galactic magnetic field. The units of B in the MKS system will have to be in teslas, where 1 T $= 10^4$ gauss. With $B_{gal} \approx 10^{-6}$ gauss $= 10^{-10}$ T and $m \approx 9 \times 10^{-31}$ kg, we have from (5.11),

$$E_\gamma \approx e \times 10^{-34} \times \frac{10^{-10}}{9 \times 10^{-31}}\,\text{V} \sim 10^{-14}\,\text{eV}.$$

To get photons with the same average energy from thermal black body radiation would require a temperature given by

$$kT \sim 10^{-14}\,\text{eV}$$

or

$$T \sim 10^{-10}\ {}^{\circ}\text{K}.$$

Thus it is not surprising that the synchrotron radiation spectrum, as obtained from galactic sources, extends down to low frequencies compared to any sensible black body spectrum. But the true synchrotron spectrum is not monochromatic; it does not contain a single-frequency photon. Instead it is clear from Fig. 5.6 that the spectrum extends to very high frequencies. Experiments show that at low velocities the electrons do radiate mainly at $\omega = qB/m$, a frequency we shall call the "cyclotron" frequency ω_0. But if electrons of higher energy are injected into the synchrotron, they begin to radiate not only at the cyclotron frequency, but at higher harmonics, at frequencies that are multiples of ω_0. This effect is more pronounced as the electron reaches relativistic speeds, with more and more energy going into higher harmonics as $v \to c$. This is illustrated in Fig. 5.7. Electron interactions broaden this spectrum into a continuum.

If we now turn back to the spectrum of the Veil or Crab Nebula, as depicted in Fig. 5.6, we see that (a) because they produce synchrotron radiation, there must be a magnetic field in the nebula about which electrons move in circular or spiral orbits; and (b) there must be a continuous source of relativistic electrons. The latter point deserves a bit of explanation. Since the electrons are radiating, they must lose energy. Since $v/R = qB/m$, v/R is a constant. Thus as v decreases, R must also decrease in order that the ratio v/R stays constant. The electrons will thus move in tighter and tighter spirals until their kinetic energy goes to zero. Of course when this happens, they no longer contribute to the synchrotron radiation spectrum. Since the Crab has been sending out synchrotron radiation for 1000 years, something must be continuously feeding in electrons to make up for the ones that spiral in and no longer radiate. But what? Indeed the Crab is presently emitting 10^{31} J of synchrotron radiation

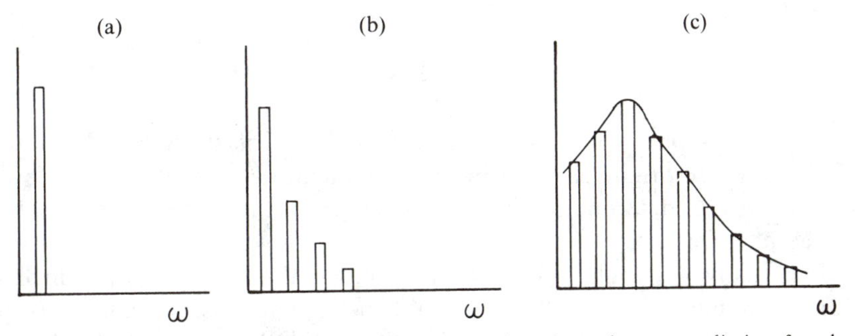

Fig. 5.7. Fundamental frequency and higher harmonics of synchrotron radiation for electrons rotating at (a) nonrelativistic, (b) slightly relativistic, and (c) very relativistic speeds.

each *second*. The electrons, therefore, must be losing energy at this rate. Clearly the nebula is being fed energy; it is being kept "alive." How? As to the magnetic field **B** of the Crab—whatever its origin—it clearly presents the mechanism by which electrons may spiral and emit synchrotron radiation. If the field is a dipole or predominantly in one direction, the electrons will effectively spiral in one plane (say x-y, as they spiral up and down along the z axis, the latter the local direction of **B**) and so emit polarized radiation.

We can also learn something if we study the rate at which the nebula is expanding. If an initial mass m_0 were ejected in the explosion at an initial velocity v_0, then as time goes on, this velocity will get smaller as the mass collides with particles at rest (i.e., $v_{ISM} < v_0$) in the interstellar medium. After the collision, the particles in the interstellar medium will pick up a radial component of velocity and will move along with the original particles; the ejected matter literally sweeps out the interstellar medium. Since within the m_0-interstellar medium system all forces are internal, momentum is conserved so that

$$P_0 = P_{\text{now}}$$

$$m_0 v_0 = v_{\text{now}}(M + m_0)$$

$$m_0 v_0 = m_0 v_{\text{now}} \left[\frac{\rho(4\pi/3)R^3}{m_0} + 1 \right],$$

where M is the matter swept out of the interstellar medium, R the radius of the nebular shell, ρ the density of the interstellar medium ($\approx 10^6 \times 1.7 \times 10^{-27}$ kg/m^3), and $M = \rho 4\pi R^3/3$. For the Crab Nebula, $v_{\text{now}} \approx 1.3 \times 10^6$ m/sec, $R_{\text{now}} \approx 2$ LY$\approx 2 \times 10^{16}$ m, and $m_0 \sim 1 M_\odot \approx 2 \times 10^{30}$ kg. Therefore, we find $\rho(4\pi/3)R^3/m_0 \ll 1$, so that $v_0 \cong v_{\text{now}}$. Thus we expect the expansion time to be

$$\Delta t = R_{\text{now}}/v_{\text{now}} \sim 1.5 \times 10^{10} \text{ sec} \sim 500 \text{ yrs},$$

which agrees pretty well with the observed age of the Crab. For other nebulae whose origins are lost in time, momentum conservation offers the opportunity of determining their age solely from the present configuration of the nebulae.

Eventually a nebula slows down so much that $v_{\text{final}} \sim v_{\text{ISM}}$, which at 100 °K would be about 10^3 m/sec. When this happens, we can no longer distinguish the nebula from the ISM; the nebula has disappeared. But by that time

$$m_0 v_0 = (M_{\text{swept}} + m_0)v_{\text{final}}$$
$$= (M_{\text{swept}} + m_0) \times 10^3.$$

Since $v_0 \sim 10^6$ m/sec, we see that

$$M_{\text{swept}} = 10^3 m_0 - m_0 \approx 10^3\, m_0.$$

Before it intermingles with the ISM, the nebula will have swept up 1000 times the initial ejected mass. You should be able to show that for the Crab, this means the nebula will disappear in about 10,000 years.

5.3.3. Sources of Supernova Explosions: Type II

It is easiest to start with Type II supernovae because $M_{\text{ej}} \sim 5\, M_\odot$ implies that the parent star is much more massive than the Sun. As we discussed in Sec. 4.4, if the mass of a star is between 3 $M_\odot$ and 10 $M_\odot$, a carbon flash will occur and either the star will disintegrate or a neutron star will remain behind. If, however, $M > 10\, M_\odot$, we are dealing with a star that has evolved fairly completely into the red giant phase with a full complement of nested rings—H, He, CSi, Fe. What would cause such a star to explode? To set the ground rules: if we are to have a supernova explosion, energy must be deposited in the envelope of the star in sufficient quantity to overcome the pull of gravity. Prior to 1966, the conventional wisdom concerning this energy production centered around the dynamics of the iron core.

Recall the average binding energy vs. nucleon number A curve, Fig. 4.1. Iron sits at the top of this curve. Thus Fe + Fe collisions should not lead to fusion. Any inelastic reaction will lead instead to a breakup of iron with a *net loss* of kinetic energy in the interaction. Thus, for example, when broken up into α-particles and neutrons,

$$M_{\text{Fe}}c^2 + M_{\text{Fe}}c^2 + \sum \text{KE}_i = \sum M_\alpha c^2 + \sum M_n c^2 + \sum \text{KE}_f$$

must result in $\sum\text{KE}_f < \sum\text{KE}_i$. We thus have new and large fluxes of αs and neutrons but their overall KE is less than that of the original iron nuclei. This suggests two things: (a) the large flux of neutrons means we can build up the heavy elements where $A > 209$ if the reactions are fast enough; and (b) the absorption of energy leads to a collapse of the core. The latter step is catastrophic. The collapsing core leads to a rapid heating throughout the entire star. There is a tremendous release of gravitational energy. This energy, however, does not go into expanding the envelope—not right away. First the increase in temperature increases the fusion rate of the unburned (or slowly burning) elements in the outer rings. Since $\epsilon_{pp} \propto T^4$, $\epsilon_{3\alpha} \propto T^{40}$ etc., these reactions go off at a tremendous rate. This in turn raises T further, increasing ϵ_{pp}, $\epsilon_{3\alpha}$, etc. The result is a fantastic increase in the luminosity at all levels of the star. Up goes $\Delta T/\Delta r$ and the star blows apart. In the process, the high densities and rapid flux of neutrons results in the synthesis of the heavier elements we spoke of above. Such is the scenario of a *thermonuclear* supernova explosion.

By 1966 serious doubts concerning this model began to creep up. Calculations showed that the reduced pressure due to the collapsed core would first lead to an inward falling envelope. The forces necessary to turn this around are clearly greater than those necessary to blow off a static envelope (remember, $F \propto \Delta\bar{v}$), with the result that thermonuclear reactions are not energetic enough to do the job. What then could transfer sufficient energy from the collapsing core to the envelope? After all, we do see supernovae and the data of Table 5.1 on Type II supernovae are consistent with their being massive objects. For example, because they are massive they evolve quickly ($\sim 10^8$ yr) and therefore must appear with young stars, which they do. Moreover, evolving quickly means that they would not deplete their hydrogen content and hence the strong hydrogen lines in their spectra. As to their being red giants, in 1970 a supernova was spotted in the galaxy M101. The peak of its black body spectrum corresponded to a surface temperature $T_s = 12{,}000$ °K. Knowing the distance to M101 and measuring the flux $\mathcal{L}$ (J/m^2 = sec) of radiation from the supernova that was incident on Earth, the absolute luminosity could be calculated from $L = \mathcal{L}\, 4\pi d^2$. From this and T_s one could deduce the radius of the star at the time it exploded via $R^2 = L/(4\pi\sigma T^4)$. It turned out to be 3×10^{12} m or $\sim\frac{1}{2}\times10^4\ R_\odot$. So we do have massive red supergiants with, very probably, iron cores that become supernovae, but most likely not via thermonuclear reactions.

In 1966 another mechanism was suggested for the explosion of massive stars. The collapsing core was generated as before through the breakup of the iron nucleus. But now the imploding core passes through the white dwarf stage of electron-degenerate stability to reach densities where $e^- + p \rightarrow n + \nu$ can proceed. This reaction is in a sense "endothermic" in that using up electrons reduces the degeneracy pressure and increases the rate of collapse. This increases the rate of neutrino generation and if the process goes rapidly enough, there will be a huge outpouring of neutrinos, which if they interact with the envelope, could blow the star apart leaving behind a neutron star. The last "if" is a rather big one, however, and only recently ($\sim$1974) have we come to sufficiently understand neutrino interactions to believe the above scenario is possible. At high energy particle accelerators, new forms of neutrino interactions called neutral-current interactions where

$\nu + p \rightarrow \nu + p$, and $\nu + e \rightarrow \nu + e$ have been detected. If so, these interactions may have sufficiently large cross section to allow for intense fluxes of neutrinos to blow off the mantle of a star. This marriage of experimental and theoretical high-energy-particle physics and astrophysics is an exciting development. Equally exciting is the fact that the remnant of such an explosion will be a neutron star. To add further intrigue, if the initial star is massive enough (say $M > 20\ M_\odot$) the imploding core may exceed $3\ M_\odot$, in which case the outward neutron-degenerate pressure would be insufficient to stop the collapse and a black hole may result. It sure beats science fiction, doesn't it?

Regardless of how the explosion takes place, it is interesting to note that the ejected matter will be rich in heavy elements. Since $M_{\text{swept}} \sim 10^3\ M_\odot$ we are in effect seeding the interstellar medium with heavy elements to $\sim 1/10^3 \sim 0.1\%$. What we see in our region of the Galaxy, however, is 3% heavy elements. Unless this has a cosmological origin, our region must have been seeded by about 30 supernovae.

Stars whose masses lie between 3 and 10 $M_\odot$ present a different problem since their cores probably do not evolve beyond the carbon stage. At these masses, the stars are probably supported by the pressure of degenerate electrons. Further compression, as we have seen earlier, will lead to a catastrophic explosion where, it is estimated, not a cinder will remain. Recently, however, attempts have been made to allow for a more controlled burning in the degenerate carbon core. This allows contraction to continue so that densities can be reached where our faithful process $e^- + p \rightarrow n + \nu$ will begin. The removal of electrons reduces the degenerate pressure which, as described above, will lead to a catastrophic collapse of the core, an outpouring of neutrinos, a neutrino-induced explosion, and a remnant neutron star. These all or nothing choices leave one a bit perplexed. Clearly the last chapter has not been written on Type II supernovae.

5.3.4. Sources of Supernova Explosions: Type I

Although the ejected mass of a Type I supernova is about $\frac{1}{2} M_\odot$, this is no guarantee that the mass of the parent star is low. If we turn to Table 5.1, however, we see that many Type I supernovae appear in regions with old stars and have no hydrogen lines in their spectra. Both of these facts point to a low mass parent star since a low mass star will evolve slowly and consume its hydrogen in the process. But how does a star with $M \sim 1\ M_\odot$ explode? We will describe one possibility: that enough mass can be added to the star so that it exceeds the Chandrasekhar limit, $1.4\ M_\odot$.

Consider a binary pair of stars A and B, with $M_A > M_B$. Because it is more massive, A will evolve more quickly and reach its red giant phase before B. As we shall see later in this chapter when we study black holes, a large red giant can transfer mass to a smaller companion. If it does so, the degenerate core of A will be left behind so that after mass transfer the binary pairing is A, now a white dwarf, and a more massive star B. It is now Bs turn to evolve quickly and become a giant star whereupon it will transfer mass back to A. But A is now a white dwarf and the added mass pushes it beyond the critical white dwarf mass limit—collapse ensues—and once again $e + p \rightarrow n + \nu$ operates to produce a supernova and a neutron star remnant. The observation that a few neutron stars are found with super giant companions (see Sec. 5.5) lends a measure of credibility to the above scenario. As we shall see in Sec. 5.4, however, most supernova explosions will disrupt the binary pairing and lead to isolated neutron stars.

The fact that the brightness of a Type I supernova falls by a factor of 2 every 50 or so days has puzzled astronomers for some time. It could be due to the decay of radioactive elements. Candidates were proposed: Be^7, Sr^{90}. However, in the 1952 atomic bomb test, Cf^{254} was detected with a 55-day half life. It was produced by intense neutron fluxes in the explosion. If, however, Cf^{254} were synthesized it needed iron to build on. Indeed, the spectrum of a Type I supernova observed in 1970 appeared to have such strong iron emission lines that its abundance of iron seemed at least to be 20 times the abundance of iron in the Sun! A consistent accounting of both of these observations would indicate intense neutron fluxes from which iron and heavy radioactive elements can be produced.

Either way, Type I or Type II supernovae would have been reasonable places to look for neutron stars—at the center of expanding nebulae. But no one did. Why? Well, what would you hope to see? How would you know something was there? Remember, the neutron star is 10 km across and has very low luminosity. How would you find something that small which has so little intrinsic brightness? In what part of the EM spectrum would you look? What kind of detector would you build? Nobody else knew either.

5.4. Pulsars

The history of the discovery of pulsars is well described in a 1968 *Scientific American* article by Hewish. For those interested in the details of this adventure, we recommend that article. We would, however, like to digress for a moment and elaborate in a little more detail on just what led up to the discovery. The story is a beautiful example of how great finds in physics are often made: through persistence, patience, and a good deal of luck.

Between the Sun and the Earth is a flux of charged particles called the solar wind. If an electric vector coming from outer space, $\mathbf{E}_{inc}$, is incident on the solar wind, then each charged particle will experience an acceleration proportional to $\mathbf{a} = -\mathbf{E}_{inc}C$ where C is a constant that takes into account the charge, mass, and other parameters involved in the Coulomb interaction. Each accelerating charge will produce scattered radiation, where $\mathbf{E}_{scat} \propto \mathbf{a} = -\mathbf{E}_{inc}C'$. Here the minus sign is important because it tells us the scattered electric vector is opposite to the incident one. The total scattered radiation will be proportional to $\mathbf{E}_{scat}$ times the number density of particles in the solar wind (represented by n), i.e.

$$\mathbf{E}_{scat,tot} = -\mathbf{E}_{inc}nC',$$

where C' is another constant of proportionality. The radiation that makes it through and reaches the Earth will be the vector sum of $\mathbf{E}_{inc}$ and $\mathbf{E}_{scat,tot}$. Calling this the transmitted radiation, we have

$$\begin{aligned}\mathbf{E}_{tran} &= \mathbf{E}_{inc} + \mathbf{E}_{scat,tot}\\ &= \mathbf{E}_{inc} - \mathbf{E}_{inc}nC'.\end{aligned}$$

What Hewish and his collaborators set out to study was the "scintillations" caused by the changing density of the solar wind, i.e., the changes in E_{tran} due to fluctuations in n. The scintillations in turn were to be used to look for quasars, using their emitted radiation in the 3-m region of the radio spectrum. Up until 1966 no detection devices had been built to study short time changes, i.e., scintillations, for wavelengths that long. As the article so beautifully tells, what Hewish and collaborators found was a new and strange object emitting pulses of radio frequency radiation, the *pulsar*. It was Hewish's good fortune to be in the right place at the right time with the right instrument even though for the "wrong" reasons. This is not uncommon in physics: Oersted's discovery of the relation between currents and magnetic fields and Davisson and Germer's discovery of the wave nature of the electron are but two more examples of similar "accidents." Nonetheless, they earned their discoveries through their hard work and persistence. In 1974 Hewish won the Nobel prize for his discovery of pulsars. It should be a lesson to us all.

5.4.1. General Considerations

Presently, there are about 300 observed pulsars. Each emits a continuous train of radio frequency pulses with an almost constant time interval between signals (the period) given by τ (see Fig. 5.8). One must not confuse the frequency of the radiation with the period of the pulses. Although the pulses are detected at, say, a frequency of 10^8 cps, their periods range from

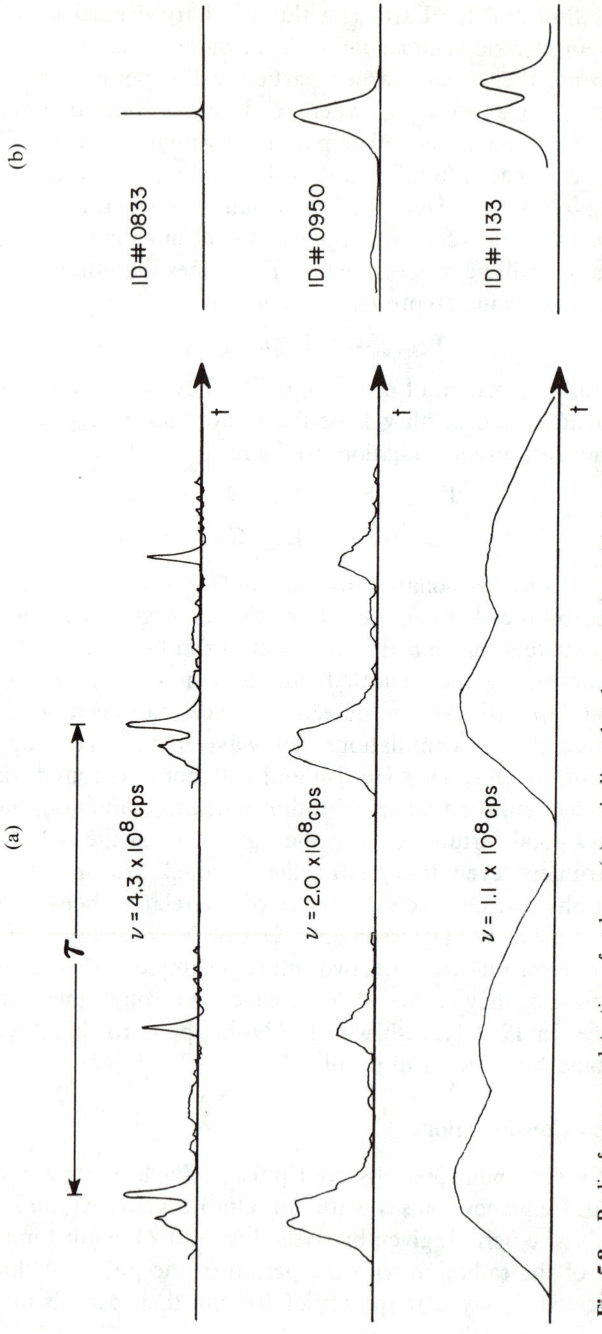

Fig. 5.8. Radio frequency detection of pulsars: (a) periodic pulse shape varying with frequency for a single pulsar; (b) integrated pulse shape for various pulsars.

0.033 sec to 3 sec. It was soon discovered that the pulses are not confined to a single frequency. Figure 5.8a shows *two* pulses from the *same* pulsar recorded at three different frequencies. Notice that even the pulse has a structure and that structure varies from frequency to frequency! Nevertheless, even though they have different frequencies, the pulses all have the same period. In Fig. 5.8b we show the shape of three different pulsars integrated over the radio frequency spectrum. As you can see, 0833 is very sharp, 0950 has a small prepulse, and 1133 is double-humped. Clearly there is a wealth of information contained in these pulses, if we can only figure out what is going on. Besides the radio frequency part of the spectrum, which is where Hewish's apparatus was tuned, it was certainly natural to look for pulsed radiation at other frequencies, especially in the visible. In 1969, scientists at the Lick Observatory took the photographs shown in Fig. 5.9. These remarkable pictures show a visible pulsar found near the center of the Crab Nebula. This pulsar also emits in the radio part of the spectrum where it was first discovered and in the X-ray region as well! This brings us to the important question: Where are pulsars found?

All known pulsars are found in the galactic plane, although a few may be heading out in highly elliptic orbits. In 1970 a study of their distribution about the galactic plane led to an interesting speculation about their masses. For stars in general, the more massive they are, the nearer to the galactic plane they can be found. The distribution of pulsars indicated that if they derived from stars, the *parent star* (not the pulsar) had a mass in the 2–10 $M_\odot$ range.

Another exciting discovery was that short-period pulsars were found close to the center of the Crab and Veil nebulae, both thought to be nebular remnants of supernova explosions. However, not all pulsars have supernova remnants surrounding them and not all supernova remnants have associated pulsars. Also it was found that most pulsars are loners, i.e., they are not part of a double star system, even though better than 50% of all stars are part of multiple star complexes.

One can begin to make a bit of sense out of this by looking at the selected pulsar data of Table 5.2. First of all, let us note that different pulsars do not have the same period. As the table shows, they vary significantly, from 33 msec to 1.96 sec, with most pulsar periods between $\frac{1}{4}$ and 2 sec. Second, the periods are changing with time. In fact, they are all getting longer. Consider the Crab pulsar (#053132). Its period is lengthening by 4.22×10^{-13} sec each sec. Therefore, in one year its period will have increased by $4.22 \times 10^{-13} \times \pi 10^7 \approx 10^{-5}$ sec. This is not very long on our time scale, but it can be appreciable over longer periods of time. In fact, if we assume this gain in period has been steady since the birth of the pulsar,

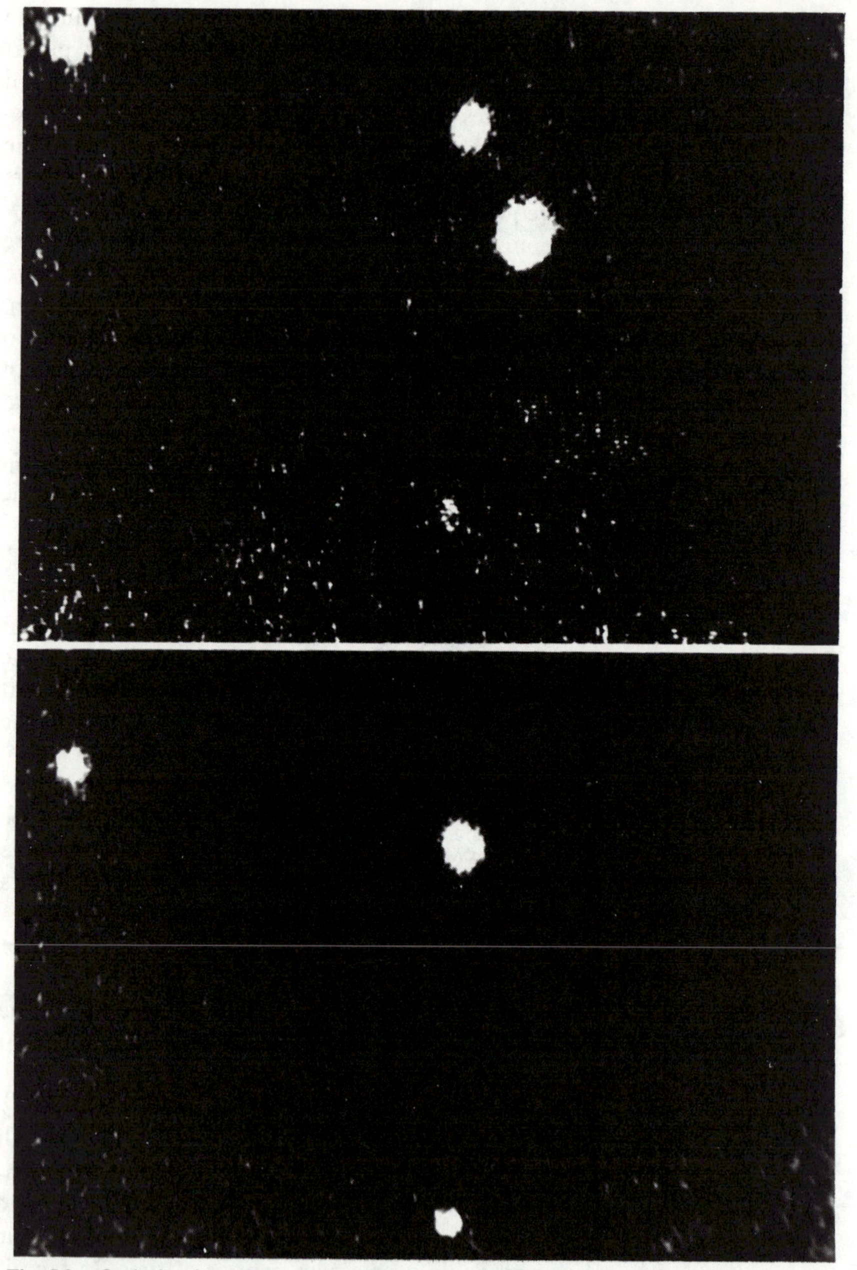

Fig. 5.9. Optical pulsar photographed in the center of the Crab Nebula (Lick Observatory photograph).

TABLE 5.2.

Pulsar parameters

Pulsar Identification No.	τ (sec)	$\Delta\tau/\Delta t$ (10^{-13})	Estimated maximum age (yr)
053132	0.033	4.22	2.5×10^3
083339	0.089	1.24	2.3×10^4
095031	0.253	0.002	3.4×10^7
174949	0.562	0.71	2.5×10^5
150803	0.739	0.05	4.6×10^6
191936	1.337	0.013	3.2×10^7
204547	1.96	0.11	5.3×10^6

then according to Fig. 5.10 the age of the pulsar is

$$\text{age} = \frac{\tau_{\text{now}} - \tau_0}{\Delta\tau/\Delta t}, \tag{5.12}$$

where τ_0 was its period when the pulsar was born. Of course we do not know what τ_0 is, but if we assume that most pulsars have been around long enough so that τ_o τ_{now}, we can simulate a maximum age from $(\text{age})_{\max} = \tau_{\text{now}}(\Delta\tau/\Delta t)$. Thus for the Crab pulsar, we have from (5.12),

$$(\text{age})_{\max} \approx 33 \times 10^{-3}/4.22 \times 10^{-13} \approx 7.8 \times 10^{10} \text{ sec} \approx 2.5 \times 10^3 \text{ yr}.$$

Not bad when compared with the known age of the Crab nebula of 930

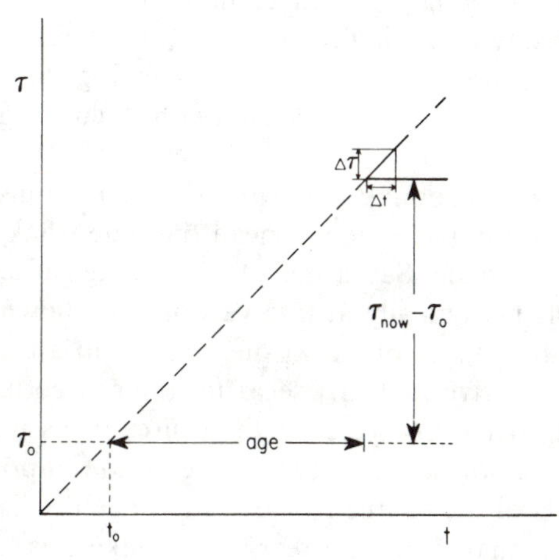

Fig. 5.10. Assumed linear relationship between pulse period and time.

years. Turning the above argument around, we could have solved for τ_0 for the Crab from its known age and the above equation (5.12). We find $\tau_0 \sim 0.02$ sec. Just why the pulsars are slowing down ($\omega = 2\pi/\tau$) is an important question that we eventually shall have to answer.

Now let us see if we can explain why only some pulsars are found at the center of supernova remnants. Let us assume all pulsars are produced in supernova explosions (not that all explosions lead to the formation of a pulsar—maybe some do not, in which case we would see a remnant without a pulsar). Then the young pulsars should still show the supernova remnant. An older one (say age $\sim 10^4$ yr) would have "lost" the nebula to the extent that it would have mixed with the ISM. Thus some pulsars have nebulae, some do not.

What about the fact that few pulsars are part of binary-star systems? Again we shall assume that the original star was one part of a double star. The force holding this star to its companion is

$$F = GMm/r^2,$$

where M is the mass of the potential supernova, m the mass of the companion. To continue in a circular orbit about a common axis of rotation (see Fig. 5.11) requires that the companion satisfy

$$mv^2/a = GMm/r^2,$$

where a is the distance from the center of the companion to the axis of rotation. Now if there is an explosion, M is reduced by some large amount so that the new mass satisfies $M' \ll M$. This reduces the gravitational pull on the companion star. If indeed we find $mv^2/a \gg GM'm/r^2$, then the companion star is no longer bound to move in a circular orbit and will move away at velocity v, leaving the remnant of the exploding star to move in the opposite direction, alone, as depicted in Fig. 5.11. Thus the infrequency of binary pulsars is consistent with their origin in supernova explosions.

We can also learn something about the size of the pulsar from the "rise time" of each pulse. By rise time we mean the time it takes the pulse to go from zero to its full value. Say a region is emitting pulses of radiation by turning on and off periodically at intervals of τ as shown in Fig. 5.12. Let us assume the region turns on all at once. Then to an observer, the light from the front will arrive first, Δt_1 sec after the source turns on. The light from the rear will arrive Δt_2 sec after the source turns on. This means the observer sees a pulse whose intensity will rise up to full power only after the light from all parts of the source has arrived. Since the light from the front arrives L/c earlier than from the rear, it will take this time, L/c, for the observer to record the full power of the pulse. The pulse will stay at full power until the source turns off. Then the front will go dark first, as will the

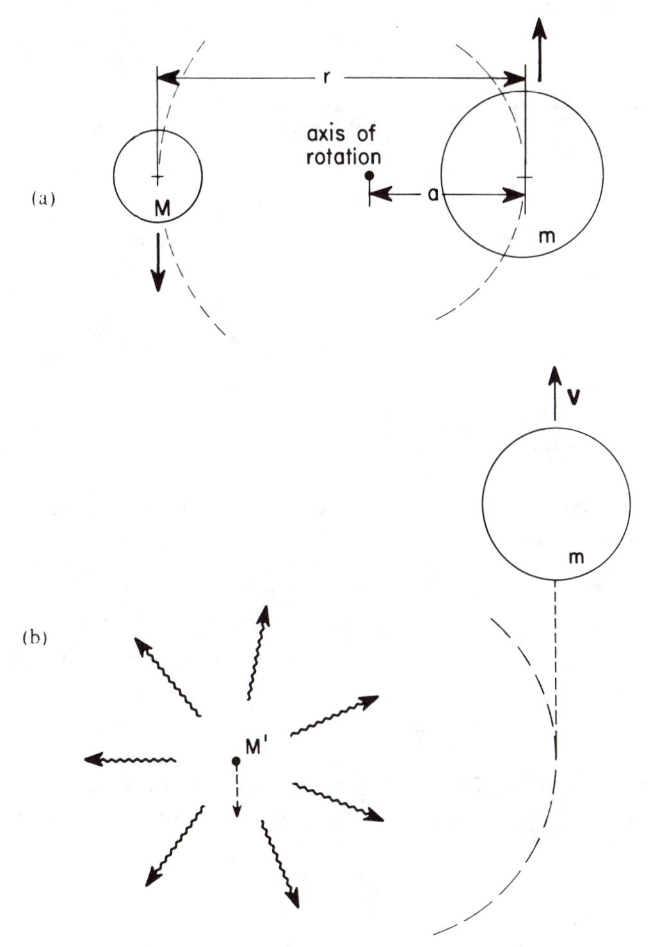

Fig. 5.11. Relative motion of binary pair M and m: (a) before and (b) after supernova explosion.

rear at a time L/c later. Then L/c is also the fall time. At a time τ sec later, the whole process will be repeated.

Thus, measuring the rise time of the pulse tells us, under the conditions described above, the size of the region emitting the EM radiation via

$$\text{rise time} = L/c. \tag{5.13}$$

It is possible that the region does not turn on all at once, but rather an internal signal propagates with velocity v that tells each part of the region when to turn on. Since this propagating signal must always have $v < c$, it will take a much smaller object to produce the same rise time, τ_R, as in the

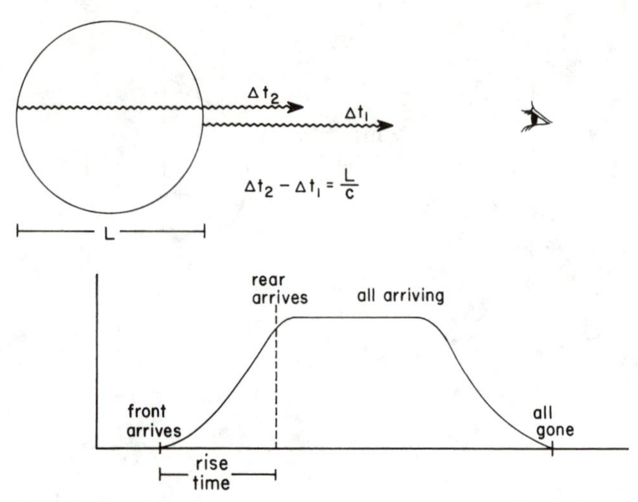

Fig. 5.12. Pulse shape arising from a source turning on at $t = 0$. Radiation from the front arrives Δt_1 sec later, from the rear Δt_2 sec later.

first case. But we do not know what conditions produced the pulse, so that all we can say for sure is that the region emitting the radiation has a maximum size of

$$L_{\max} = \tau_R c.$$

From Fig. 5.13, we have for the fine structure of the pulsar pulse, a rise time of about 10^{-4} sec, which implies an overall size for the emitting region

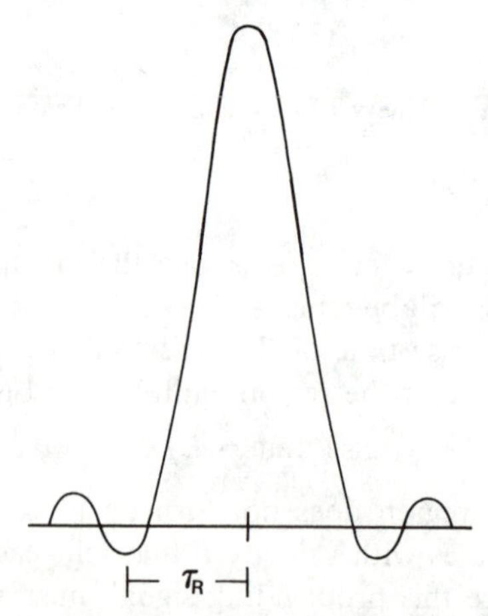

Fig. 5.13. Fine structure of a pulsar pulse indicating the rise time τ_R.

of

$$L_{\max} \approx 10^{-4} \times 3 \times 10^{8} = 3 \times 10^{4} \text{ m} = 30 \text{ km(!)}$$

This is amazingly close to the size of a neutron star if we assume that the entire star participates in the formation of the pulse.

Now that we know the approximate size of a pulsar, let us ask what its period of rotation would be if it developed from a star approximately the size of the Sun. The link, of course, is the conservation of angular momentum, $\omega r^2 = \text{const}$. Then, as we have seen earlier, with $\tau_\odot \approx 2 \times 10^6$ sec, or $\omega_\odot = 2\pi/\tau_\odot \sim 3 \times 10^{-6}\ \text{sec}^{-1}$ and $R_i = R_\odot$, $R_f = R_{\text{pulsar}} = \frac{1}{2} L_{\max} \sim 15$ km, we are led to

$$\omega_{\text{pulsar}} = \omega_\odot \left(\frac{R_i}{R_f} \right)^2 \sim 3 \times 10^{-6} \left(\frac{7 \times 10^8}{15 \times 10^3} \right)^2 \sim 10^4\ \text{sec}^{-1},$$

or

$$\tau_{\text{pulsar}} = 2\pi/\omega_{\text{pulsar}} \sim 10^{-3}\ \text{sec}.$$

This is a reasonable number because it is fairly close to the measured pulse periods of pulsars, $\frac{1}{3}10^{-2}$–10^0 sec. Indeed, the fact that $\tau_{\text{pulsar}} \leftrightarrow \tau_0 < \tau_{\text{now}}$, is compatible with some breaking action on the pulsar—if the period of rotation and the period of pulsation are in fact related.

5.4.2. Is the Pulsar a Neutron Star?

It should be obvious by now that a natural explanation of a pulsar would be a neutron star because (a) they both result from supernova explosions; (b) the size of each is around 10 km; and (c) the rotational period of a pulsar is compatible with that of a neutron star.

Perhaps the most stunning link between the two comes from considering the lengthening of the period τ. Assuming we are dealing with a *rotating* neutron star, this means that not only is its period lengthening, but it is losing rotational energy as well. To see this, consider the rotational KE of a spinning ball, $\text{KE}_{\text{rot}} = \frac{1}{2} I\omega^2$. Then as τ increases or ω decreases, KE_{rot} decreases at a rate

$$\frac{\Delta \text{KE}_{\text{rot}}}{\Delta t} = \frac{1}{2} \frac{I\, \Delta(\omega^2)}{\Delta t}.$$

But $\Delta(\omega^2) = (\omega + \Delta\omega)^2 - \omega^2 \approx 2\omega\Delta\omega$ for small $\Delta\omega$, so that

$$\frac{\Delta(\omega^2)}{\Delta t} \approx 2\omega \frac{\Delta\omega}{\Delta t}.$$

To find the rate of change of the angular velocity (angular acceleration

$\alpha = \Delta\omega/\Delta t$), we use $\omega = 2\pi/\tau$ so that

$$\frac{\Delta\omega}{\Delta t} = 2\pi\frac{\Delta(1/\tau)}{\Delta t} = 2\pi\frac{\left[(1/\tau) - (1/(\tau + \Delta\tau))\right]}{\Delta t}$$

$$\approx 2\pi\frac{\Delta\tau/\tau^2}{\Delta t}\,.$$

Putting the last three equations together, we have

$$\frac{\Delta \mathrm{KE_{rot}}}{\Delta t} = \frac{(2\pi)^2 I}{\tau^3}\cdot\frac{\Delta\tau}{\Delta t}\,. \tag{5.14}$$

Then applying this equation to the Crab pulsar with $\tau \sim (1/30)$ sec and $\Delta\tau/\Delta t \sim 4\times10^{-13}$ and a moment of inertia for a solid sphere $I = \frac{2}{5}MR^2$, the size of a neutron star, $M \sim M_\odot$, $R \sim 10$ km, we find from (5.14),

$$\frac{\Delta(\mathrm{KE_{rot}})}{\Delta t} \approx \frac{2}{5}\,\frac{4\pi^2(2\times10^{30})(10^4)^2}{(1/30)^3}(4\times10^{-13}) \approx 3\times10^{31}\ \mathrm{J/sec}. \tag{5.15}$$

But this is almost exactly the same energy being put out by the Crab nebula in the form of synchrotron radiation! We said earlier that something had to supply the 10^{31} J/sec being radiated by the Crab and that it could not be left over from the explosion. Notice also the sensitivity of the energy loss to the radius of the rotating object. The moment of inertia I is proportional to R^2. If R were 100 km, we would be off by a factor of 100. Thus we see that a rotating neutron star slowing up could supply the energy of the Crab Nebula.

What about the pulsed energy? This is different from the continuous synchrotron radiation and in fact is 1000 times *smaller* at 10^{28} J/sec. Therefore the production of pulses is not important in slowing up the pulsar-neutron star. In a way this is fortunate because we can decouple the two problems and worry separately about (a) how the nebula is supplied with energy from the neutron star, and (b) how the pulsed radiation is produced. Remember the pulsed radiation is coming from the surface or very near the surface of the neutron star, while the continuous radiation is coming from the nebula which is 2 LY away from the centrally located neutron star.

5.4.3. One Way of Producing Continuous Radiation

Since the region emitting the synchrotron radiation coincides with the Crab Nebula, we must look for a mechanism for injecting high-speed electrons into that region. The mechanism turns out to be the corotating magnetic

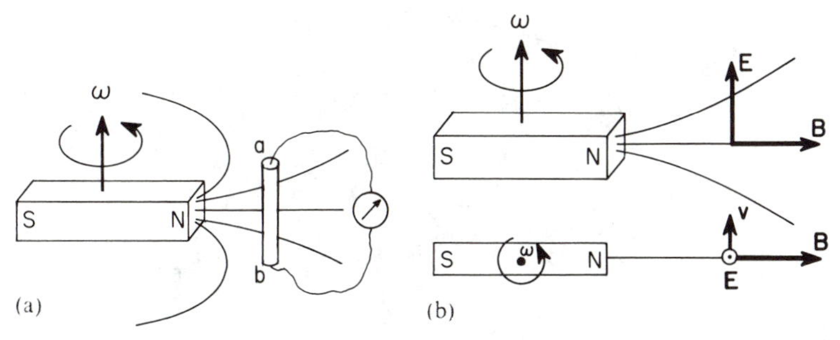

Fig. 5.14. (a) Magnetic field of a rotating bar magnet producing a current in a stationary wire *ab*; (b) electric field induced by this magnetic field which moves at velocity v across the wire.

field of the neutron star. Consider first a bar magnet that is rotating so that its dipole axis is perpendicular to the axis of rotation. If we allow the moving field to cross a wire ab, as in Fig. 5.14a, there will be a current generated. One way of explaining this current is to postulate that the moving magnetic field is accompanied by a moving electric field and that this **E** field drives the electrons in the wire. Experimentally we see that the direction of **E** must be related to the direction of **B** and **v** via $\mathbf{E} \times \mathbf{B}$, which is a vector that points in the direction of **v**, the velocity of propagation of the **B** field. If we remove the wire, we do not affect the relationship between **E** and **B**—we just have no way of detecting **E**. So we can still imagine **E** being there as we show in Fig. 5.14b. Experimentally one finds that in terms of their magnitudes

$$E = vB.$$

Now as you may recall, Maxwell predicted another way of producing a moving electric and magnetic field. If a charge is oscillating and thereby accelerating, it produces an electromagnetic field with $E = cB$. This radiation, as you well know, propagates with the velocity of light c.

There is one major difference between the EM fields produced by the spinning bar magnet and the EM fields produced by the oscillating charge. In the latter case the fields will continue to propagate even after the charge has stopped oscillating, but once the bar magnet has stopped rotating, the electric fields will immediately die out. In order to be self-sustaining, as Maxwell first pointed out, the fields must move with the velocity c.

At first glance it would appear to be impossible to achieve a self-sustaining field for a spinning bar magnet. But if we carry the analogy of the moving magnetic field a bit further, we see that the velocity of **B** at a distance R from the axis of rotation is $v = R\omega$. There is clearly a radius

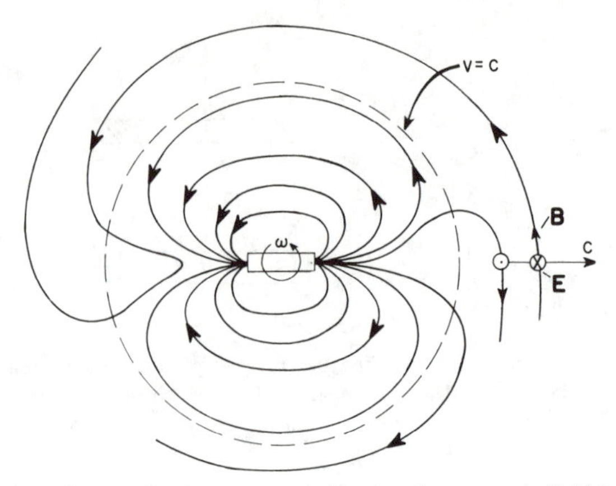

Fig. 5.15. Top view of a rotating bar magnet indicating the magnetic field lines as they cross the light cylinder where $v = c$.

beyond which $v = c$, i.e., when

$$R \geqslant \frac{c}{\omega} = \frac{c\tau}{2\pi} .$$

Inside the radius $R = c/\omega$ it will be business as usual, an electric field of intensity $E = vB$ accompanying the dipole magnetic field as long as the magnet rotates. But at $R = c/\omega$, $v = c$, and therefore $E = cB$ we have satisfied the conditions for self-sustained EM radiation. And indeed beyond this point (remember, nothing can move faster than c) the dipole field will break up into the **E** and **B** field of a propagating EM wave!! We try to illustrate this in Fig. 5.15.

As the corotating dipole field turns into an EM wave outside of what some people call the "light cylinder," we begin to see how the neutron star feeds energy into the nebula. Recall that a corotating magnetic field can drag charged particles with it. If such particles exist, say by ejection through a crack in the surface of the neutron star due to accumulated strain, or by being pulled out of the surface by the **E** field that accompanies the corotating **B** field, they will be dragged in the spin direction, pulling energy and angular momentum out of the neutron star as they are accelerated outward. Beyond the light cylinder, **B** opens up and the electrons that cross the cylinder will now ride the EM wave radially outward and into the nebula. This latter trip is achieved as follows:

(a) After crossing the light cylinder such an electron will see crossed electric and magnetic fields.

(b) The electric field will act on the electron and for the direction shown in Fig. 5.16, produce an upward force

$$\mathbf{F} = -e\mathbf{E}_y .$$

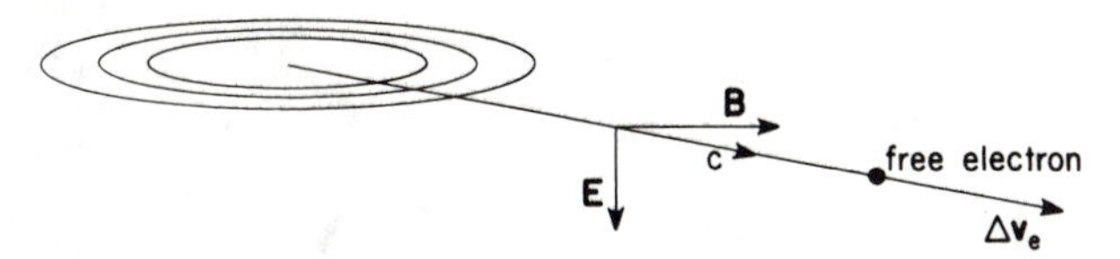

Fig. 5.16. Electromagnetic radiation interacting with a free electron.

(c) The electron now has an upward component of velocity which will interact with **B** via

$$\mathbf{F} = q\mathbf{v} \times \mathbf{B}.$$

(d) Again for the EM vectors shown in Fig. 5.16, the cross product of **v** and **B** indicates that this force will be in the *forward* direction.

(e) Thus the electron gets a radial push from the EM wave. The electron will ride this wave, feeling a steady forward force or pressure and so can in principle reach very high velocities ($\Delta v_e = a\Delta t = F_{\text{mag}}\Delta t/m_e$) if,
  (i) the magnitude of **B** beyond the light cylinder is of reasonable size; and
  (ii) there is enough distance between the light cylinder and the nebula so that with a steady acceleration the electron can reach high velocities.

As to point (i), the field $\mathbf{B}_s$ at the surface of the neutron star is about 10^{12} gauss. A *static* dipole field falls off as $1/r^3$. This would give us a magnetic field at the light cylinder B_c of magnitude

$$B_c = B_s\left(\frac{R_s}{R_c}\right)^3.$$

For the Crab pulsar, $\tau = 1/30$ sec implies that the radius of the light cylinder R_c is

$$R_c = c/\omega \approx 3 \times 10^8/(2\pi/1/30) \sim 10^6 \text{ m}.$$

This would give us a field at the light cylinder of magnitude

$$B_c = 10^{12}\left(\frac{10^4}{10^6}\right)^3 \sim 10^6 \text{ gauss!}$$

This is a substantial field and with the nebula 2 LY from the pulsar, there is plenty of room between the light cylinder ($R_c \sim 10^6$ m) and the nebula ($R_{\text{neb}} \approx 2 \times 10^{16}$ m) for the electrons to reach the necessary speeds.

As the electrons accelerate beyond the light cylinder, the propagating dynamic magnetic field falls off as $1/r$. This weakening magnetic field

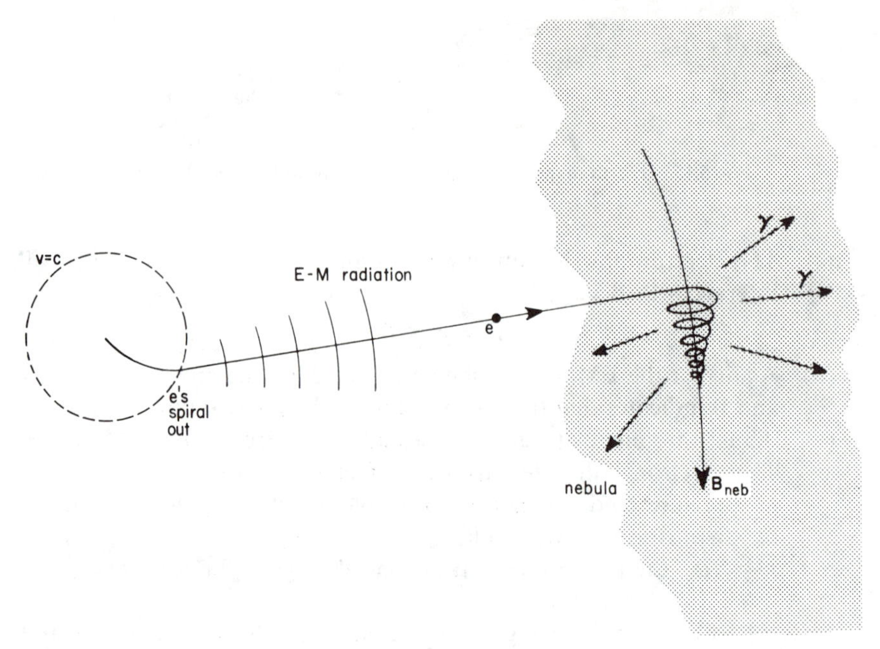

Fig. 5.17. The communication of energy from the pulsar to the nebula by the acceleration of electrons.

allows the electrons to eventually decouple from the EM wave. When they reach the nebula, the electrons will spiral about the magnetic field lines as shown in Fig. 5.17 that thread this region of space and emit synchrotron radiation whose polarization and nonthermal spectrum we observe. Thus the electrons gain energy on the way out at the expense of the neutron star and lose that energy in the form of synchrotron radiation in the nebula. This is the nature of the link (maybe) that we surmised earlier must exist because the $\Delta(\mathrm{KE})/\Delta t$ of the pulsar-neutron star (5.15) matched the power output of the nebula.

Based on the model of a corotating magnetic field, we can produce another connection, this time between the rotational energy loss of the neutron star and EM radiation produced by the corotating magnetic field. The link here is the energy per unit volume, $\mathcal{E}/V$, carried by the photons in the EM field. This is given by the equation $\mathcal{E}/V = B^2/4\pi k_m$ where $k_m = k_0/c^2 = 10^{-7}$ in MKS units. Since the EM field decouples from the corotating field at the light cylinder, the energy density flowing out through the cylinder at that point is given by (where $\mathcal{E}$ represents energy, not electric field)

$$\frac{\mathcal{E}}{V} = \frac{B_c^2}{4\pi k_m}, \tag{5.16}$$

where B_c is the field at $r = R_c$. From conservation of energy, any energy that flows across the light cylinder must have originally come from the neutron star. Thus if we can calculate the flux of EM energy at $r = R_c$, this should be the same as the rate at which energy is lost by the neutron star.

To make this calculation, all we have to note is that the energy flowing across the boundary in time Δt is just the energy contained in a spherical shell of thickness $c\Delta t$. This is given by

$$\Delta\mathcal{E} = \left(\frac{\mathcal{E}}{V}\right)\delta V,$$

where δV is the volume of the shell. At the light cylinder, this volume will be $\delta V = (4\pi R_c^2)c\Delta t$ and the rate of energy flow across the boundary will be

$$\frac{\Delta\mathcal{E}}{\Delta t} = \left(\frac{B_c^2}{4\pi k_m}\right)4\pi R_c^2\frac{c\Delta t}{\Delta t} = \frac{cB_c^2R_c^2}{k_m}. \tag{5.17}$$

As we mentioned above, this should also be the rate at which energy is lost by the neutron star. Using $B_c \sim 10^6$ gauss $= 10^2$ T, $R_c = 10^6$ m, and $k_m = 10^{-7}$ we have from (5.17)

$$\frac{\Delta\mathcal{E}}{\Delta t} \approx \left(\frac{10^4}{10^{-7}}\right)(10^{12}) \times 3 \times 10^8 = 3 \times 10^{31}\ \text{J/sec}. \tag{5.18}$$

Just great! This matches $\Delta\text{KE}/\Delta t$ calculated in (5.15); a rotating magnetic field must be directly linked to the energy loss of the pulsar. Moreover, since we calculated B_c from our estimate of the surface field B_s, the above results are a convincing indication that the surface field is indeed 10^{12} gauss! In fact it is instructive to rewrite $\Delta\mathcal{E}/\Delta t$ in terms of the surface field and surface radius since these are the most direct links with the properties of the neutron star. Recalling that the dipole field falls off as $1/r^3$, so that $B_c = B_s(R_s/R_c)^3$ and using $c = R_c\omega$, we have from (5.17)

$$\frac{\Delta\mathcal{E}}{\Delta t} = \frac{B_s^2}{k_m}R_s^6\frac{\omega^4}{c^3}, \tag{5.19}$$

which gives us the energy loss directly in terms of the surface properties of the pulsar.

We can, however, have even more fun. We showed in Sec. 5.4.2 that the loss in rotational KE of the neutron star is given by $(2\pi)^2 I(\Delta\tau/\Delta t)\tau^3$ (J/sec). Equating $\Delta\text{KE}/\Delta t$ to $\Delta\text{E}/\Delta t$, from (5.14) and (5.19), we find

$$\frac{(2\pi)^2\cdot\frac{2}{5}MR_s^2}{\tau^3}\cdot\frac{\Delta\tau}{\Delta t} = \frac{B_s^2}{k_m}R_s^6\frac{(2\pi)^4}{\tau^4}\frac{1}{c^3}.$$

After cancelling and rearranging, we have an expression for the surface magnetic field of a pulsar in terms of its mass, radius, period, and rate at which the period is changing,

$$B_s^2 = \tfrac{2}{5} k_m \frac{M}{R_s^4} \frac{c^3}{(2\pi)^2} \tau\left(\frac{\Delta\tau}{\Delta t}\right). \tag{5.20}$$

If you look at Table 5.2 you will notice that the product of $\tau \times \Delta\tau/\Delta t$ is approximately a constant! Since $M \sim M_\odot$ and $R_s \sim 10$ km, it would appear from (5.20) that all neutron stars or pulsars have surface fields of approximately the same magnitude, 10^{12} gauss. This is a great triumph for the magnetic dipole model of a pulsar. This model implies the pulsars are all formed in a similar manner and therefore, outside of age, should have very similar properties. The fact that B_s is determined to be 10^{12} gauss is a striking confirmation of this prediction.

5.4.4. The Pulsed Radiation

The production mechanism for the pulsed radiation is even less well understood than that for the continuous synchrotron radiation. As a reflection of this fact there are at least a dozen theories purporting to explain the phenomenon. One of the first, proposed by Gold at Cornell in 1969, is called the lighthouse model. Since it is a natural extension of our previous discussion, we will attempt to describe it here.

We have already indicated that the corotating magnetic field will drag any electrons that might be in the vicinity of the neutron star around and outward. Since these electrons are curving as they move outward, they too will radiate as shown in Fig. 5.18. Close to the neutron star, where their velocity is much less than c, this radiation will be emitted in all directions. But as the electrons move out and pick up tangential velocity from the corotating field, the photons will begin to be projected in a forward cone. This idea can be made quantitative with the aid of the relativistic velocity transformation equations that are obtained from the special theory of

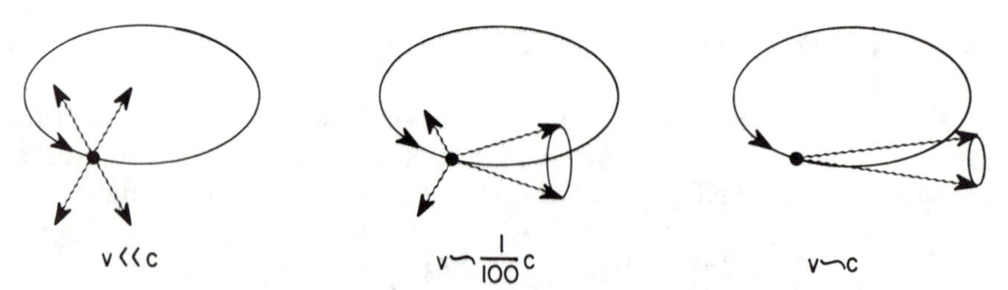

Fig. 5.18. The dependence of the direction of synchrotron radiation on the velocity of the moving source.

relativity with $c_{0x}^2 + c_{0y}^2 = c_x^2 + c_y^2 = c^2$:

$$c_x = \frac{c_{0x} + u}{1 + uc_{0x}/c^2}\ ; \qquad c_y = \frac{c_{0y}\sqrt{1 - u^2/c^2}}{1 + uc_{0x}/c^2}\ , \tag{5.21}$$

where c_{0x} and c_{0y} are the components of the photon velocity in the rest frame of the electron (i.e., as an observer riding along with the electron would see the emission) and c_x, c_y are the components of the photon velocity as observed in an inertial frame where the electron appears to be moving with the velocity u in the x-direction. For example, a photon emitted at 90° in the rest frame of the electron would have $c_{0x} = 0$ and $c_{0y} = c$. In the lab or inertial frame, the above equations give us

$$c_x = u \qquad \text{and} \qquad c_y = c\sqrt{1 - u^2/c^2}\ ,$$

with the result that the angle at which the photon appears would be

$$\tan\theta = \frac{c_y}{c_x} = \frac{c\sqrt{1 - u^2/c^2}}{u} = \sqrt{\frac{c^2}{u^2} - 1}\ .$$

As $u \to c$, $\tan\theta$ gets smaller and smaller which means that the photons are projected into a smaller and smaller forward cone. As the electrons approach the light cylinder, the photon beam is reduced to a small pencil of radiation, the equivalent of a well-collimated light house beacon—hence the name, the lighthouse model.

Although the beam of radiation is pencil thin, the radiating particles may be spread out over some fraction of the light cylinder as shown in Fig. 5.19. This will give the beam an apparent angular spread of $\Delta\theta$, so that in the vicinity of the Earth, the beam will spread out over a distance $\Delta S = d\Delta\theta$, with ΔS much greater than the diameter of the Earth and where d is the distance between the pulsar and the Earth. It will therefore take a time $\Delta t = \Delta S/v$ for the beam to sweep by. But v, the tangential velocity of the beam at d, is equal to $v = d\omega$. Therefore we obtain

$$\Delta t = \frac{\Delta s}{v} = \frac{d\Delta\theta}{d\omega} = \frac{\Delta\theta}{\omega}\ .$$

Of course, Δt is nothing but the duration of a single pulse. For the Crab pulsar, $\Delta t = \frac{1}{2} \times 10^{-3}$ sec, and with $\tau = 1/30$ sec we have

$$\Delta\theta = \omega\Delta t \approx \tfrac{1}{2} \times 10^{-3} \times 2\pi \times \left(\frac{1}{1/30}\right)$$

$$\sim \frac{1}{10}\,\text{radian} \sim 6°.$$

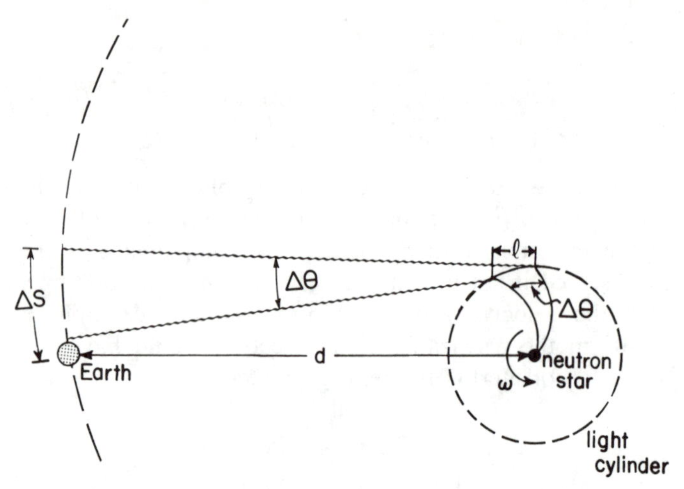

Fig. 5.19. Electrons spiral out of the neutron star and occupy a distance l along the light cylinder.

From this we can actually measure the linear size of the emitting region, l, at the light cylinder with $R_c \approx 10^6$ m:

$$l = R_c \Delta\theta \sim 10^6 \times (1/10) \sim 10^5 \text{ m}.$$

As nice as the lighthouse model is, it also raises a few difficult questions:

(a) Why is the pulse so stable if it is due to matter ejected through a break in the neutron star's surface?
(b) What happens to the radiation that should be produced while the electrons are in transit to the light cylinder?
(c) Some pulsars exhibit a pulse structure which contains a second component located about $\frac{1}{2}$ period from the main pulse. Does this imply a second distinct source of electrons? If two, why not ten?

Because of objections like these, other models have gained favor, especially one that says the radiation is due to "hot spots" located on the surface of the neutron star at the magnetic poles as shown in Fig. 5.20. If electrons are trapped in the strong dipole field, they will spiral around the field lines and reach their maximum density at the magnetic poles. The conjunction of a strong magnetic field and a dense plasma of electrons can lead to the emission of radiation. If the magnetic dipole does not line up with the spin axis, the magnetic poles will rotate once again, giving us a lighthouse-type effect but one whose origin is far different from the beaming at the light cylinder. Since there are two poles, there will be two

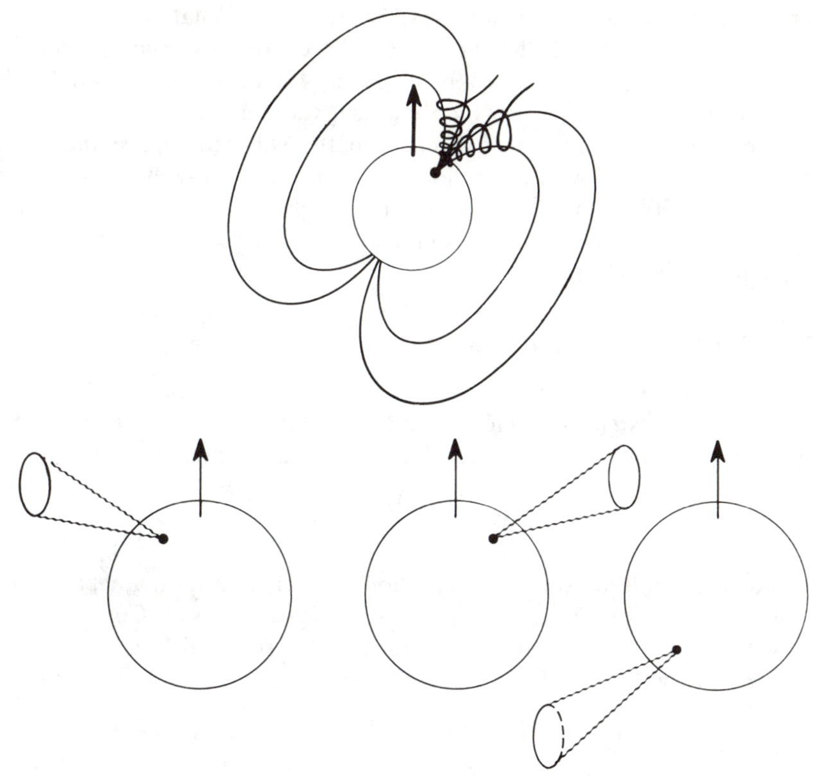

Fig. 5.20. Hot spots generated at the magnetic poles of a neutron star where the electron density is highest. Each pole produces a hot spot although an observer can see only one.

pulses per one complete revolution of the pulsar although we will only see one pulse unless the poles are located on the equator, as with the Crab pulsar. This is illustrated in Fig. 5.20.

In concluding this section, we should point out that although we have not been able to come up with a definitive model for either the continuous or pulsed radiation, the marriage between the neutron star and the pulsar is on a reasonably firm basis.

5.5. Black Holes

What happens when the degeneracy pressure has built up to relativistic levels and stellar masses exceed 4 $M_{\odot}$? At this point we can no longer inhibit the gravitational collapse by any known physical means. Gravity wins out. But to what does the object collapse? That is indeed the question.

To nothing? To a "singular point" in space and time? What laws of physics would apply to matter as it approaches such conditions? Such objects in gravitational collapse are called black holes. Unfortunately to do black holes justice, one should talk a bit about general and special relativity and what we expect the behavior of compact matter will be in space and time. Instead, we shall deal with black holes in a more qualitative way. We shall concentrate on that point in space beyond which the condensing star is no longer visible and then turn to compact X-ray sources and the possible detection of a black hole.

5.5.1. The Schwarzschild Radius

Consider a photon leaving the surface of a body of mass M and radius R. According to Einstein, the photon has a gravitational "mass" $m = h\nu/c^2$. Thus it has not only kinetic energy $\mathrm{KE} = h\nu$, but potential energy

$$\mathrm{PE} = -\frac{GM}{R}\left(\frac{h\nu}{c^2}\right).$$

We should be able to apply conservation of energy to its flight path and follow explicitly its decreasing KE as it moves away from M. Consider Fig. 5.21 with the photon initially at the surface of a star and finally escaping out at infinity with $\mathrm{PE}_f = 0$:

$$\mathrm{KE}_i + \mathrm{PE}_i = \mathrm{KE}_f + \mathrm{PE}_f$$

$$h\nu_i - \frac{GMh\nu_i}{Rc^2} = h\nu_f + 0.$$

Interestingly enough, Planck's constant h cancels out and we can solve for

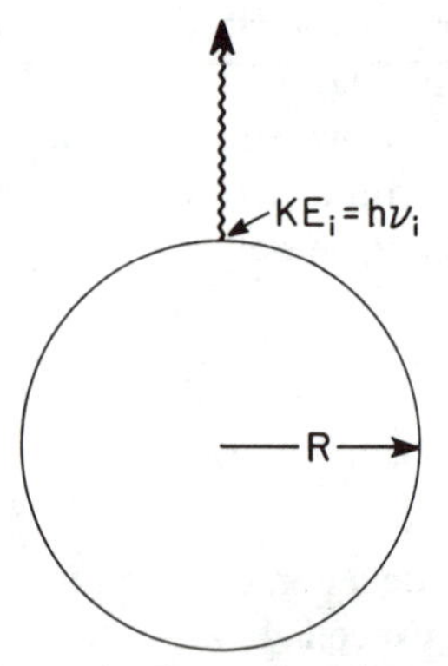

Fig. 5.21. Photon emitted normal to the surface of a star.

the frequency shift,

$$\nu_f = \nu_i\left(1 - \frac{GM}{Rc^2}\right).$$

It should be obvious from this equation that $\nu_f \neq \nu_i$; we have a gravitational red shift. This relationship has a very interesting consequence. When the term in parenthesis goes to zero, ν_f goes to zero and equivalently the wavelength of the photon, λ_f, goes to infinity; the photon has neither energy nor momentum. It is undetectable. We can say that it ceases to exist. But to set $(1 - GM/Rc^2)$ equal to zero requires $GM/Rc^2 = 1$, or

$$R = GM/c^2.$$

Thus a photon shooting straight up out from a star of mass M whose radius has reached the above critical value will never be detected.

Or consider the case shown in Fig. 5.22, a photon trying to leave tangential to the star's surface. Since it has gravitational mass $m = h\nu/c^2$, it will be bent into a curved path by the gravitational force. Usually such a photon will escape. However, if M is big enough and/or R small enough, the photon may feel such a large gravitational pull that it will circle the star and of course never be detected. For this to happen we must have for a centripetal acceleration of v^2/R,

$$m\frac{c^2}{R} = \frac{GMm}{R^2}$$

or

$$R = GM/c^2.$$

Both types of photon motion give the same result: a photon leaving a star where $M/R = c^2/G$ will not be detected. It turns out that general relativity modifies this result only by a factor of 2 and R is then called the

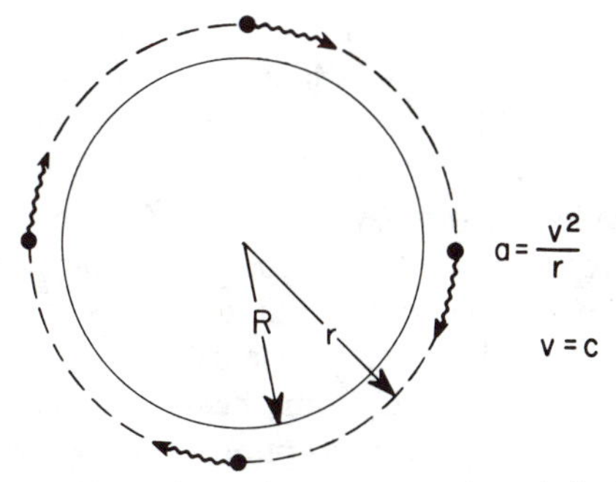

Fig. 5.22. Conditions for a photon to move in a circle about a star.

Schwarzschild radius

$$R_{\text{Sch}} = 2GM/c^2. \tag{5.22}$$

Now clearly in the course of the collapse of a star whose mass exceeds 4 $M_\odot$, its radius will become small enough to satisfy the Schwarzschild condition. At that moment *any* radiation emitted by the star (visible radiation, X-rays, γ-rays, whatever) will no longer be "visible." The star will thus simply disappear as if down a black hole. This does not mean, however, that the star stops contracting. It continues past this point, going to who knows what state of matter. It is just that once past what we call the Schwarzschild radius, photons are no longer received from the star. It is now a black hole. For a star with $M \sim 4\ M_\odot$, we have

$$R_{\text{Sch}} \approx 2 \times 7 \times 10^{-11} \frac{4 \times 2 \times 10^{30}}{(3 \times 10^8)^2} \approx 12 \text{ km},$$

which is about the size of a neutron star. Now R_{Sch} will be larger for more massive stars, so at the *point of extinction* black hole stars are not that small. Even their density at this point is not outrageous,

$$\rho_{\text{Sch}} = \frac{M}{(4\pi/3) R_{\text{Sch}}^3} \sim \frac{8 \times 10^{30}.}{4(12 \times 10^3)^3} \sim 10^{18} \text{ kg/m}^3,$$

which is only slightly more dense than a relativistic neutron star. Of course as the black hole continues to contract, it will reach much smaller radii and much higher densities. Indeed, when the densities reach values between 10^{55} kg/m^3 and 10^{96} kg/m^3, the black holes have intriguing properties which we shall discuss in the closing section of this book.

5.5.2. Compact X-Ray Sources and Cygnus X-1

In 1964, satellites flying X-ray detectors (detecting photons with energies between 1 and 20 keV) found intense X-ray sources generating about 10^{30} J/sec of power. These first detectors had such poor angular resolution that one could not tell whether the sources were extended objects like hot clouds or more compact objects like stars. This problem was "resolved" in 1971 when a satellite called Uhuru was launched with detectors of vastly improved angular resolution. It soon became clear that many of these X-ray sources were extremely compact, and what is more, some eight have been found to be part of binary star systems.

This latter point was deduced from the periodic nature of the X-ray reception. A typical plot of X-ray reception vs. time is shown in Fig. 5.23. We can deduce from Fig. 5.23 that two periods are involved: a short term τ of the order of seconds and a long term τ_b of 2 to 10 days. Since these latter

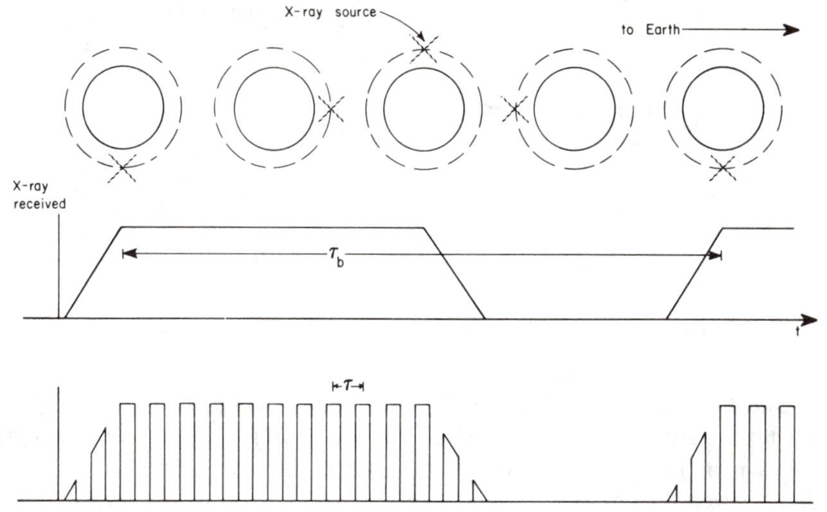

Fig. 5.23. Eclipsing binary period τ_b and pulse period τ for a rotating X-ray source.

times are typical of the period of a close companion orbiting another star and since the pulse was "on" four times longer than it was "off," it was not hard to deduce that the periodicity of the pulse was due to an X-ray-emitting star going behind its companion and thus having its X-rays blocked for a short time by the other star.

What was even more surprising was that two of these X-ray "eclipsing" binaries had even shorter period pulses superimposed on the long pulse:

(a) One source, called Hercules X-1 (Her X-1), had an eclipsing period of 1.7 days and a pulse period of $\tau = 1.24$ sec.
(b) The other, called Centaurus X-3 (Cen X-3), had an eclipsing period of 2 days and a pulse period of $\tau = 4.84$ sec.

Could these latter two compact X-ray sources mean that we were seeing an X-ray-emitting neutron star and/or white dwarf? The mechanism of the pulsed period would have to be something like that illustrated in Fig. 5.24, where two X-ray hot spots produce the short period radiation. Setting aside for a moment the question of how this radiation might be produced, can we use any of the parameters of these binary systems to tell more about the nature of the compact X-ray source?

Here is where the fact that we were dealing with a binary system proved to be a major breakthrough. In the case of most of the X-ray binaries (Hercules X-1 is an exception), the large central star was determined to be a blue supergiant. (Why a blue supergiant and not a red giant? We shall get to

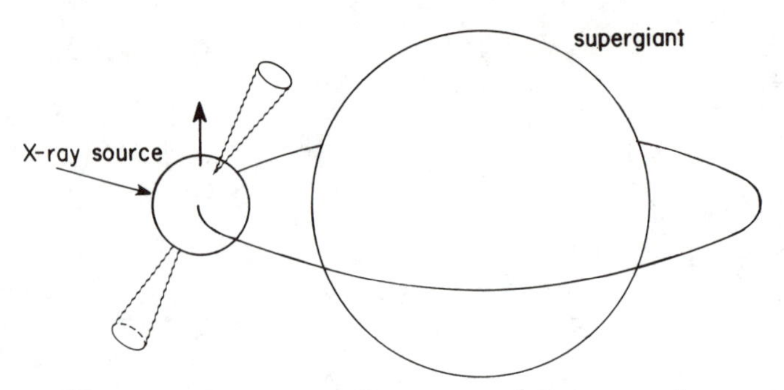

Fig. 5.24. Illustration of an X-ray source circling a supergiant.

that shortly.) Consider then a binary-star system composed of a compact star and a giant companion revolving about a common center of mass (cm) as shown in Fig. 5.25. The distance between the stars is related to their distance from the cm as follows:

(a) Since both objects rotate about a common cm, they have the same angular frequency ω so that

$$v_1 = r_1\omega \qquad \text{and} \qquad v_2 = r_2\omega.$$

(b) Since the momenta are equal (but opposite in direction) as seen from the cm, we have $Mv_1 = mv_2$ or $Mr_1\omega = mr_2\omega$ and with the distance a between the stars, $a = r_1 + r_2$, we have

$$a = \frac{mr_2}{M} + r_2 = r_2\left(\frac{m + M}{M}\right).$$

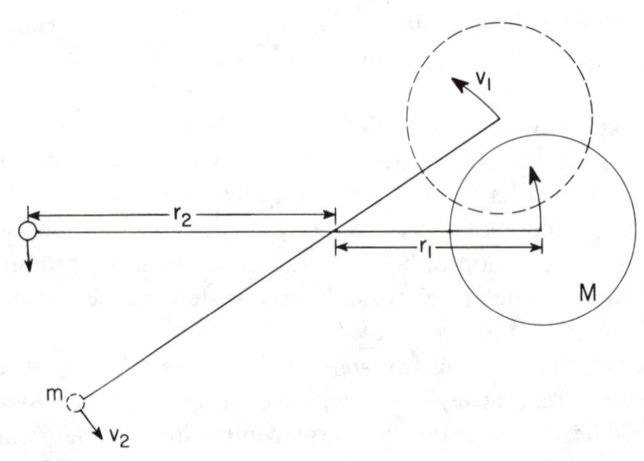

Fig. 5.25. A binary star system rotating about their common c.m.

The force between the stars is $F = GMm/a^2$; this gravitational force must supply the acceleration v_2^2/r_2 of the compact object, so that

$$\frac{mv_2^2}{r_2} = \frac{GMm}{a^2} = \frac{GMm}{r_2^2(m + M/M)^2},$$

or

$$v_2^2 \cdot r_2 = \frac{GM^3}{(m+M)^2}$$

and solving for m we have

$$m = \sqrt{\frac{GM^3}{r_2 v_2^2}} - M. \tag{5.23}$$

If, and it is a big if, we can find r_2, v_2, and M, then we can solve for the mass of the compact source. Clearly the mass is a crucial parameter in determining whether we have a white dwarf, a neutron star, or even a black hole. How one goes about finding m is somewhat as follows: (a) from the eclipsing period we get τ' (1.7 days for Her X-1); (b) from the Doppler shift we get v_2; and (c) since $v = r(2\pi/\tau)$, we can solve for $r_2 = v_2\tau'/2\pi$. All we need then is M. One method is to find first the luminosity of the blue supergiant and then place it on an H–R diagram. Since it will be short-lived, one can then trace its evolutionary track back to the main sequence. On the main sequence we know that for heavier stars, $L \propto M^4$. Thus knowing L on the main sequence, we can find M. Using techniques like this, one estimates for

Cen X-3: $M \sim 20\ M_\odot$ and $m \sim 1\ M_\odot$,

Her X-1: $M \sim 2\ M_\odot$ and $m \sim 0.7\ M_\odot$.

Quite obviously Cen X-3 and Her X-1 are not necessarily black holes, but they can be white dwarfs or neutron stars. If we recall, however, our arguments concerning the maximum rotational velocity of neutron stars (Sec. 5.2.4), we see that the minimum allowed period of any gravitationally bound rotating object can be found from (5.10) as

$$\tau_{\min} = 2\pi\sqrt{\frac{R^3}{Gm}}.$$

The masses of Cen X-3 and Her X-1 are given above. As potential white dwarfs, their radii would have to be $10^{-2}\ R_\odot$. Plugging these values into the above equation, we find that $\tau_{\min} \sim 9$ sec. The pulse periods of 1.24 sec and 4.84 sec would thus seem to exclude the possibility that either is a white dwarf; Her X-1 and Cen X-3 are solid choices for X-ray-emitting neutron stars!

And then there is the strange case of Cygnus X-1 (Cyg X-1). It is an X-ray-emitting object that has a blue supergiant companion, but it does not produce pulsed radiation. If it is a neutron star or white dwarf, then it might produce both the eclipsing and pulsed X-ray radiation, just as Her X-1 and Cen X-3 do. If not a white dwarf or neutron star, then what? Dare we say a black hole? Here the mass determination is crucial. The results are exciting indeed. For Cygnus X-1 and its companion, the masses deduced from (5.23) are

$$\text{Cyg X-1} \qquad M \sim 20\, M_\odot \qquad \text{and} \qquad m \sim 8\, M_\odot\,.$$

If these numbers hold up, then when we are looking at Cyg X-1 we are looking at a black hole.

But what about the source of X-rays for all of these objects? How do you explain X-ray-emitting white dwarfs and neutron stars? And after all that we have said about photons not able to leave a black hole, how do we explain the X-ray radiation of Cygnus X-1? How does such a binary system evolve? To resolve these questions and to gain confidence in our interpretation of the nature of the various compact X-ray sources, we turn now to the production of X-rays.

5.5.3. The X-Ray Production Process

To produce 20-keV X-rays from a thermal black body spectrum would require an energy of $E_\gamma^{peak} \sim 3kT \sim 20$ keV, corresponding to a temperature of $T \sim 5 \times 10^7$ °K. This is an unheard of temperature for the surface of any star, much less a neutron star or white dwarf, so the X-rays will not be generated by any effect internal to either of these objects. Nor will the known pulsation methods of a neutron star produce X-ray radiation of an intensity ($\sim 10^{28}$ J/sec) comparable to that of a compact X-ray source ($\sim 10^{30}$ J/sec).

Thus if they are alone, white dwarfs and neutron stars will not be strong sources of X-ray radiation. They need help and that help is certainly close at hand in the massive form of the blue supergiant. Like our sun—but probably more so because of the weaker gravitational field at its surface—the blue supergiant will possess a "solar" wind. As the small companion star sweeps through the solar wind, it can accrete matter, in this case protons, as shown in Fig. 5.26. The protons will be gravitationally attracted to the compact star and go crashing into its surface. If we consider a proton falling onto the surface of a neutron star or white dwarf, we can easily calculate the energy it picks up on the way in. Conservation of energy then

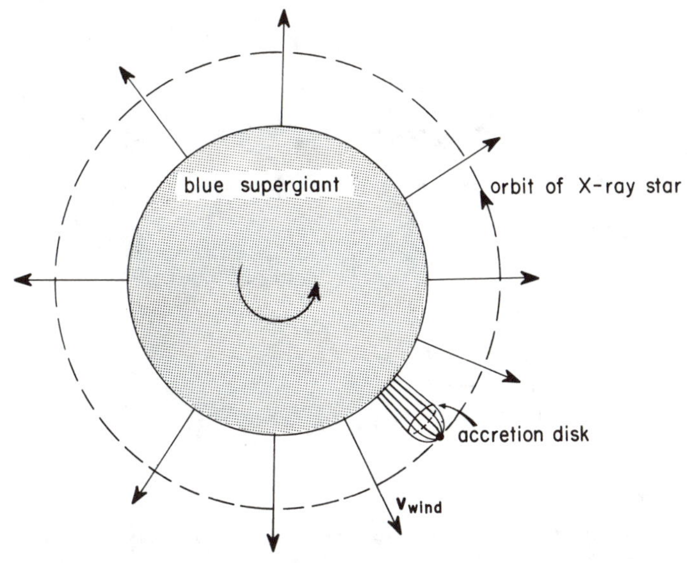

Fig. 5.26. X-ray source moving through the solar wind emitted from a blue supergiant.

gives, for a single proton with $PE_i = 0$ and $KE_i = \frac{1}{2} m_p v_{\text{wind}}^2$,

$$\tfrac{1}{2} m_p v_{\text{wind}}^2 + 0 = KE_f - \frac{Gmm_p}{R},$$

where R is the radius, say, of the neutron star and m its mass. The solar wind velocity is 10^6 m/sec, so we can solve for KE_f, using $m \sim M_\odot$ and $R \sim 10^4$ m. Therefore we find,

$$\begin{aligned} KE_f &= Gmm_p/R + \tfrac{1}{2} m_p v_{\text{wind}}^2 \\ &\approx \left[\frac{6.7 \times 10^{-11} \times 2 \times 10^{30} \times 1.7 \times 10^{-27}}{10^4} + \tfrac{1}{2} 1.7 \times 10^{-27} \times (10^6)^2 \right] \\ &\sim 2 \times 10^{-11} J \sim 10^8 \text{ eV} = 100 \text{ MeV}. \end{aligned}$$

Thus a proton need only convert a tiny fraction of its KE to produce a single 20-keV X-ray photon. Of course a freely falling particle, no matter what its energy, cannot produce X-rays. It needs large accelerations in order to shake off photons. By crashing into the surface of the neutron star and producing a "hot spot" of thermal radiation, that is exactly what the proton will accomplish. A typical X-ray source generates about 10^{30} J/sec or $\sim 3 \times 10^{37}$ J/yr. Assuming full conversion of each 2×10^{-11} J/proton

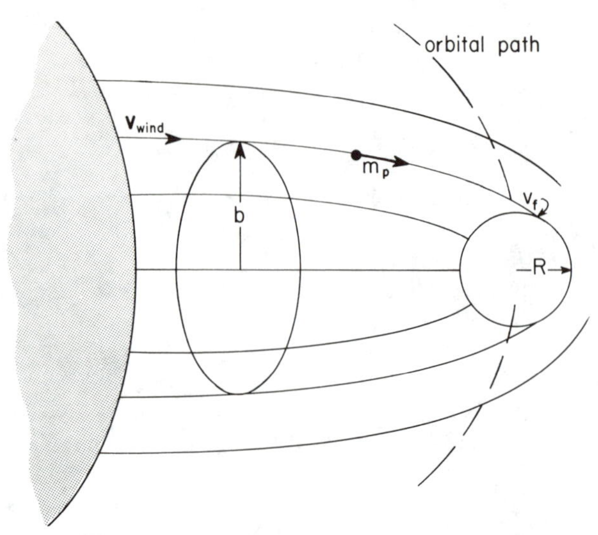

Fig. 5.27. Component of solar wind captured by X-ray source.

into X-rays, we would need to accrete N protons/year where

$$N/\text{year} \approx 3 \times 10^{37}\ \text{J/yr}/2 \times 10^{-11}\ \text{J/proton} \sim 10^{48}\ \text{protons/yr}.$$

With about 10^{57} protons in one solar mass we would need to use the equivalent of $10^{-9}\ M_\odot/\text{yr}$ to supply the X-ray source of a neutron star (what would it be for a white dwarf?).

Is $10^{-9}\ M_\odot/\text{yr}$ a reasonable number? At first glance it would seem so; a massive blue giant could supply that mass flux for a very long time. But remember, the solar wind is not very dense, nor is the conversion from solar wind to accreted matter 100% efficient. To indicate just how inefficient the accretion process can be, consider how matter from the stellar wind will be pulled onto the compact star. If we imagine we are sitting on the neutron star, then the stellar wind will appear to be streaming by, pretty much at right angles to our orbital path. This last point is due to the fact that $v_{\text{wind}} > v_{\text{orbit}}$. Because of gravity, these particles will be deflected toward the neutron star, as shown in Fig. 5.27, from distances that are quite a bit greater than the geometric size of the neutron star. Some, of course, will miss the neutron star altogether, but all of those within the area πb^2 should make contact. We can find this cross section by using conservation of energy and angular momentum for those particles that make a grazing collision with the star. The initial and final angular momentum of one of these particles is $L_i = m_p v_{\text{wind}} b$ and $L_f = m_p v_f R$, respectively. From conservation of L, we find

$$m_p v_{\text{wind}} b = m_p v_f R.$$

As we have already seen, from conservation of energy we get

$$\tfrac{1}{2} m_p v_{\text{wind}}^2 = \tfrac{1}{2} m_p v_f^2 - \frac{Gmm_p}{R}.$$

If we solve these two equations for b and eliminating v_f, we find

$$b = \frac{R}{v_{\text{wind}}} \sqrt{v_{\text{wind}}^2 + \frac{2Gm}{R}}.$$

For a neutron star $2Gm/R \sim 10^{16} m^2/\text{sec}^2$, which is much greater than v_{wind}^2. The radius of the accretion disk therefore is about

$$b = \frac{R}{v_{\text{wind}}} \times 10^8 \text{ m} \sim 100\, R,$$

which is really very small; only protons in the stellar wind that are close to the neutron star will be accreted. Detailed calculations show that about 0.1% of the stellar wind makes it onto the compact X-ray source. Since 10^{-9} solar masses per year must be accreted in order to produce the correct X-ray intensity for a neutron star, the stellar wind will have to be strong enough to carry away about $10^{-6}\ M_\odot/\text{yr}$ from the massive companion. After all this, it is gratifying to note that the *observed* mass loss from a blue supergiant is just about $10^{-6}\ M_\odot/\text{yr}$!

5.5.4. Hercules X-1 Anomaly: Roche Lobe

Her X-1 is an interesting anomaly, however. It is one of the few compact X-ray sources whose companion star has a mass of $< 20\ M_\odot$. With a mass of $2\ M_\odot$, calculations show that the associated stellar wind is far too small ($< 10^{-13}\ M_\odot/\text{yr}$) to account for the X-ray intensity of Her X-1. (The Sun has a similar mass loss which should not be too surprising since its mass is about the same.) The fact that the rate of mass loss decreases as the mass of the star decreases is in part due to the radiation pressure which drives the solar wind. Recall from Sec. 1.4.1 that $P_{\text{rad}} \propto T^4 \propto L$ (where L is the luminosity of the star), and since $L \propto M^4$, the intensity of the solar wind will depend strongly upon the mass of the star; the smaller the mass, the less intense the stellar wind. With $M = 20\ M_\odot$, Cen X-3's companion is massive enough to produce an adequate stellar wind: with $M = 2\ M_\odot$, Her X-1's companion is not. So, if the stellar wind will not account for the X-ray intensity of Her X-1, then what will?

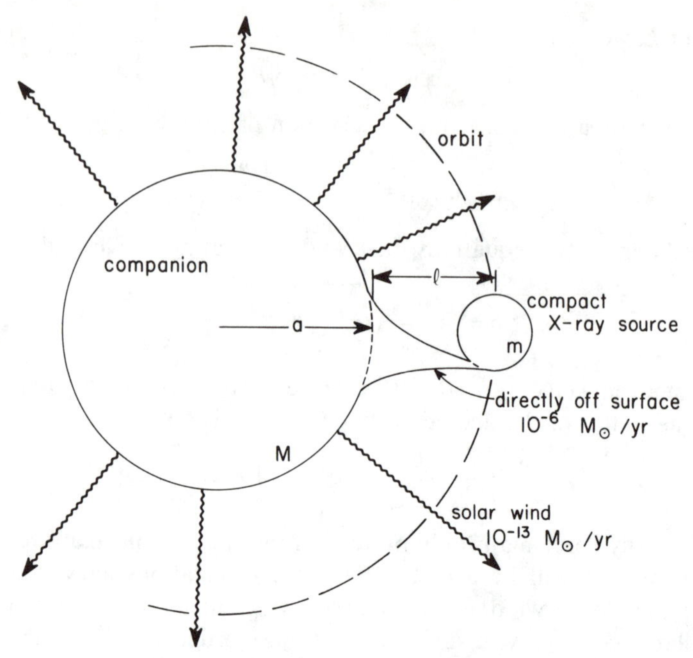

Fig. 5.28. Illustration of mass being pulled from companion star M to compact X-ray source m.

Consider what will happen when the *companion* to the compact source becomes so large that the particles on its surface feel a *net* force in the direction of the compact star, i.e., when

$$F_{\text{net}} = \frac{Gmm_p}{l^2} - \frac{GMm}{a^2} > 0,$$

where the various lengths are illustrated in Fig. 5.28. When this happens, mass will be pulled directly off the companion and channeled right onto the surface of the compact star! The rate of mass transfer will far exceed that of the accompanying solar wind ($10^{-6}\ M_{\odot}$/yr for accretion to $10^{-13}\ M_{\odot}$/yr for the solar wind). Indeed, at $10^{-6}\ M_{\odot}$/yr, the influx of matter is sufficient to drive Her X-1's X-ray source! The surface of radius a, that the companion M must reach before mass transfer can take place, is called the Roche surface or Roche lobe. The details of the mass transfer through the Roche lobe are more complicated than we have let on. The above force argument is a little too simplistic, but it will do to give us a feel for the way a low-mass star can feed a compact X-ray source.

So we now have a continuous flow of protons from the companion to the compact X-ray source. The protons crash into the surface, heat it, and produce X-rays. If the flow of matter is continuous, why is the radiation pulsed? The reason has to do with the intense corotating magnetic field of the neutron star. Because the infalling protons are charged, they will be bent by the magnetic force, $\mathbf{F} = q\,\mathbf{v} \times \mathbf{B}$. In the very same fashion, the solar wind emitted from the Sun and the cosmic-ray protons coming from beyond the solar system are deflected by the Earth's dipole field. In the case of a neutron star, the picture is further complicated by the intense gravitational force which accelerates the protons at the same time the magnetic force is bending them. As you can imagine, it is a complicated problem to solve. In all of these cases, a proton can only penetrate so far before the magnetic force turns its momentum vector around and the particle heads back out. Of course, if the particle's energy is high enough and/or the magnetic field weak enough, the particle can break through this "barrier" and crash into the surface of the object. In the case of the Earth, high-energy cosmic rays do just this. The magnetic field of a neutron star is so intense, however, that there is a region around the star that is essentially forbidden to all particles in the stellar wind. The result is that because of the relative motion of the two stars, the stellar wind forms into a flat disk with an inner region 2000 km across that is almost devoid of particles. Almost, that is, because there are two spots where protons can be funneled onto the surface of the neutron star. These are at the magnetic poles. Any particles heading in these directions, parallel to the magnetic field, will not feel a magnetic force since the cross product $\mathbf{v} \times \mathbf{B}$ vanishes. Thus, spinning out of the disk, protons will get a magnetic free ride right onto the surface of the star as illustrated in Fig. 5.29. The two magnetic poles become X-ray

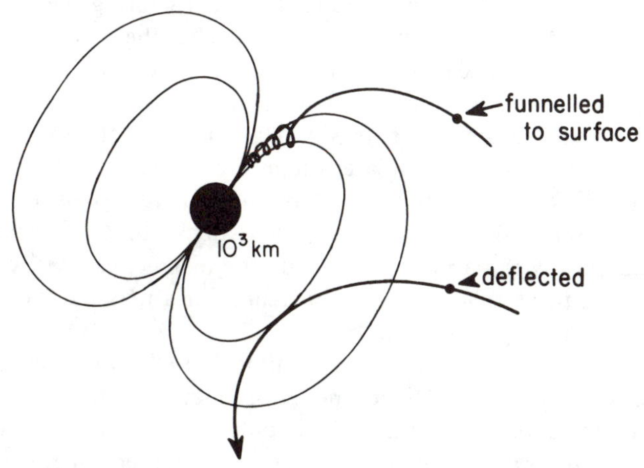

Fig. 5.29. Intense magnetic field around a neutron star funnels protons onto the magnetic poles.

hot spots that will rotate as the magnetic pole rotates about the spin axis. This will give us an X-ray-flashing mechanism in the same spirit as the radio pulse model we spoke of earlier in Sec. 5.4.4. Note that with a single neutron star, there is no companion to feed in protons and thus there is no X-ray component.

This very nicely explains why we see short-period X-ray pulses from neutron stars. But why only continuous eclipsing radiation from a black hole? Indeed, why any radiation at all?

First of all, any X-ray production would have to take place outside the Schwarzschild radius. This turns out not to be a drawback so long as the density of particles is so high at this point that they interact among themselves. Consider then the stellar wind and a black hole. As we noted above, because of the relative motion between the normal star and in this case the black hole, the stellar wind will spiral in toward the Schwarzschild radius in what is called an accretion column. As the column approaches the Schwarzschild radius, it will become denser and denser and flatten into an accretion disk. As the particles in the accretion disk approach to within 1000 km of R_{Sch}, their density becomes so great that they begin to collide amongst themselves and emit X-ray radiation (and no doubt γ radiation, too) via the bremsstrahlung process. Since the inflow of matter is continuous and the interactions are around the entire disk, the X-ray source will be continuous. This type of self-generation of X-rays is also possible for the neutron star but is overwhelmed by the pulsed radiation produced by the particles crashing into the star's surface. For black holes, the latter is not possible.

One question remains. Why are most complex X-ray sources associated with blue supergiants? The answer comes from considering the Roche lobe accretion column. When the companion star reaches the blue giant stage, it overflows the Roche surface and matter is drawn directly onto the compact star. Hence we have a Roche accretion column as well as the accretion column from the solar wind. The density of the Roche column, however, is several orders of magnitude greater than the density of the accretion column. And this is the problem. Subsequently, the Roche accretion column becomes a disk that is so thick and dense that it blocks off all X-ray radiation; any radiation produced by the disk is absorbed within the disk. Thus, once the Roche surface is breached for a massive star, the rush of matter shuts off the X-ray source.

Unfortunately there is only a narrow time limit between the onset of X-ray generation by the solar wind and the extinction of the X-rays by the dense Roche accretion disk. Indeed, it is only during the blue supergiant phase of a binary system's evolution that the compact X-ray source is visible. This means that X-ray sources will be rare—which seems to be the

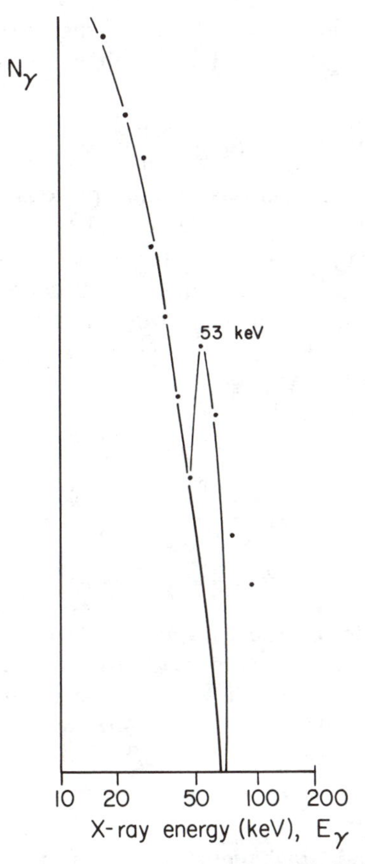

Fig. 5.30. X-ray spectrum from Her X-1.

case. Before the blue supergiant phase, while one of the binary components evolves into a neutron star or white dwarf, there will be no X-rays. Once the compact star has evolved, the solar wind will produce X-rays, and then the overflow of the Roche surface will lead to extinction. There must be many millions of binary stars in the pre-X-ray stage and many millions in the post stage. Thus our seeing only a few X-ray binaries should not be surprising.

Finally we would like to end this section by discussing the exciting discovery of a bump in the X-ray spectrum of Her X-1. The sharp increase in intensity, shown in Fig. 5.30, is centered around 53 keV. Now how does one obtain a monochromatic spike superimposed on a continuous spectrum? Protons crashing into the magnetic pole will not do it. But low energy electrons spiraling around the magnetic field lines might. You will recall from Sec. 5.3.2 that a nonrelativistic charged particle moving in a circle in a uniform magnetic field will emit monochromatic radiation of energy

$$E_\gamma = \hbar \frac{eB}{m} . \tag{5.11}$$

Why consider low-energy electrons and not protons? Because of the mass term in the denominator of (5.11); the mass of the proton is 2000 times the mass of the electron and this makes the emission of 53-keV photons by protons unlikely even in fields as strong as 10^{12} gauss ($= 10^8$ T).

Indeed, let us assume that it is electrons that produce the synchrotron radiation bump and use that fact to calculate the neutron star's magnetic field. Solving for B in (5.11) we get

$$\begin{aligned} B &= E_\gamma \frac{m}{\hbar e} \\ &\approx \frac{(53 \times 10^3 \times 1.6 \times 10^{-19})9 \times 10^{-31}}{10^{-34} \times 1.6 \times 10^{-19}} \\ &\approx 5 \times 10^8 T = 5 \times 10^{12} \text{ gauss!} \end{aligned} \tag{5.24}$$

Thus, turning the tables on the data and using the considerable evidence that Her X-1 is a neutron star, Fig. 5.30 provides the first direct measurement of the magnetic field on the surface of such an object. Such surface fields are consistent with B_s as inferred from the slowdown of pulsar periods in (5.20).

5.6. Neutron Stars and Gravitational Radiation

Neutron star pulsars are now such a "familiar" part of astrophysical observations that they are beginning to be used to probe even more exotic physical constructs. In a recent and exciting discovery, "gravitational radiation" appears to have been "seen" from a binary pulsar system designated as PSR 1913 + 16.

5.6.1. The Binary Pulsar System

Consider two stars, each having mass M, tumbling about one another according to the law of gravity. As shown in Fig. 5.25, they rotate about their common center-of-mass, which for equal-mass stars is halfway between their separation a. Recall that for circular orbits the acceleration is v^2/r, where r is the distance to the center-of-mass, so that we have from $F = \text{mass} \times \text{acceleration}$

$$GM^2/a^2 = Mv^2/(a/2). \tag{5.25}$$

Equation (5.25) then requires the total kinetic energy ($= 2 \times \frac{1}{2} Mv^2$) to be one-half the potential energy, with the net total energy given by

$$E_{\text{tot}} = \text{KE}_1 + \text{KE}_2 + \text{PE} = -GM^2/2a. \tag{5.26}$$

We recognize the factor of $\frac{1}{2}$ in Eq. (5.26) as the virial theorem, which in fact insures that (5.26) is valid even for *noncircular* orbits. As we shall see, this will be an important fact because the binary system in question PSR 1913 + 16 is highly elliptical.

Proceeding on, we convert v to the angular velocity $\omega = v/r$ to obtain from (5.25),

$$\omega^2 a^3 = 2GM. \tag{5.27}$$

Now one of the two stars in the binary system PSR 1913 + 16 is a pulsar, and using the pulsed-time techniques discussed in Sec. 5.4.1, the period of *orbital motion* of the pair about their center-of-mass is measured to fantastic accuracy, $\tau_b = 27906.98172 \pm 0.00005$ sec, so that the observed angular frequency is

$$\omega = 2\pi/\tau \approx 2.25 \times 10^{-4}\ \text{sec}^{-1}. \tag{5.28}$$

While it is not an easy business to pin down both masses of a binary pulsar system, the observational data are most consistent with both stars being neutron stars with (typical) masses 1.4 $M_\odot$. Then substituting this mass and frequency (5.28) into (5.27), we find that the separation between the stars is $a \approx 1.95 \times 10^9$ m. Since we know that our sun is about this size, clearly the radius of each star in this binary system must be much less than that of a main sequence star.

5.6.2. Observed Rotational Rate Increase

Because the observed orbital period or frequency is known to such amazing accuracy, it was not too difficult for the radio astronomers at Arecibo, Puerto Rico to detect that in fact the angular velocity was speeding up at a rate corresponding to

$$\frac{\Delta\omega}{\Delta t} \approx 2.6 \times 10^{-20}\ \text{sec}^{-2}. \tag{5.29}$$

If we then eliminate a in Eq. (5.26) in favor of ω, the increase in ω corresponds to a total energy loss of (where, again, we use $\mathcal{E}$ as energy)

$$\frac{\Delta\mathcal{E}}{\Delta t} = \frac{GM^2}{2} \frac{\Delta}{\Delta t}\left(\frac{\omega^2}{2GM}\right)^{1/3} = \frac{GM^2}{3(2GM\omega)^{1/3}} \frac{\Delta\omega}{\Delta t}, \tag{5.30}$$

where we have used $\Delta\omega^{2/3}/\Delta t = \frac{2}{3}\omega^{-1/3}\Delta\omega/\Delta t$. Then plugging in $M =$

1.4 $M_\odot$, and Eq. (5.28) and (5.29) into (5.30), we find that

$$\frac{\Delta \mathcal{E}}{\Delta t} \approx \frac{6.7\times 10^{-11}(1.4\times 2\times 10^{30})^2 \times 2.6\times 10^{-20}}{3(2\times 6.7\times 10^{-11}\times 1.4\times 2\times 10^{30}\times 2.25\times 10^{-4})^{1/3}}$$
$$\approx 1.0\times 10^{25}\ \mathrm{J/sec}. \tag{5.31}$$

Can such a power loss be accounted for by a corotating magnetic field that is generated by the *orbital* motion of the pulsars? Recall that for the Crab pulsar, with a *rotational* frequency of 10^2 sec^{-1}, the power loss $\Delta\mathcal{E}/\Delta t$ as found from (5.17) and (5.19) is 10^{31} J/sec. For the above *binary orbital frequency* of 2×10^{-4} sec^{-1}, however, the radiation loss from (5.19) is

$$\frac{\Delta \mathcal{E}}{\Delta t} = \frac{B_s^2}{k_m} R_s^6 \frac{\omega^4}{c^3} \approx \frac{(10^8)^2}{10^{-7}}(10^4)^6 \frac{(2\times 10^{-4})^4}{(3\times 10^8)^3} \sim 10^7\ \mathrm{J/sec}.$$

Clearly this small power loss cannot account for the observed 10^{25} J/sec calculated in (5.31). But what then is the cause of this latter loss of energy?

5.6.3. Gravitational Radiation

A possible candidate is *gravitational* radiation. Such radiation is a logical consequence of relativistic theories of gravity, including the presently accepted general theory of relativity proposed by Einstein. Unfortunately it has proved too difficult to observe this radiation with Earthbound detection devices. Nevertheless the orbital rotation of binary neutron stars generates gravitational radiation of the "quadrupole" type with a large enough angular frequency (5.28) so that this energy-loss mechanism should be given serious consideration as supplying the observed loss (5.31).

More specifically, we can understand the gravitational energy loss from a quadrupole source distribution in much the same way that we found the EM energy loss from a dipole source in Sec. 5.4.2. Just as monopole (charge) fields fall off like r^{-2} and dipole fields like r^{-3}, it can be shown (see exercise 32) that quadrupole fields fall off like r^{-4}. Then an analogous magnetic-quadrupole distribution would give energy densities $B_c^2 R_c^2$ [see (5.17)], with now $B_c = B_s(R_s/R_c)^4$ so that with $R_c \sim \omega^{-1}$ we have $\Delta\mathcal{E}/\Delta t \sim \omega^6$.

This characteristic ω^6 dependence also holds for the gravitational mass-quadrupole distribution case. A detailed analysis based upon general relativity for our binary system configuration leads to

$$\frac{\Delta \mathcal{E}}{\Delta t} = \frac{8}{5}\frac{GM^2 a^4 \omega^6}{c^5} f(e), \tag{5.32}$$

where $f(e)$ is a geometric factor measuring the eccentricity of the binary orbit ($e \to 0$ and $f(e) \to 1$ for a circular orbit). In the case of the binary pulsar PSR 1913+16, the measured eccentricity is $e = 0.617$ and $f(e) \approx 12$. Then for the observed $\omega \approx 2.25 \times 10^{-4}$ sec^{-1}, $a \sim 1.95 \times 10^9$ m and $M \sim 1.4\, M_\odot$, we find from (5.32) the radiated power loss

$$\frac{\Delta \mathcal{E}}{\Delta t} \approx \frac{8}{5} \frac{(6.7 \times 10^{-11})(1.4 \times 2 \times 10^{30})^2 (1.95 \times 10^9)^4 (2.25 \times 10^{-4})^6}{(3 \times 10^8)^5} \times 12$$

$$\sim 0.6 \times 10^{25} \text{ J/sec.} \tag{5.33}$$

The agreement between (5.31) and (5.33) is too close to be mere chance; we perhaps for the first time, are detecting gravitational radiation!

"I DON'T CARE WHAT IT LOOKS LIKE — THEY'RE PULSARS."

Problems

1. Assume the interior of a white dwarf is at 10^7 °K.
   a. Find the density ρ that makes the white dwarf degenerate.
   b. If the core of the white dwarf is Si ($A = 28$, $Z = 14$), would these atoms be completely ionized?
   c. Would the Si atoms be degenerate?
   d. What pressure would the degenerate gas exert?
2. What would the average mass density ρ be in a white dwarf with a carbon core in order that the carbon atoms are completely ionized?
3. If a white dwarf were composed of Si, what internal mass density ρ would mark the division between a solid interior and a gas-liquid envelope? Assume $T = 10^7$ °K.
   Compare the location of this phase transition point with the transition point in a white dwarf composed of helium, assuming both white dwarfs have the same density profile.
4. Show that (5.4) follows from (5.3).
5. A white dwarf need not be relativistic in order to be stable. The relativistic condition only determines the *maximum* mass for which the white dwarf will be stable.
   a. Write down an expression for the nonrelativistic degenerate pressure. Show that $\langle P_{\text{nonrel}} \rangle \sim \rho^{5/3}$.
   b. Find the nonrelativistic stability condition between M and R when $M = 1.4\ M_\odot$. Show that
   $$M^{1/3}R = \text{constant.}$$
   c. Try and evaluate the constant in the above expression.
6. Explain what happens in a condensing star when the average kinetic energy of the electrons reaches 0.8 MeV.
   At this time are the nuclei in the star degenerate? Explain quantitatively.
7. In the life of a neutron star there are two crucial moments, one when ρ becomes $\approx 10^{11}$ kg/m^3 and another when $\rho \approx 10^{17}$ kg/m^3. Discuss what happens at each of these densities and show why these particular values of ρ occur.
8. Why is the critical mass of a neutron star greater than that of a white dwarf?
9. Infer from the charges of the proton, neutron, and $\Sigma^{\pm}$ hyperons that a consistent set of charges for the u, d, and s quark are the nonintegral values $2/3$, $-1/3$, $-1/3$, respectively.
10. In the text we noted that if a neutron star could further decompose into a sea of quarks its maximum mass would be much higher than $4\ M_\odot$. Assuming a relativistic quark distribution and recalling that

each neutron is composed of 3 quarks, explain why the maximum mass is larger than 4 $M_\odot$ and give a rough estimate of what this mass should be.

11. If nonrelativistic *protons* happen to be spiraling around the magnetic field lines of the Crab Nebula where $B_{\text{crab}} \sim 10^{-4}$ gauss, what will be the energy of the radiated photons?
    What temperature would a thermal black body radiator have to have in order to produce photons with the same average energy found above?
12. In the text we noted that the Crab Nebula will disappear in 10,000 years. Derive this result from the information given in Sec. 5.3.
13. In some supernovae the initial velocity of the ejected material is $\sim 10^7$ m/sec. How long would we be able to "see" such a supernova remnant?
14. Discuss the difference between thermal radiation and synchrotron radiation.
15. How old would pulsar #083339 be if it were "born" with a period of rotation of 29 msec?
16. If we find a pulsar with a period of rotation of 6 msec and if the period is increasing at the rate of $\frac{1}{2} \times 10^{-14}$ sec/sec, what would be the upper limit on the age of the pulsar?
17. Say the radiation received from an object in space has the pulse shape shown below.

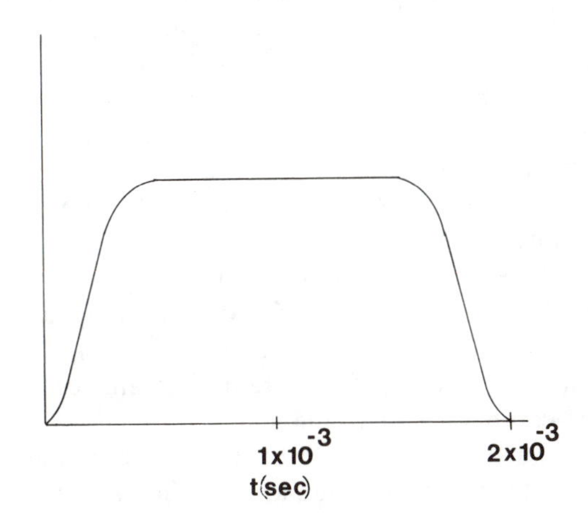

   a. Estimate the size of the object.
   b. Assuming its mass is 2 $M_\odot$, what is its minimum period of rotation?
   c. If it had a magnetic field, how big would **B** be at its surface?
   d. Is it likely that the object is a neutron star?

18. In Table 5.1 consider pulsar #150803. Using reasonable assumptions for its mass and radius, find out how much rotational kinetic energy it is losing each second.
19. It is believed that a pulsar is a neutron star. List and discuss as many reasons as you can why we believe the first statement to be true.
20. What is the radius of the light cylinder for pulsar #174949?
    Assuming a surface magnetic field of 10^{12} gauss, what is the value of dipole field at the light cylinder?
    At what rate is electromagnetic radiation flowing out of the light cylinder?
21. Show that the magnetic field in the Crab Nebula would be $\sim 10^{-4}$ gauss if the fields origin is the co-rotating magnetic field of a neutron star.
22. Using Eq. (5.20), find the magnitude of the surface magnetic field for pulsar #174949.
23. Show that $c_{0x}^2 + c_{0y}^2 = c_x^2 + c_y^2$ follows from (5.21).
24. As described in the text, an electron moving in a circle at velocity u will radiate photons into a cone of angle θ.

    a. If $u = c/2$, what is θ?
    b. If $u = 9c/10$, what is θ?
25. Estimate the size of the Schwarzschild radius for a black hole with $M = 80\ M_\odot$.
    Indicate what must be determined experimentally in order to find M and also how these quantities are determined.
26. It has been conjectured that the central parsec (3 LY!) of our galaxy contains a black hole of 6 million solar masses. What would be the size of this black hole's Schwarzschild radius?
27. If the compact source revolving about a blue giant is a white dwarf, how many 20-keV photons can be produced by 100% conversion of a single infalling proton?
    How much of the blue giant's mass would have to be used up in one year to produce 10^{29} J/sec of X rays assuming a 1% conversion efficiency of the energy of the infalling proton?
28. What would be the approximate radius of an accretion disk for a white dwarf companion to a blue giant?
29. Assume the Earth is moving through the interstellar medium at a velocity equal to its orbital speed around the Sun.

    a. What would be the radius of the accretion disk for the Earth as it sweeps up interstellar material?
    b. How long would it take the Earth to double its mass this way? (Take $n_{\text{ISM}} \sim 1$ atom/cc, $M_E \approx 6 \times 10^{24}$ kg and $R_E \approx 6 \times 10^6$ m.)

30. What is a compact X-ray binary?
    How are the X rays produced? What would be the maximum X-ray energy in the vicinity of a black hole?
    Why do some X-ray binaries pulsate with periods of a few seconds while others do not pulsate at all?
    Why do the large companion stars always appear to be blue giants?
31. Show that (5.30) follows from (5.26).
32. Given the electric *quadrupole* configuration indicated below, show that the electric field along the x axis falls off as r^{-4} for r much greater than the separation a_0 between opposite charges.

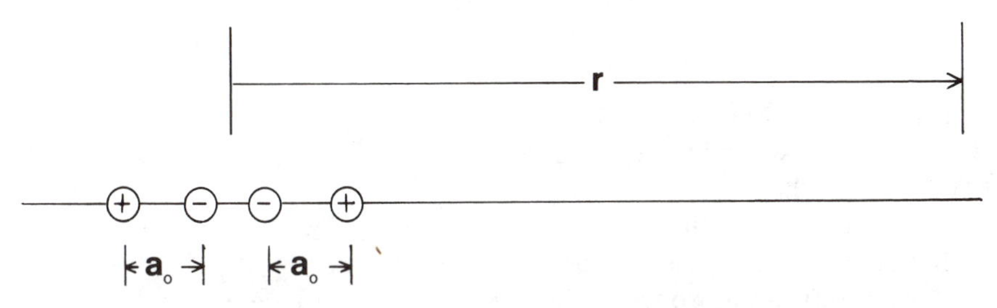

33. In Sec. 5.6 we indicated that the masses of the binary pulsar system could be obtained from observational data. To do this we assume that $M_1 = M_2$ and employ the expression derived from Einstein's theory of general relativity

$$\frac{\omega_{\text{prec}}}{\omega} \approx \frac{6GM}{(1-e^2)ac^2},$$

where ω_{prec} corresponds to the precession of the oblate binary pulsar orbit and e is the eccentricity.

  a. Given the measured values $\omega_{\text{prec}} \approx 4$ deg/yr, $\omega = 2.25 \times 10^{-4}$ sec^{-1} and $e \approx 0.617$, eliminate a from the equation using (5.27). Then solve for the mass, to find $M \approx 1.4\ M_{\odot}$.
  b. Returning to (5.27), input this M and ω to find a.

CHAPTER 6

Cosmology I—Basic Facts

But the stars throng out in their glory,
    And they sing of the God in man;
They sing of the Mighty Master,
    Of the loom his fingers span,
Where a star or a soul is part of the whole,
    And weft in the wondrous plan.

Robert Service, *The Three Voices*

6.1. Simple Extragalactic Observations

Now that the evolution of individual stars within a galaxy has been explored, we turn to the dynamics between galaxies, leading to an overall picture of the universe itself. This subject is referred to as cosmology. In this chapter we shall consider the general features of cosmology, simple observations of sizes and densities of galactic and extragalactic objects, the cosmic distance ladder, the kinematic expansion and Hubble's constant, the dynamics of the now suspect model of a steady-state universe, and the detection of the microwave background radiation.

6.1.1. Theoretical Generalities

We begin by making four general observations about the macroscopic universe around us. First, while there are four fundamental forces in nature —strong (nuclear fusion), electromagnetic, weak (radioactivity), and gravity —only the weakest of these forces, gravity, appears to control the dynamical interactions between planets, stars and galaxies. Part of the explanation is that the gravitational force is *long range*, falling off like r^{-2}, while the strong and weak forces are extremely short range, falling off essentially to zero over microscopic nuclear dimensions of order 10^{-15} m. On the other hand, the electrostatic Coulomb force is also long range with an r^{-2} dependence. But the gravity force is much weaker in strength than the Coulomb force as measured by the ratio of these forces as they act between two protons separated by a distance r,

$$\frac{F_G}{F_{em}} = \frac{Gm_p^2/r^2}{k_0 e^2/r^2} \approx \frac{7 \times 10^{-11}(1.7 \times 10^{-27})^2}{9 \times 10^9 (1.6 \times 10^{-19})^2} \sim 10^{-36}!$$

Why then does gravity dominate over electrostatics in the macroscopic and cosmological world? Put simply, the electrostatic force is *screened* because there are two kinds of electric charge producing attractive as well as repulsive forces. When protons and electrons of opposite charge form atoms, the superposition of the long range electric field due to each charge reduces the force outside the dimensions of the atom (10^{-10} m) to zero. The electrostatic force is then short range and screened provided all the positive and negative charges in the universe form atoms. If any charge would be left over, say more + than −, then the excess + charge would not be screened and we would feel a very strong long-range force that would drastically alter our picture of a gravity-dominated macroscopic world. We therefore assume that the universe is electrically neutral and since charge is known to be conserved, the universe has always had, and will always have, equal numbers of positively and negatively charged particles which effectively screen the electrostatic force on a macroscopic scale to zero. Even with this neutral universe, however, one can imagine a macroscopic separation of positive and negative charge into a dipole. In this case the resulting dipole force falls off like r^{-3} and again the monopole gravitation force dominates at large distances. This is because the gravitational force is *unscreened*; that is, both the matter–matter and matter–antimatter gravitational interactions are attractive, so that the superposition of all gravitational forces serves to reinforce one another rather than cancel. Even if antimatter forces are somehow different from matter forces, the observed dominance of matter over antimatter (at least in our part of the universe) means that the gravitational force wins out over long distances. We shall also see that the stronger but shorter range strong and weak forces played a role in the early cosmological universe when the interaction distances were smaller and the interaction times shorter.

There is, however, one electromagnetic effect which influences long range macroscopic interactions. It is light—electromagnetic radiation due to the bremsstrahlung of accelerated charges. This radiation is propagated at $v = 3 \times 10^8$ m/sec in free space with electric and magnetic fields that fall off like r^{-1}, enabling one part of the universe to communicate with another part.

As a final observation, we note that our discussion of gravity so far neglects corrections due to the "curvature" of space and time. The theory of general relativity predicts these corrections, but they are important only near very strong gravitational sources such as black holes or the universe itself. Since the parameters associated with the universe in the large are not that well measured, if at all, we shall continue to neglect quantitative effects of general relativity even at this level. Surprisingly, in fact Euclidean geometry with spherical volumes $V = 4\pi R^3/3$ having "edges," coupled with Newtonian dynamics driven by $F = ma$, $F_G = Gm_1m_2/r^2$ and energy conservation, suffice to give a reasonable description of the kinematical and dynamical consequences of cosmology. We shall tacitly assume, however, the validity of one of the cornerstones of general relativity (and of a Newtonian world), called the "equivalence principle," which requires the equivalence of gravitational and inertial mass. We shall return to a brief discussion of general relativity only in the closing section of this book.

6.1.2. The Cosmic Distance Ladder

If we are going to build a picture of the cosmos we shall have to determine the distances to the seeming infinity of galaxies that appear around us. Unfortunately, the trigonometric parallax technique that we discussed in Sec. 1.5 is accurate only to about 100 LY, a distance which hardly gets us out of the corner of our galaxy. If we are going to probe deeper into the universe we shall have to build on the information contained in those few stars whose distances from us we do know.

The technique for finding distances beyond 100 LY centers on the fundamental relationship

$$L_0(\text{actual luminosity}) = 4\pi d^2 \mathcal{L}_{\text{obs}}(\text{observed energy flux}). \qquad (6.1)$$

If L_0(joules/sec) and $\mathcal{L}_{\text{obs}}$(joules/m^2-sec) can be determined, then we can infer the distance d from (6.1). The measurement of $\mathcal{L}_{\text{obs}}$, of course, is just a function of our experimental equipment; the difficult parameter to determine will be L_0, the actual luminosity of the stellar object. To probe deeper into the universe we shall need brighter and brighter objects whose fluxes we can determine and whose absolute luminosities we can somehow infer. We can imagine a cosmic distance ladder where each rung is a different cosmic object whose brightness allows us to climb deeper into space.

To step beyond the first rung of trigonometric measurement, astronomers used an H–R diagram of nearby main sequence stars. The absolute luminosity of these stars is known from $\mathcal{L}_{\text{obs}}$ and d, where d is determined by parallax measurements. One then looks for a cluster of stars

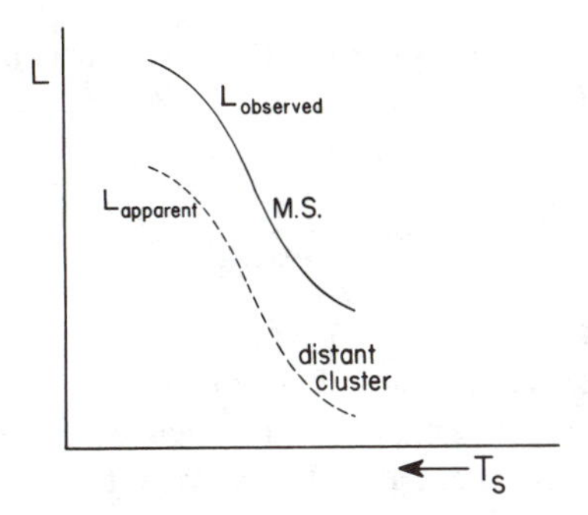

Fig. 6.1. Observed and apparent luminosities for our main sequence and a distant cluster of stars.

such as open clusters, associations, or globular clusters, since the stars in a given cluster will all be about the same distance from us. If we plot the apparent luminosity ($\mathfrak{L}_{obs} \times \Delta A$) vs. temperature of the stars in the cluster (where ΔA is the area of the detector) and if the shape of the resultant curve parallels that of the main-sequence stars (as indicated in Fig. 6.1), then one can safely assume that the absolute luminosities of the stars in the cluster are the same as the absolute luminosity of the nearby stars. Thus deducing L_0 for a given star and measuring its $\mathfrak{L}_{obs}$ we can determine d for the group. Based on the strength of current telescopes and the rather low brightness of main-sequence stars, this main-sequence photometric rung on our ladder can be safely used out to 3×10^5 LY, taking us out to the edge of our galaxy.

To proceed to the third rung of the ladder, it is first necessary to find another astronomical "signal" which, like main-sequence photometry is observed not only at great distances from us, but can be accurately calibrated at closer distances. Such "signals" come from the variable stars discussed in chapter 2. Presently nine Cepheids, whose distances are determined by main-sequence photometry, are used to produce an accurate curve of period vs. luminosity for this class of stars. Using this curve, which we hope is universal, we can determine the luminosity of distant Cepheids by merely measuring their periods. This L_0 along with $\mathfrak{L}_{obs}$ is sufficient to find d. Cepheid periods in the great galaxy M31 (Andromeda), place it 2×10^6 LY from us. These variable star techniques now take us out beyond our local group of galaxies to 10^7 LY on the cosmic distance ladder.

To measure out to further distances we need brighter objects, such as novae, brightest stars, and globular clusters. For example, brightest star luminosities peak at roughly $10^6\ L_\odot$—at least those do in nearby galaxies whose distances we have measured. The assumption, and it is a crucial one, is that the peak brightness of *all* brightest stars in any galaxy is $10^6\ L_\odot$.

Therefore, if $\mathcal{L}_{obs}$ is the energy flux that we measure from a galaxy a distance d away, then the usual relation $L_0 = 4\pi d^2 \mathcal{L}_{obs}$ leads to the estimate

$$d \approx \sqrt{\frac{10^6 L_\odot}{4\pi \mathcal{L}_{obs}}} .$$

This technique extends d out to 10^8 LY.

Finally we can use the galaxies themselves as a reference. We are in a cluster of galaxies (the local group) surrounded by other clusters of galaxies (the nearest being the Virgo cluster). For those clusters less than 3×10^8 LY away, we see a variety of galactic sizes and shapes with each cluster having a giant elliptical galaxy with a peak brightness of about $10^{10}\ L_\odot$. Again making a crucial assumption that all *galactic clusters*, even those some 10^{10} LY away (and in the past) have a giant elliptical galaxy with a peak luminosity of $10^{10}\ L_\odot$, the apparent measured energy flux $\mathcal{L}_{obs}$ can be converted to the distance

$$d \approx \sqrt{\frac{10^{10} L_\odot}{4\pi \mathcal{L}_{obs}}} ,$$

which extends d to about 10^{10} LY. Brighter galactic clusters such as the Virgo cluster can be used at shorter distances as well, being 6×10^7 LY away. To date, we have not seen cosmological objects further than 10^{10} LY from us.

The cosmic distance ladder is not completely on firm footing, however, and as closer distances become more accurately known, greater distances must be recalibrated. Furthermore the whole method is fraught with uncertainty because as we look deeper into space we also are looking back in time. Thus to assume that the peak luminosity of a nearby galaxy is the same as the peak luminosity of a distant galaxy is to have unbounded faith in the evolutionary invariance of luminosity vs. time. Astronomers have in fact tried very hard to account for possible corrections to this assumption.

6.1.3. Other Experimental Tools

Luminosity measurements are, of course, appropriate only for optical wavelengths. *Radio telescopes* have been used to measure the energy flux of extragalactic objects in the long wavelength region. Radio galaxies and quasistellar objects (quasars—having very small optical diameters) have been detected in this manner. It is thought that some quasistellar objects are at a distance of 10^9 LY to 10^{10} LY. *Doppler shifts* are of central importance for extragalactic observations and we shall devote all of Sec. 6.2 to them. Particular spectra, such as the 21-cm *hyperfine emission* line and the *Lyman absorption* line of hydrogen, are also important tools in the search for nonluminous hydrogen in the region between the galaxies. Finally *binary systems of galaxies* are used to estimate the mass of the galaxies via $v^2 = GM/R$ for circular orbits in a similar manner as applied to binary systems of stars within our galaxy.

6.1.4. Cosmic Distance and Density Estimates

We have seen that our galaxy contains about $1.7 \times 10^{11}\ M_\odot$ and has a diameter of 10^5 LY. While the (spiral) Andromeda galaxy is even bigger ($M \sim 7 \times 10^{11}\ M_\odot$, $D \sim 1.3 \times 10^5$ LY), most other spiral galaxies vary in mass between $10^9\ M_\odot$ and $4 \times 10^{11}\ M_\odot$ and in diameter between 2×10^4 and 1.5×10^5 LY. Elliptical galaxies are even more wide-ranging with $M \sim 10^6$–$10^{13}\ M_\odot$ and $D \sim 2 \times 10^4$–5×10^5 LY. As a *very crude* estimate, let us take our own galaxy as an average. To compute the average galactic mass density ρ_{gal}, we return to the arguments of Sec. 2.1 and express the central mass of our galaxy, obeying $M = v^2 r/G$, as $4\pi r^3 \rho_{gal}/3$ with $v = r\omega$. This leads to a galactic mass density

$$\rho_{gal} \sim \omega^2/4G.$$

Since the angular velocity of the galaxy is $\frac{1}{2} \times 10^{-15}\ \text{sec}^{-1}$, we have

$$\rho_{gal} \sim \frac{\frac{1}{4} \times 10^{-30}}{4 \cdot 6.7 \times 10^{-11}} \sim 10^{-21}\ \text{kg/m}^3,$$

equivalent to roughly one proton per cubic centimeter!

One other galactic parameter is of general interest, a measure of the luminosity of the galaxies. As with stars, galactic luminosities vary with size but the ratio of luminosity to mass is roughly constant. For the galaxies the present measured average ratio of observable galaxies is

$$(L/M)_{gal} \approx \tfrac{1}{15}(L/M)_\odot \approx \tfrac{1}{8} \times 10^{-4}\ \text{J/sec-kg}. \tag{6.2}$$

We will make use of this number shortly.

Two intergalactic parameters are useful: the amount of (nonluminous) mass between the galaxies in the intergalactic medium (IGM), and the average distance between the galaxies. In the former case, almost all measurements to date detecting the 21-cm or Lyman-α lines of existing hydrogen in the IGM conclude that there is little, if any, mass between the galaxies. More specifically, the mass could be in neutral hydrogen, but its absorption of 21-cm radiation emitted by quasars has been measured to require $\rho_H < 6 \times 10^{-28}\ \text{kg/m}^3$. The IGM mass could also reside in molecular hydrogen not absorbing 21-cm radiation, but the measured absorption of the Lyman-α line emitted by quasars indicates that $\rho_{H_2} < 10^{-29}\ \text{kg/m}^3$.

An IGM composed mainly of ionized hydrogen is invisible with respect to spectral-absorption measurments, but ionized electrons in such a hot medium will emit bremsstrahlung in the X-ray region. While it is difficult to distinguish this X-ray radiation from that caused by other sources, it is now estimated that the density of ionized hydrogen ρ_{H^+} is no more, and probably a lot less, than

$$\rho_{H^+} < 10^{-27} \text{ kg/m}^3.$$

At present, conventional wisdom ignores such a possibility and assumes that the mass in the IGM contributes little to the mass of the universe.

As to the distance between galaxies, Fig. 6.2a shows that the average spacing between nearby galaxies is about $R_{IG} \sim 10^6$ LY $\approx 10^{22}$ m. This determines a volume associated with each galaxy and its surrounding IGM of $V_{IG} \sim R_{IG}^3 / \sim 10^{66}$ m^3. Again assuming that all the mass is concentrated in the galaxies, the average mass density within a galactic cluster (GC) is

$$\rho_{GC} \sim M_{gal}/V_{IG} \sim 10^{41}/10^{66} \sim 10^{-25} \text{ kg/m}^3.$$

At the next distance level, Fig. 6.2b indicates that galactic clusters themselves are roughly uniformly spaced every 10^8 LY$\sim 10^{24}$ m. This fact of uniformity is usually taken as the initial ansatz in any cosmological theory and is referred to as the *cosmological principle*. The average volume of a galactic cluster is then $V_{GC} \sim 10^6\, V_{IG}$. Typical galactic clusters contain $\sim 10^3$ galaxies so that the average mass density ρ_{cos} of each galactic cluster within a cosmological cell of 10^8 LY on a side is

$$\rho_{cos} = M_{GC}/V_{GC} \sim 10^3 \text{ M}_{gal}/10^6\, V_{IG} \sim 10^{-28} \text{ kg/m}^3. \tag{6.3}$$

This cosmic mass density is our crude geometrical estimate of the mass density within the entire universe, a mass density which ws shall refer to as ρ_0 at present.

To obtain another estimate of ρ_0, we invoke the cosmological principle to give significance to the present observed luminosity density in the universe,

$$(L/V)_{cos} \approx 4.6 \times 10^{-60}\, L_\odot/\text{m}^3 \sim 2 \times 10^{-33} \text{ J/sec-m}^3.$$

Combining this number with $(L/M)_{gal}$ in (6.2), we see that the luminous mass density ρ_{lum} in the universe can be calculated as

$$\rho_{lum} = (L/V)_{cos}/(L/M)_{gal} \approx 2 \times 10^{-28} \text{ kg/m}^3. \tag{6.4}$$

What about nonluminous sources of ρ? They should be added to ρ_{lum} and then compared with ρ_{cos}.

With little nonluminous mass in the IGM, it might appear that the only other candidate for a significant mass density is due to radiated light itself, with an effective mass $m = E/c^2$. For the Sun, we recall from Sec. 4.1.1

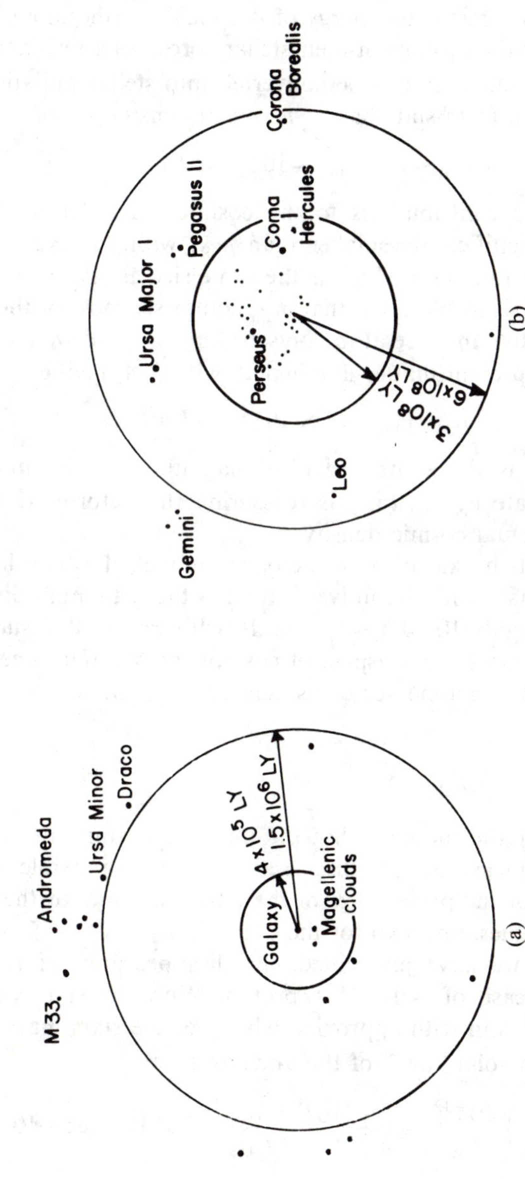

Fig. 6.2. (a) Our local group of galaxies projected on the plane of the Milky Way; (b) relationship between clusters of galaxies.

that each fusion reaction involves the disappearance of four protons and the release of 25 MeV of energy or about 8 MeV/proton. This represents $\sim 1\%$ of the proton's rest mass energy of ~ 1 GeV. Furthermore, most stars convert only $\frac{1}{10}$ of the protons in their stellar cores. As a result, at the most 10^{-3} of the luminous mass can be converted into stellar radiation, so that the stellar radiation from stars has a *maximum* density ρ_{SR} of

$$\rho_{SR} \sim 10^{-3} \rho_{lum} \sim 10^{-31} \text{ kg/m}^3. \tag{6.5}$$

Other more exotic contributions to the cosmic mass density have been estimated to be significantly smaller than ρ_{SR}, with the exception of the cosmic microwave radiation which is the same size as ρ_{SR} (to be discussed later). In any case, it would seem that ρ_{lum} comprises most of the mass and energy density in the universe at the present time, ρ_0, and we conclude that the universe is at present matter-dominated, with (6.4) giving

$$\rho_0 \approx \rho_{lum} \approx 2 \times 10^{-28} \text{ kg/m}^3. \tag{6.6}$$

The fact that ρ_0 is the same order of magnitude as the much cruder geometrical estimate ρ_{cos} in (6.3) is reassuring; henceforth we accept the above ρ_0 as the actual cosmic density.

Finally, without the aid of a cosmological model, the only length scale that we can associate with the universe itself is the maximum distance that we can detect, roughly 10^{10} LY$\sim 10^{26}$ m. It will turn out that such a length has meaning in almost every aspect of cosmology. We reinforce its significance by investigating time scales associated with the size or age of the universe.

6.1.5. Cosmic Time Scales

Let us assume for the moment that the universe is finite in extent. Then given the simple measurements discussed so far, it is possible to compare four time scales for the present age of the universe—one for the lifetime of stars, one for galaxies, and two for the universe.

For a star, as we have just noted, the disappearance of four protons results in the release of $\sim 10^{-12}$ J/proton. With the star containing $N = M/m_H$ protons and with approximately $\frac{1}{10}$ of the stars mass consumed, we can establish a solar "age" of the universe from

$$t_{star} \approx \frac{1}{10}\left(\frac{M}{m_H}\right)\frac{10^{-12}}{L_\odot} \sim \frac{10^{44}\text{ J}}{4 \times 10^{26}\text{ J/sec}} \sim 3 \times 10^{17}\text{ sec} \sim 10^{10}\text{ yr}. \tag{6.7}$$

where $N \sim 10^{57}$ protons/star and $L_\odot \approx 4 \times 10^{26}$ J/sec.

For the Galaxy, the measured uranium content (U^{235} and U^{238}) allows one to infer, using the theory of radioactive decay, that $t_{gal} \sim 10^{10}$ yr as well.

For the universe, one cosmic time scale is the time it takes for light to travel across the farthest distances so far detected, 10^{10} LY, also giving

$$t_{\cos} \sim \frac{R_u}{c} \sim \frac{10^{10}\,\text{LY}}{c} \approx 10^{10}\,\text{yr}. \tag{6.8}$$

A second cosmic time scale for the universe is its expansion time. This can be calculated rather easily if we assume the universe is in an early stage of expansion. In the brief time after such an "explosion" takes place, it is the attractive force of gravity which "pulls" the particles together and sets a time scale. We may obtain an expansion time in much the same way we deduced contraction or free-fall times for condensing stars or black holes. Namely, over a short enough period the acceleration a is constant with $a = F/m = GM/R^2 \sim G\rho R$. Thus we find again that for $R \sim at^2$,

$$t_{\exp} \sim 1/\sqrt{G\rho} \tag{6.9}$$

is the characteristic time scale for any system expanding or contracting under the sole influence of gravity. For the estimate $\rho_{\cos} \sim 10^{-28}$ kg/m^3, (6.9) gives

$$t_{\cos} \sim 10^{11}\,\text{yr},$$

roughly consistent with the other time scales. Of course, this scale is an upper bound, because the attractive gravitational force slows down the expansion, i.e., the acceleration a is not constant.

If the universe is of finite size, then the length scale of $R_u \sim 10^{10}$ LY$\sim 10^{26}$ m leads to a volume $V_u \sim 4\pi R_u^3/3 \sim 10^{79}$ m^3, not altered significantly by general relatively. The total mass in such a universe of constant density is

$$M_u = \rho_0 V_u \sim 10^{-28} \times 10^{79} \sim 10^{51}\,\text{kg},$$

corresponding to 10^{78} protons or 10^{21} stars or 10^{10} galaxies, give or take a few orders of magnitude!

6.1.6. Olber's Paradox

A final simple cosmological observation is that our night sky appears uniformly dark, save for a slight "sprinkling" of stars and galaxies. According to 18th century physics and assuming that the universe is eternal and infinite in extent, the sky ought to have the uniform intensity of our sun—a paradox indeed! It has become traditional to escape this paradox in expanding universe models by invoking the ideas of general relativity. However quite recently Harrison has argued that even in the extreme case of a static universe the paradox is resolved by applying the simple kinetic theory notion of a mean free path.

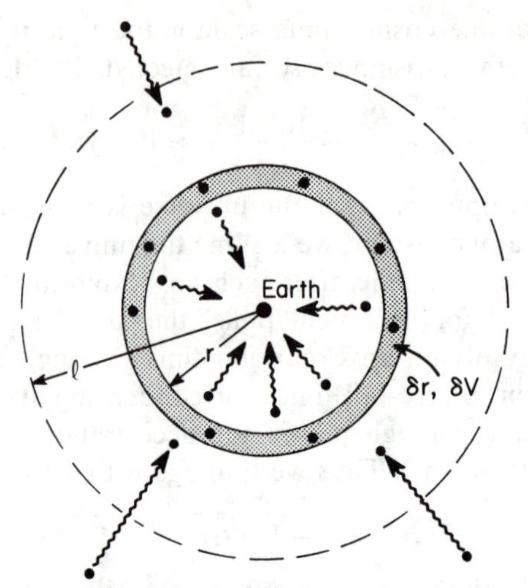

Fig. 6.3. Contribution to the energy density that we receive from stars in the spherical shell of radius r and thickness δr. Also depicted are stars just beyond a mean free path length whose radiation becomes absorbed before reaching the Earth.

Consider a thin spherical shell with thickness δr at distance r from the Earth as depicted in Fig. 6.3. For a number density n of *point* sources per unit volume having luminosity L(J/sec), the energy density of radiation δu that we receive from this shell in time δt is

$$\delta u = \frac{(n\delta V)(L\delta t)}{\delta V} = nL\,\delta t. \tag{6.10}$$

Since $\delta t = \delta r/c$ for light, if we take n and L constant in (6.10), the energy density of the night sky becomes infinite for an infinitely large universe:

$$u = \sum uL\,\delta t = \frac{nL}{c}\sum \delta r \to \infty. \tag{6.11}$$

Stars of course are not point sources and will intercept some of this energy as it heads towards the Earth. Radiation from stars beyond the distance of the mean free path length of $l = 1/n\sigma$ from us will in effect "miss" us because these photons will be absorbed by stars in the sphere $r < l$ surrounding the Earth. Thus we do not see stars for $r > l$ so that the sum $\sum \delta r$ in (6.11) is cut off by l. As a result we find

$$u = \frac{nLl}{c}, \tag{6.12}$$

which is no longer infinite. But alas the paradox is not avoided because the energy density is still finite and equal to

$$u = \frac{nL}{c}\frac{1}{n\sigma} = \frac{L}{c\sigma}.$$

How large is u? If we assume the stars are on the average as bright and as big as the Sun, then we have $L = L_\odot$ and $\sigma = \pi R_\odot^2$ and the last equation becomes

$$u = \frac{L_\odot}{c\pi R_\odot^2}. \tag{6.13}$$

Recall from Sec. 1.4.1 that the solar flux is given by $\mathfrak{L}_\odot = L_\odot / 4\pi R_\odot^2$ and that $u_\odot = 4\mathfrak{L}_\odot / c$. As a result we find for (6.13) that $u = u_\odot$. Thus the paradox is not resolved and we are led to agree with the 18th century physicists that the night sky *is* uniformly as bright as the Sun.

The loophole in the argument is an implicit assumption in (6.11) and (6.12) that the stars radiate forever. From the previous discussion we know that the time scale for stars is 10^{10} years, after which they no longer radiate with luminosities $L_\odot$. However, the mean free path $l = 1/n\sigma$ in Fig. 6.3 is roughly of size

$$n = \frac{\rho_0}{M_\odot} \sim \frac{10^{-28}\ \text{kg/m}^3}{10^{30}\ \text{kg/star}} \sim 10^{-58}\ \text{stars/m}^3$$

$$\sigma = \pi R_\odot^2 \sim (10^9\ \text{m})^2 \sim 10^{18}\ \text{m}^2$$

$$l = \frac{1}{n\sigma} \sim \frac{1}{10^{-58}10^{18}} \sim 10^{40}\ \text{m} \sim 10^{24}\ \text{LY}.$$

That is, radiation from stars beyond the distance $r > l$ from us roughly take $\tau > l/c \sim 10^{24}$ years to hit us, while such stars live only 10^{10} years! Thus the right hand side of (6.10)–(6.13) should be further reduced by the fraction t_{star}/τ:

$$u = \frac{t_{\text{star}}}{\tau} u_\odot \sim \frac{10^{10}}{10^{24}} u_\odot \sim 10^{-14} u_\odot. \tag{6.14}$$

So we learn from (6.14) that although the night sky is uniformly bright in a static universe, its brightness is greatly reduced from that at the surface of the Sun. Only if stars lived for $\sim 10^{24}$ years would our sky be ablaze in such a static universe.

The moral of this story should therefore be "burned" into us for future cosmological calculations. Before concluding that a possible dynamical process can occur according to the laws of physics, we must compare the length or time scale of our star or universe with the interaction length or time scale $l = 1/n\sigma$ or $t_{\text{int}} = l/v$. Only if $l \ll R_u$ or for $t_{\text{int}} \ll t_{\text{exp}}$ can processes proceed in the "normal" fashion.

6.2. Hubble's Constant and Its Implications

Recall that velocities of distant stars and galaxies are determined by the Doppler shifts of their absorption spectra. It is customary to describe such line shifts by $z = \Delta\lambda/\lambda_0$ where $\Delta\lambda = \lambda - \lambda_0$, λ being the measured line of a distant star and λ_0 being the line of the relevant element on Earth. For stars moving away from us at nonrelativistic velocities v, the Doppler shift to the red, $\lambda = \lambda_0(1 + v/c)$, implies $z = v/c$. On the other hand, for relativistic velocities, $\lambda = \lambda_0(1 + v/c)^{1/2}(1 - v/c)^{-1/2}$ from (1.17) leads to

$$z \equiv \frac{\Delta\lambda}{\lambda_0} = \sqrt{\frac{1 + v/c}{1 - v/c}} - 1. \tag{6.15}$$

While $z \ll 1$ in the former case, a v of 0.9 c in the latter case corresponds to $z = \sqrt{19} - 1 \approx 3.35$, which is about the maximum z detected for a quasi-stellar source. Gravitational red shifts are significant only for very strong local gravitational fields, such as for black holes (recall the discussion in section 5.5). Gravitational red shifts for the entire universe will be discussed in chapter 7.

6.2.1. The Red Shift Curve

As early as 1912 it was discovered that most of the galaxies near us have red-shifted spectral features. By 1922, 41 spiral galaxies detected had red-shifted absorption lines and only 5 such galaxies had slightly blue-shifted lines. Neglecting the small shifts from the random drift velocities of the galaxies, this indicated that on the average the galaxies were moving away from us. As further distance measurements became feasible, larger values of z were detected. Ruling out the values of z smaller in magnitude than 10^{-2}, those values being obscured by the random motions of the galaxies $\sim 10^{-2}$ c, Hubble, in 1929, made the exciting discovery that distant galaxies were receding faster than nearby galaxies, with roughly a linear relation between velocities and distances.

Needless to say, Hubble's observations were based upon original distance measurements which have since been improved along the lines discussed in Sec. 6.1.2. At that time Hubble's linear relation indicated a straight line curve with slope in units of time of 2×10^9 LY. Now the Hubble slope is believed to be about 2×10^{10} LY. More specifically, Fig. 6.4 is a plot of the recessional velocities of 45 galaxies (with $v > 10^{-2}$ c) vs. distance from us.

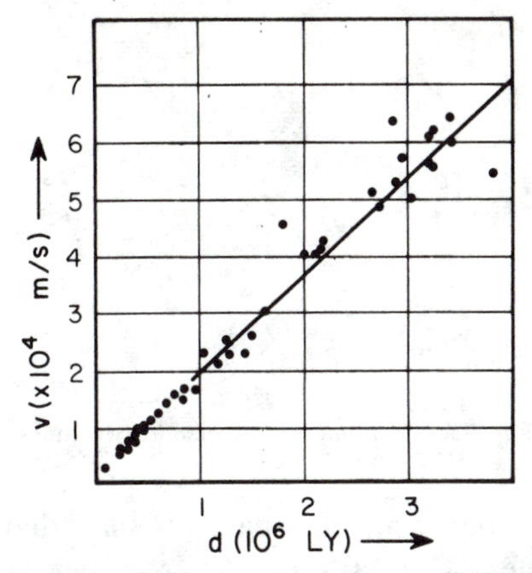

Fig. 6.4. Recessional velocity of galaxies vs. distance from Earth.

Through this series of points one can try to fit a curve. Certainly to first order a straight line suffices so that

$$v = H_0 d, \tag{6.16}$$

with H_0 the slope v/r having units of $\sec^{-1}$. This constant is called Hubble's constant and the current best straight line fit gives

$$H_0 \approx 17 \text{ km/sec}/10^6 \text{ LY} \approx 55 \text{ km/sec}/10^6 \text{ pc}, \tag{6.17a}$$

which says that, for every 10^6 LY away from the Earth, the velocity of a galaxy increases by 17 km/sec. Alternatively, in units of time the inverse of H_0 is

$$\begin{aligned} H_0^{-1} &\approx 10^6 \text{ LY}/17 \text{ km/sec} \times 10^{13} \text{ km/LY} \times 1 \text{ yr}/3 \times 10^7 \text{ sec} \\ &\approx 1.8 \times 10^{10} \text{ yr}. \end{aligned} \tag{6.17b}$$

It is significant that this "measured" time scale of the universe is essentially the same as the rough estimate suggested in the last section; we proceed to explain why this should be the case. (Very recently a few astronomers at Kitt Peak have come to the conclusion that the motion of the Earth was not properly accounted for in previous determinations of H_0 and that (6.17a) should be a factor of 2 larger! The scramble is on to see if they are correct.)

6.2.2. H_0 and an Expanding Universe

Consider the universe to be a system of particles forming a gas cloud of finite radius R (remember, we are neglecting the effects of general relativity here). The cloud then expands, with two points A and B at "$t = 0$" at distances $r_A = \frac{1}{2} r_B = \frac{1}{2} R$ from the center changing as shown in Fig. 6.5.

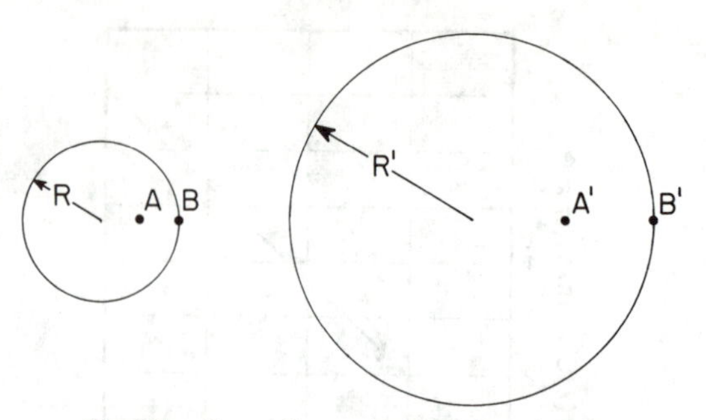

Fig. 6.5. Expanding universe at two times.

Then choose a particular kind of expansion such that this relationship is preserved at *all* times. Thus at time Δt later, we should find $r'_A = \frac{1}{2} r'_B = \frac{1}{2} R'$. Indeed, for any radial point where $r = fR$, at $t = 0$ we should find at $t = \Delta t$ that $r' = fR'$. In this way we are assured that the particles keep their same relative positions as the cloud expands. Since $r = fR$, we have $\Delta r = f\Delta R$ and therefore

$$\frac{\Delta r}{\Delta t} = f\frac{\Delta R}{\Delta t} .$$

Solving for $f = r/R$ and plugging into the above we can rearrange this equation to read

$$\frac{\Delta r}{\Delta t} \cdot \frac{1}{r} = \frac{\Delta R}{\Delta t} \frac{1}{R} .$$

If this relationship holds for any arbitrary r—anywhere inside the cloud—as measured from the origin, we must have

$$\frac{\Delta r}{\Delta t} \cdot \frac{1}{r} = \text{constant},$$

i.e., $\Delta r_1/r_1\Delta t = \Delta r_2/r_2\Delta t$, etc. But $\Delta r/\Delta t$ is the velocity v of that particular particle in the cloud. Thus our very special cloud has the following relationship between the vs of the particles and their respective distances from the center:

$$v/r = \text{constant}.$$

If we call this constant H_0, we have Hubble's law, $v = H_0 r$. But what if our origin is not at the center of the cloud but off to one side, say on particle A as shown in Fig. 6.6? In that case the distance d to particle B as measured from this origin will be related to its distance from the original center by the vector equation $\mathbf{d} = \mathbf{r}_b - \mathbf{r}_a$. Then the recessional velocity with respect to the offset origin A is

$$\mathbf{v}_a = \frac{\Delta \mathbf{d}}{\Delta t} = \frac{\Delta \mathbf{r}_b}{\Delta t} - \frac{\Delta \mathbf{r}_a}{\Delta t} ,$$

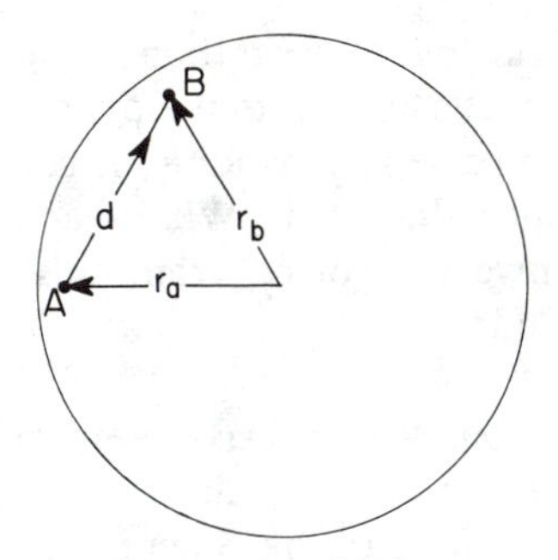

Fig. 6.6. Separation between A and B in an expanding universe.

and since $H_0 r_a = \Delta r_a/\Delta t$, $H_0 r_b = \Delta r_b/\Delta t$ we see that

$$\mathbf{v}_d = H_0(\mathbf{r}_b - \mathbf{r}_a) = H_0\mathbf{d}.$$

Thus Hubble's law holds for any origin in the cloud providing the cloud expands in the way we have described. This picture is sometimes referred to as the "raisin bread model" for the expansion of the universe.

If we now replace the particles in the cloud (or the raisins in the bread) by clusters of galaxies according to the cosmological principle (but not individual galaxies which move randomly within the cluster) and the offset origin by the Earth in Fig. 6.5, we see that our discovery of Hubble's law implies an expanding universe where each cluster retains its relative position but follows the lead of the outermost clusters of galaxies. The cosmological principle is then preserved in time because if the clusters are uniformly distributed, they will stay uniformly distributed in the future; expansion will not affect the homogeneity of space. The present average spacing between clusters of galaxies of 3×10^7 LY should increase uniformly in the "near future."

6.2.3. Formulating Dynamical Models

Some pertinent and fundamental questions about this expansion of the universe still remain:

(a) If the universe is expanding, will it continue to do so indefinitely or will it at some later time, begin to contract (blue shift)?

(b) How is the mass density of the universe changing with time? If matter + energy is fixed, then an expanding universe will obviously lead to a decreasing density. If the mass density remains constant, then matter must somehow be created! But out of what? How can we check this experimentally?

(c) Is Hubble's constant changing with time, i.e., is $H = H(t)$? If so, how can we detect this "acceleration or deceleration"? How is this kinematical parameter related to the dynamical mass density of (b)?

(d) Finally, if our universe is expanding, it could imply a beginning at $t = 0$. What happened at $t = 0$? What reactions took place? In what way has the universe evolved from this point? How will it all end—or will it end?

To answer these questions we must assume a model constrained by the measured parameters of Sec. 6.1 and the present measured value of the Hubble constant, $H_0^{-1} \approx 1.8 \times 10^{10}$ yr. Such models will tell us how R, the outer radius of the universe, changes with time, $R = R(t)$, while describing the motion of entire galactic clusters treated as particles in a large gas cloud according to the cosmological principle. Three examples of now discarded universes are depicted in Fig. 6.7. A′ is the Einstein static universe with $R = \text{const}$, which is ruled out by the observed Hubble expansion. (Actually, general relativity in its simplest form predicts a nonstatic universe—a prediction which Einstein circumvented by modifying the theory with an ad hoc "cosmological constant." After the Hubble discovery Einstein is said to have admitted that this was "the biggest blunder" of his life.) B′ is a uniformly expanding universe with $v = \text{const}$ at any R. This universe is excluded because the gravitational force on a cluster at radius r is proportional to $M_r r^{-2}$ which implies that v cannot be constant. C′ is the Lemaître universe in which the initial expansion was halted while the galaxies and perhaps the quasistellar objects were formed. This model, while not excluded by observation, appears to be somewhat artificial.

In Fig. 6.8 the evolution of three contemporary models, Friedmann "big bang" models, is traced with t_0 the present time. A is a forever expanding universe, with $v_{\text{univ}}^2 > 0$ at $R = \infty$, $t = \infty$. B is a forever expanding universe, with $v_{\text{univ}}^2 = 0$ at $R = \infty$, $t = \infty$ (called the Einstein–deSitter model). In these two cases, v_{univ} represents the velocity of a particle (star) near the

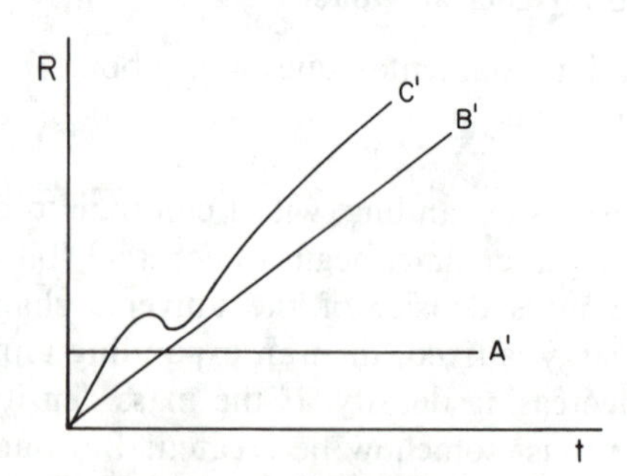

Fig. 6.7. The radius of the universe vs. time for the Einstein static universe (A′), a uniformly expanding universe (B′) and the Lemaître universe (C′).

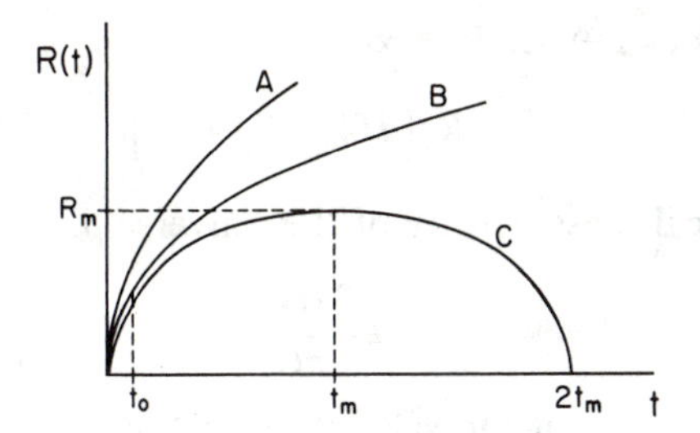

Fig. 6.8. The radius of the universe vs. time for (A) $v^2_{\text{univ}} > 0$, (B) $v^2_{\text{univ}} = 0$, (C) "v^2_{univ}" < 0, all at $R = \infty$.

"edge" of the universe. For case C, $v^2_{\text{univ}} < 0$ at $R = \infty$, $t = \infty$, which means that at some intermediate time t_m, the expansion halts with $v^2_{\text{univ}} = 0$ and then contraction begins. At time $t = 2t_m$, the universe will contract to a point, the "big crunch." In all three models, the expansion (and contraction for case C) must be consistent with energy conservation.

6.2.4. Energy Conservation and the Critical Mass Density

How are we to decide between the big bang universes A, B, and C of Fig. 6.8? The key point is the behavior of the velocity of a galactic cluster on the outer rim as $R \to \infty$. We can determine this behavior and what parameters affect v by invoking, of all things, conservation of energy. As shown in Fig. 6.9, we consider a galactic cluster on the "edge" of the universe and calculate its energy at any two times, say now at t_0 and anytime in the future, t. Then the equality of the total kinetic and gravitational potential energy of the cluster, at the two different times,

$$\tfrac{1}{2}mv^2 - \frac{GMm}{R} = \tfrac{1}{2}mv_0^2 - \frac{GMm}{R_0}, \qquad (6.18)$$

when combined with the present Hubble law $v_0 = H_0 R_0$ and M

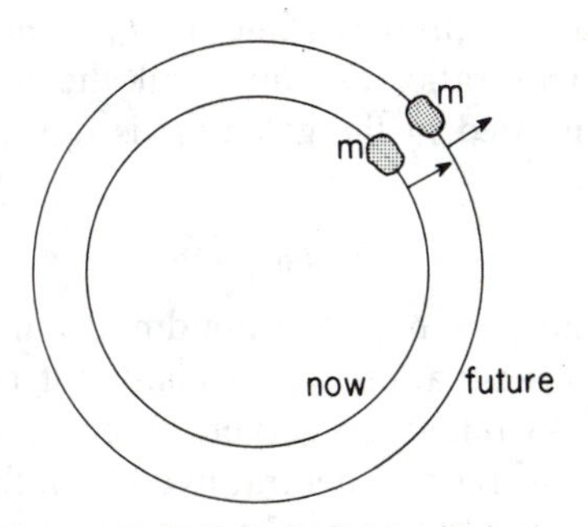

Fig. 6.9. Galactic cluster on the "edge" of the universe now and in the future.

$= \rho_0 4\pi R_0^3/3$ gives as $t \to \infty$, $R \to \infty$,

$$v^2 \to R_0^2\left(H_0^2 - \frac{8\pi}{3} G\rho_0\right). \tag{6.19}$$

Defining a critical mass density of the universe ρ_c as

$$\rho_c \equiv \frac{3H_0^2}{8\pi G}, \tag{6.20}$$

the above limiting form for the velocity on the outer rim (6.19) leads to an explicit constraint on the present mass density ρ_0 for universes A, B, and C of Fig. 6.8:

$$\text{Universe A with } v^2 > 0 \text{ requires } \rho_0 < \rho_c$$

$$\text{Universe B with } v^2 = 0 \text{ requires } \rho_0 = \rho_c$$

$$\text{Universe C with ``} v^2 < 0\text{'' requires } \rho_0 > \rho_c.$$

Amazingly enough, a detailed analysis of the theory of general relativity leads to the same constraints and values for ρ_c as the classical description given above. Since ρ_c is independent of R_0, we can compute it directly, using $G \approx 6.7 \times 10^{-11}$ m^3/kg-sec^2 and the present value $H_0^{-1} \approx 1.8 \times 10^{10}$ yr $\approx 5.7 \times 10^{17}$ sec. Then (6.20) becomes

$$\rho_c \approx \frac{3}{8\pi} \frac{1}{6.7 \times 10^{-11}(5.7 \times 10^{17})^2} \approx 6 \times 10^{-27} \text{ kg/m}^3. \tag{6.21}$$

It has been a major quest of experimental cosmologists to measure parameters which would infer the present mass density ρ_0 so that a comparison with ρ_c would determine the ultimate course of our expansion. In Sec. 6.1.4 we demonstrated that a rough estimate of ρ_0 based on the mass and size of typical galaxies (6.3) is quite near to the above value of ρ_c. Moreover we also showed that a more accurate determination of ρ_0 based upon luminosity measurements of galaxies gave $\rho_{lum} \approx \rho_0 \approx 2 \times 10^{-28}$ kg/m^3, a factor of only 30 to 40 shy from ρ_c! It turns out that while both ρ_c and ρ_{lum} depend upon the present value of H_0, their ratio is *independent* of the Hubble value. This ratio, assuming all the mass in the universe is luminous and concentrated in the galaxies, is found from (6.6) and (6.21) to be

$$\rho_c/\rho_0 \approx 36. \tag{6.22}$$

Thus we conclude that $\rho_0 < \rho_c$, corresponding to an "open universe" A in Fig. 6.8. Of course this situation would change if enough "missing mass" resided in the IGM. As remarked earlier, based upon the measurement of the 21-cm and Lyman-α lines of neutral hydrogen, there is no missing mass in the IGM. It is possible but somewhat improbable that the missing mass is in hot, ionized, hydrogen in the IGM. As we have noted in Sec. 6.1.4, the upper bound for ρ_{H^+} is 10^{-27} kg/m^3, now intriguingly close to the value of ρ_c. It will be interesting to see if this bound improves in the future. It is also possible that some of the missing mass is in black hole stars or dust, but this is thought to be insufficient to make up the factor of 36 in mass needed to have a closed universe (Case C). Lastly, this "missing mass" could reside in neutrinos; we shall return to this possibility in Sec. 7.1.5.

6.2.5. Hubble's Constant vs. Time

We may also view the evolution of universes A, B, and C in terms of Newton's second law, $F = ma$. A galactic cluster on the "edge" of the universe experiences a gravitational acceleration toward the center of the universe with $a_0 = -GM/R_0^2 = -4\pi G\rho_0 R_0/3$ (the minus sign indicates that the acceleration vector points toward the center). Let us define a dimensionless *deceleration parameter* q_0 as $-a_0/R_0H_0^2$, so that

$$q_0 = -a_0/R_0H_0^2 = 4\pi G\rho_0/3H_0^2 = \rho_0/2\rho_c. \tag{6.23}$$

Then for an open universe $\rho_0 < \rho_c$, we expect $q_0 < \frac{1}{2}$; conversely for a closed universe, $q_0 > \frac{1}{2}$. This parameter q_0 measures the time dependence of Hubble's constant, as we now show.

How do we measure q_0 directly? Since the expansion velocity is changing according to $v = HR$, we expect the rate of change of velocity, i.e., acceleration, to be related to the rate of change of H and of R. In quantitative terms we may write

$$\begin{aligned}\frac{\Delta v}{\Delta t} &= \frac{\Delta H}{\Delta t}R + H\frac{\Delta R}{\Delta t}\\ &= \frac{\Delta H}{\Delta t}R + H^2R,\end{aligned} \tag{6.24a}$$

since $\Delta R/\Delta t = v = HR$. Solving for $\Delta H/\Delta t$ and using $\Delta v/\Delta t = a$, we find that at present

$$-\frac{1}{H_0^2}\frac{\Delta H_0}{\Delta t_0} = 1 + q_0. \tag{6.24b}$$

Thus q_0 is determined by the rate of change of Hubble's constant. Now we know that the deceleration parameter must be positive, $q_0 = \rho_0/2\rho_c > 0$, so that the above relation indicates that $\Delta H_0/\Delta t_0 < -H_0^2$. It should be noted that general relativity again leads to the same results.

To see if Hubble's constant is actually decreasing with time, we take a detailed look at the Hubble curve for distant galaxies in Fig. 6.10a as in this case we are looking back in time. Thus the slope of the v vs. d curve should

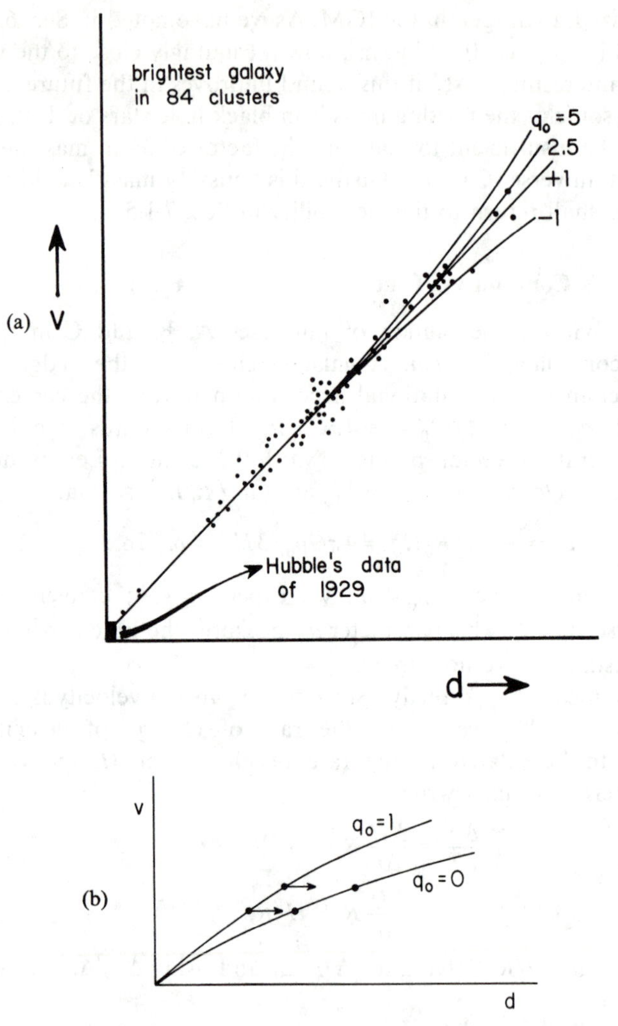

Fig. 6.10. Recessional velocity of galaxies vs. distance (a) parametrized according to q_0 and (b) indicating how the curve shifts if d has been underestimated.

not be constant but increase as d increases since at these much *earlier* times the slope H must have been larger than at present ($q > q_0$). In Fig. 6.10a we plot a family of curves all having $\Delta H/\Delta t < 0$, but with various values of q_0. The data points of distant galaxies are based upon the brightness of local galaxies (a tricky business, as noted in Sec. 6.1.2). In principle, then it is possible to fit for q_0. While this is difficult to do in practice, it would appear that indeed $\Delta H_0/\Delta t_0 < H_0^2$; in fact one finds $q_0^{\text{fit}} \sim 1$. Without further corrections this in turn would indicate that $\rho_0 \sim 2\rho_c$, implying that the universe is closed and that there is a missing nonluminous mass 70 times the observed luminous mass in the universe yet to be found!

Unfortunately, it is not clear that q_0^{fit} is really the theoretical q_0 we are seeking. This is because looking back in time means we should also correct for the evolution of the galaxies. For example, if the galaxies were brighter in the past, our present estimate of their distances from us ($d^2 = L_{\text{GC}}/4\pi\mathcal{L}_{\text{obs}}$) is too small. In terms of Fig. 6.10a the data points would have to be moved for a given measured v to larger values of d, therefore falling on a lower q_0 curve, as illustrated in Fig. 6.10b. It is now believed that such galactic evolution corrections reduce q_0 from $q_0^{\text{fit}} \sim 1$ down to a value closer to 0 than to $\frac{1}{2}$, implying once again $\rho_0 < \rho_c$ and an open universe! But even here we must be careful. It may be possible that certain galaxies in a cluster could gobble up smaller galaxies by accretion and thus balance the evolutionary diminution in brightness. Then no correction, or even a partial correction in the other direction for q_0, may be necessary (help) . . . . In the specific context of the big bang model there are other ways to infer ρ_0 and q_0 and so in chapter 7 we shall return to this fascinating question of the "missing mass."

6.3. The Steady-State Universe

Extrapolating the Hubble curve to $v = d = 0$, one is led to a big bang origin of the universe where, roughly 10^{10} years ago, there was a huge energy source concentrated at a single "point"! The principle of energy conservation makes such an explosive beginning hard to fathom, and so for many years alternative cosmologies were taken seriously. These alternatives were given further impetus because the initial measured H_0^{-1} was ten times smaller, $H_0^{-1} \sim 2 \times 10^9$ yr, which meant that the age of a big bang universe was less than the estimated age of our sun and even less than the known geological age of the Earth! While this rationale no longer can be used, it is instructive to put the big bang model into proper context by considering a reasonable alternative, the steady-state cosmology proposed by Bondi and Hoyle.

In their model the universe was allowed to expand in a manner consistent with Hubble's observations, with matter continuously created so that the density $\rho(t) = \text{const}$. Thus, new galaxies would be formed but with densities the same now as in the past and as they would be in the future. In this case the age of the universe had no meaning; any discrepancies with geological time resulted because our galaxy was created sometime well into the evolution of other galaxies.

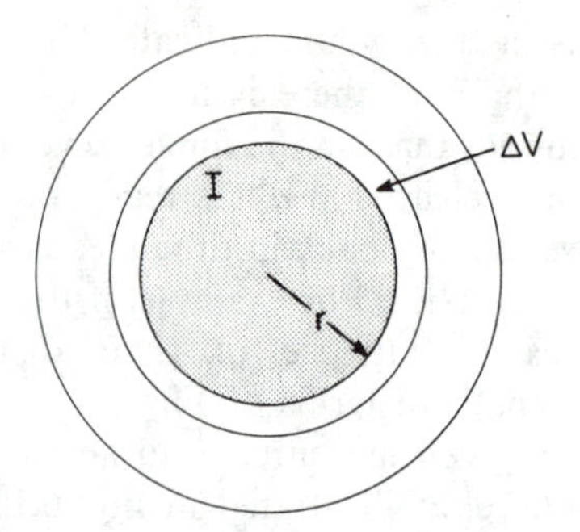

Fig. 6.11. The interior of a spherical universe.

6.3.1. Creation of Matter

Since creation of matter is the major ingredient in the steady state theory, we shall examine it a bit more closely. Again, invoking the cosmological principle, we envision a spherical universe uniformly filled with galactic clusters as in Fig. 6.11. Imagine a thin shell of volume $\Delta V = 4\pi r^2 \Delta r$ into which the mass in region I has expanded. Assume for the moment that matter and energy are conserved, so that if $\rho = M/V$ then $\Delta\rho = M\Delta(1/V)$, where

$$\Delta \frac{1}{V} = \frac{1}{V_1} - \frac{1}{V_2} = \frac{V_2 - V_1}{V_1 V_2} \approx \frac{\Delta V}{V^2} = \frac{3\Delta r}{Vr},$$

since $V = 4\pi r^3/3$ up to radius r. Then we find

$$\frac{\Delta\rho}{\Delta t} = \frac{3M}{rV}\frac{\Delta r}{\Delta t} = 3\rho H,$$

where we have used $\Delta r/\Delta t = v = Hr$. This is the rate at which a universe which conserves matter and energy loses density; therefore in a steady state universe, we must gain density at precisely this same rate, $3\rho_0 H_0$ at present. With $H_0^{-1} \sim 2 \times 10^{10}$ yr and $\rho_0 \approx 2 \times 10^{-28}$ kg/m^3, we see that

$$\frac{\Delta\rho_0}{\Delta t_0} \sim \frac{3 \cdot 2 \times 10^{-28}\ \mathrm{kg/m^3}}{2 \times 10^{10}\ \mathrm{yr}} \sim 3 \times 10^{-38} \frac{\mathrm{kg}}{\mathrm{m^3 yr}},$$

which amounts to the creation of two protons per cubic meter every 10^{10} years. This is such a low production rate that one could not hope to eliminate the steady state cosmology on these grounds alone.

6.3.2. Number Counts

An alternative approach is a probe of the number density n of thousands of recently discovered faint radio sources (presumably distant galaxies) whose red shifts are not measured. One assumes that these galaxies all have the same intrinsic luminosity L, with the number N of such galaxies inside a given volume up to a radius r from the Earth given by

$$N = \frac{4\pi}{3} r^3 n.$$

Although each galaxy has the same luminosity, our optical instruments will measure a diversity of luminosities due to the distance dependence $L = 4\pi r^2 \mathcal{L}_{min}$, where $\mathcal{L}_{min}$ is the smallest luminosity per unit area (energy flux) received within the volume $4\pi r^3/3$. If our detectors can detect all $\mathcal{L}$ above $\mathcal{L}_{min}$, then we are in principle detecting all radio galaxies up to radius r. Eliminating $r \propto (L/\mathcal{L}_{min})^{1/2}$ in N we see that

$$N \propto L^{3/2} n / \mathcal{L}_{min}^{3/2}.$$

Since one can alter the detector threshold, and therefore $\mathcal{L}_{min}$, it is possible to plot N vs. $\mathcal{L}_{min}$ in order to see if the density n is indeed a constant, as assumed in the steady-state cosmology. It turns out to be more convenient to plot the $\log N$ vs. $\log \mathcal{L}_{min}$, in which case

$$\log N = \text{const} + \log L^{3/2} n - \tfrac{3}{2} \log \mathcal{L}_{min}.$$

For n a constant, this curve should be a straight line with slope $-\frac{3}{2}$. From Fig. 6.12, we see that the data does not fall on a straight line and so n is not a constant. Since lower thresholds mean we are looking deeper in space and

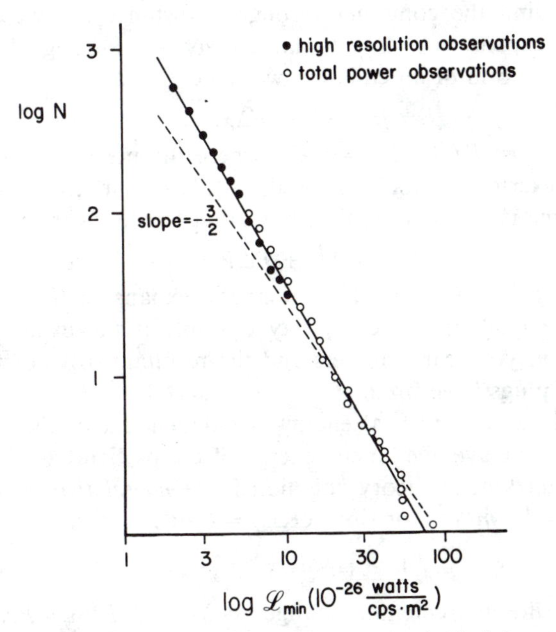

Fig. 6.12. Number of radio sources vs. threshold of detector.

earlier in time, the rising $\log N$ curve for small $\mathfrak{L}_{min}$ indicates an increasing density, as we would expect in a conserved mass (and energy) expanding universe. Unfortunately this result is not definitive because this rise might also be caused by an increasing L at early times, rather than any increasing n. Perhaps early on the galaxies were brighter but not more dense. Although it is generally believed that it is n and not L that is the culprit, this measurement alone, intriguing as it is, is not enough to prove or disprove the steady state theory.

6.3.3. Adiabatic Expansion of Particle and Photon Gases

In order to proceed further, it is worthwhile to investigate the thermodynamics of an expanding universe. We shall treat the galactic clusters as particles comprising a "particle gas" and the radiation as a "photon gas" and ask how the volume of either gas constrains its temperature. To answer the question we need to consider the first law of thermodynamics which is a statement of conservation of energy. The first law asserts that the heat energy, Q, that flows into a system must equal the work done, W, by the system against external forces plus any change in the system's internal energy ΔU, that is, $Q = W + \Delta U$. For the system under consideration the container is the universe and hence no heat can enter or leave and $Q = 0$. If we allow such a gas to expand, then the external work done by the gas, $F\Delta x$, in expanding the container through a distance Δx via a force F, comes at the expense of its internal energy U. Putting the last two statements in the form of an equation we write

$$0 = \Delta U + F\Delta x,$$

Noting that $F\Delta x = (F/A)A\Delta x = P\Delta V$, where the pressure P is the force per unit area exerted by the gas on the walls of the container and the displaced volume is $\Delta V = A\Delta x$, the above first law takes the form

$$0 = \Delta U + P\Delta V.$$

This kind of expansion is called an adiabatic expansion (from the Greek *adiabatos*: not passable). An every day example is the expansion of the gasoline vapor in your car cylinders and the resultant drive of the pistons after the spark plugs have fired.

For an ideal gas the internal energy is independent of the volume but dependent upon the average kinetic energy of the particles as $U = N\langle \mathrm{KE} \rangle$. The fundamental kinetic theory equation for a *nonrelativistic* particle gas then gives $PV = \frac{1}{3}Nm\langle v^2 \rangle$, or since $\langle \mathrm{KE} \rangle = \frac{1}{2}m\langle v^2 \rangle$,

$$PV = \tfrac{2}{3}N\langle \mathrm{KE} \rangle = \tfrac{2}{3}U, \tag{6.25a}$$

which implies the incremental changes $\frac{2}{3}\Delta U = \Delta(PV) = P\Delta V + V\Delta P$.

Combining this result with the first law above, we see that

$$0 = \tfrac{3}{2}(P\Delta V + V\Delta P) + P\Delta V = \tfrac{5}{2}P\Delta V + \tfrac{3}{2}V\Delta P,$$

which is equivalent to $(\Delta P/P) = -\tfrac{5}{3}(\Delta V/V)$. Without proof we note that this latter relation can be written as

$$PV^{5/3} = \text{const}, \tag{6.25b}$$

the adiabatic expansion law for a nonrelativistic ideal gas. Since the equation of state for such a gas is $PV = NkT$, constant particle number allows us to rewrite the adiabatic expansion law as $TV^{2/3} = \text{const}$, or assuming $V = 4\pi R^3/3$,

$$T \propto 1/R^2. \tag{6.25c}$$

Thus, as the *particle* gas expands adiabatically, the temperature decreases like the *inverse square* of the radius of the gas by (6.25c)

For a photon gas the first law relation remains unchanged, $0 = \Delta U + P\Delta V$. On the other hand, the photons are relativistic, and as shown in Sec. 1.4, $P = \tfrac{1}{3}aT^4$ and $E/V = aT^4$. Therefore since the internal energy in this case is the total photon energy, we have

$$PV = \tfrac{1}{3}U, \tag{6.26a}$$

so that now $\tfrac{1}{3}\Delta U = P\Delta V + V\Delta P$. Combining these two incremental relations leads to

$$0 = 3(P\Delta V + V\Delta P) + P\Delta V = 4P\Delta V + 3V\Delta P.$$

Again this can be stated as $(\Delta P/P) = -\tfrac{4}{3}(\Delta V/V)$, which is equivalent to

$$PV^{4/3} = \text{const}, \tag{6.26b}$$

the adiabatic expansion law for a relativistic ideal gas. Since the equation of state for a photon gas is $P = \tfrac{1}{3}aT^4$, the adiabatic law can be rephrased as $TV^{1/3} = \text{const}$, or in terms of the radius R,

$$T \propto 1/R. \tag{6.26c}$$

Note that the temperature of the *photon* gas falls off like $1/R$ whereas the nonrelativistic particle gas behaves as $1/R^2$. It can be shown that a *relativistic* particle gas has the same temperature behavior as the photon gas (also relativistic), i.e., $T \propto 1/R$. One way of justifying this is to note that a relativistic particle having a kinetic energy much greater than its rest mass energy mc^2 has $E \approx pc$ and consequently $PV = \tfrac{1}{3}U$ and $T \propto 1/R$ then follow.

By way of summary, we might ask: An adiabatically expanding gas in a closed cylinder loses energy by doing work against a piston—but what is the "piston" against which an apparently freely expanding cosmic fireball does work? The answer is that a particle or photon gas does work against the other particles or photons in the gas by virtue of their mutual gravitational attraction. Thus as we picture an expanding universe (i.e., R increasing), the density ρ decreases as $\rho \propto R^{-3}$, the gas temperature T decreases initially as $T \propto R^{-1}$, and t_{exp} and ρ are related through gravity such that $t_{\text{exp}} \sim (G\rho)^{-1/2}$.

6.3.4. The Cosmic Fireball—A Blow to the Steady-State Theory

The greatest blow against the steady-state theory was struck in 1964 with the accidental discovery of what has since been interpreted as cosmic microwave background radiation or in brief, the "cosmic fireball." This radiation now appears to follow a black body spectrum with a characteristic temperature of 2.7 °K as shown in Fig. 6.13.

In 1964 Penzias and Wilson had modified a ground-based radiometer that had been used to detect signals from Echo satellites, into a special low-noise radio antenna of length 7.35 cm. They were bothered by an excess of noise which they could not pinpoint—it did not vary with time of day nor with the season. Assuming it was due to radiation in thermal equilibrium, they calculated the "antenna temperature" of this black body radiation. More specifically, recall that a black body frequency spectrum of the energy flux $\mathfrak{L}_\nu$ peaks at a wavelength ($\nu = c/\lambda$),

$$\lambda_{\text{peak}} = 0.51 \text{ cm}/T(°\text{K}).$$

If there were one data point, at 7.35 cm, which happened to be at the peak, then the antenna temperature would be $T \sim 5/73 \sim 0.07$ °K. This was unreasonably low, so they assumed they were well below the peak, on the classical Raleigh–Jeans long-wavelength or low-frequency tail of the spec-

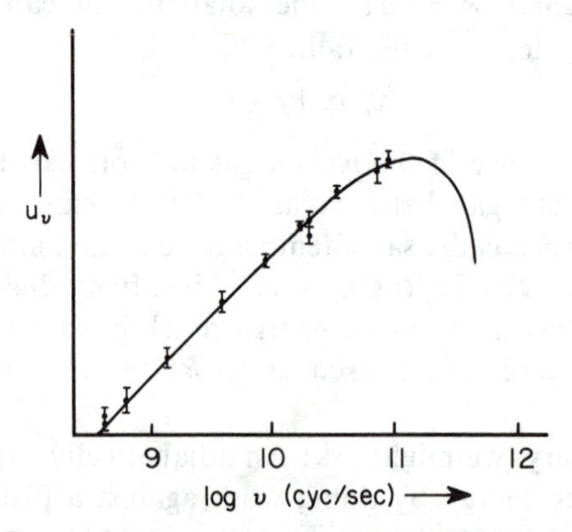

Fig. 6.13. Energy density vs. frequency for 2.7 °K black body radiation.

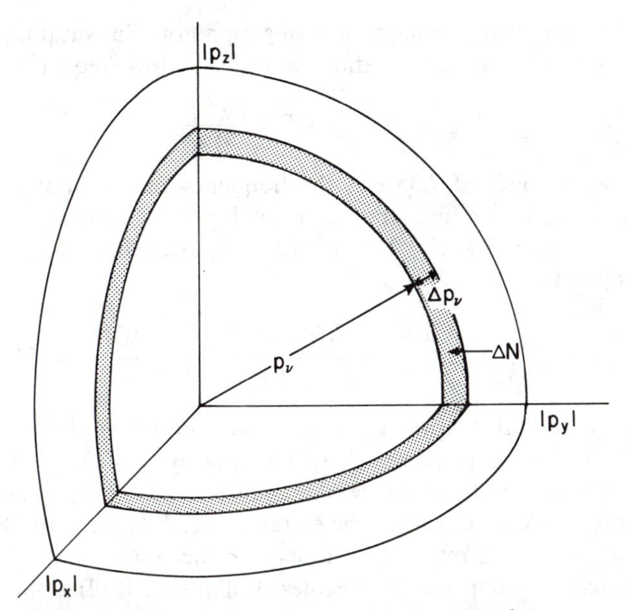

Fig. 6.14. Volume in momentum space of photons with momentum between p_ν and $p_\nu + \Delta p_\nu$.

trum in Fig. 6.13. This last assumption, together with the measured value of the energy density per unit frequency, $\Delta(E/V)/\Delta\nu \equiv u_\nu$, at $\lambda = 7.35$ cm is sufficient information for us to determine the black body temperature if we can relate the temperature to u_ν. To this end we use arguments similar to those presented in Sec. 4.3.3 related to the uncertainty principle and phase space.

Referring to Fig. 6.14 and employing an analysis similar to that of Sec. 4.3.3, we must first count the number ΔN of photon states between momentum radius p and $p + \Delta p$, i.e., in the spherical shell $4\pi p^2 \Delta p$. Borrowing the result (4.17) but for photon particles, we have

$$\Delta N = 2 \cdot 4\pi p^2 \Delta p V / h^3 \tag{6.27}$$

where V is the container volume, the factor 2 counts the two photon spin states (i.e., the E and B fields), and h^3 is the smallest "phase space" volume. Switching to "frequency space" with $p_\gamma = h\nu/c$ for photons, the quantum factor of h^3 cancels out of (6.27) and we obtain the purely classical result

$$\frac{\Delta N}{V} = \frac{8\pi\nu^2 \Delta\nu}{c^3}. \tag{6.28}$$

Also the photon energy density at frequency ν is

$$u_\nu = \frac{\Delta E/V}{\Delta\nu} = \left(\frac{\Delta N/V}{\Delta\nu}\right) kT, \tag{6.29}$$

where kT is the average energy of a single photon. Substituting (6.28) into (6.29) we have our desired equation, valid in the low frequency region

$$u_\nu = \frac{8\pi\nu^2}{c^3} kT = \frac{8\pi kT}{c\lambda^2} . \tag{6.30}$$

At the wavelength of 7.35 cm, the frequency of the photon is $\nu = c/\lambda \approx 4 \times 10^9$ cps. If we use the data in Fig. 6.13, then $u_\nu \approx 7 \times 10^{-27}$ erg/cm^3–10^9 cps $= 7 \times 10^{-28}$ J/m^3–10^9 cps, therefore the predicted antenna temperature is

$$T = \frac{c\lambda^2}{8\pi k} u_\nu \approx \frac{3 \times 10^8 (7.35 \times 10^{-2})^2 7 \times 10^{-28}}{8\pi(1.4 \times 10^{-23})} \sim 3\ °\text{K}.$$

(We might note that the present average value of T from all measurements is $T \sim 2.7$ °K with a peak wavelength in Fig. 6.13 of $\lambda_{\text{peak}} \sim 0.51/2.7 \sim 2$ mm.) Convinced that this noise was not caused by their instrument, Penzias and Wilson conjectured it might be extraterrestrial. It was with this in mind that they spoke to a group of physicists at Princeton.

The Princeton group (Dicke, Peebles, Wilkinson, Roll) at that time was discussing the possibility that one might be able to detect the leftover radiation from a cosmological big bang. Briefly stated (we shall devote the next section to a more detailed discussion), in the early stages following the big bang, the high density of charged particles would result in a bremsstrahlung black body spectrum characteristic of the temperature at that time. This primordial fireball was initially very hot, but as the universe expanded it would cool down according to the thermodynamics of a photon gas as derived in the last section, namely $T \propto 1/R$. Thus Penzias and Wilson presumably were detecting a mere shadow of the original big bang. Nonetheless the radiation should be there, it should be isotropic (i.e., the same in all directions), its spectrum should follow a *unique* black body curve (note the experimental points on the theoretical curve in Fig. 7.7 with a present temperature of about 3 °K). Amazingly enough, this fireball temperature had been roughly estimated in the late 1940s by Gamow and collaborators, but more about that in Sec. 7.2.2.

In any event, other cosmologies such as the steady-state theory do *not* predict such a cosmic fireball (except under highly unlikely circumstances) because the density of matter is never large enough to allow matter to be in equilibrium with radiation. Thus, there is no opportunity for the bremsstrahling interactions which would lead to a unique black body spectrum in the past or at present.

Even though the steady state theory is now out of favor, it had two beneficial effects. One, it was a stimulus to the development of the theory of stellar nucleosynthesis, for in this picture one could not fall back on element formation in a big bang. Secondly, "it was an important stimulus also to observational work, much of it to prove the cosmology wrong" (Peebles, 1971). In any event, it is noteworthy that Penzias and Wilson were recently awarded the Nobel Prize for their discovery. As we shall see in the next chapter, this microwave background radiation is crucial to making detailed predictions in the big bang cosmology. Without it we might justifiably end the book at this point.

"THAT WRAPS IT UP — THE MASS OF THE UNIVERSE."

Problems

1. The range R of each of the fundamental forces is determined by the mass of the particle exchanged between the interacting objects,

$$R \approx h/mc$$

   a. Show that the above relationship follows from the uncertainty principle $\Delta p \Delta x \approx h$ when $\Delta x \approx R$ and the exchanged particle moves with $v \approx c$.

b. Why is the range of the electromagnetic and gravitational force infinite?

c. The particle responsible for the strong force is the π meson whose mass is ≈ 140 MeV. What is the range of the strong force?

d. It is believed that the range of the weak force is less than 10^{-16} m. What does this imply about the mass of the exchanged particle?

2. The electric field from a point charge is $E_1 = k_0 q/r^2$. The electric field component of the EM radiation from the same charge with acceleration a_0 is $E_2 = (k_0 q a_0)/c^2 r$. Assume the charged particle is in a galaxy 10^{22} m from us.

   a. What would be the direction of E_1 and E_2 if the charge accelerates along a line perpendicular to our line of sight with the galaxy.

   b. If the acceleration of the charge is 10 m/sec^2 (similar to a freely falling object on Earth), what would be the ratio of E_2/E_1?

3. If we measure an energy flux of 10^{-14} J/m^2-sec from a giant elliptical galaxy, how far away is the galaxy?

4. In the text we noted that the brightest star luminosities are $10^6\ L_\odot$ and that they can be detected out to $\sim 10^8$ LY.

   a. What does this imply about the threshold value of $\mathcal{L}_{obs}$?

   b. Assuming that the radiation comes in the form of 3 eV photons, how many photons/sec can we detect per square meter at threshold?

5. Using ρ_{gal} find the number of interactions between stars if two galaxies collide and pass through one another.

6. Estimate how long it would take the radius of an expanding universe to exceed the range of the strong force. (See problem 1.)

7. Calculate the total potential energy of the universe. Calculate the total rest mass energy of the universe.

8. For an *expanding* universe, show that $t_{exp} \sim (G\rho)^{-1/2}$. In obtaining your answer, what assumptions did you make besides the assumption that the acceleration a is a constant?

9. Show that Eq. (6.13) yields the result for the energy density $u = u_\odot$.

10. Concerning Olber's paradox, what would be the energy density of the night sky as found from Eq. (6.14), if the average star had a radius of $100\ R_\odot$? How would increasing the luminosity of the average star to $10\ L_\odot$ change the result of Eq. (6.14)?

11. Within the last few years several quasistellar objects have been found that have values of z around 2.

    a. What are their recessional velocities?

  b. How far are they from the Earth according to Hubble's Law?
  c. How long ago was the light we received from them emitted?
12. Based on the current value of Hubble's constant, how fast is a galaxy receeding that is 10^{24} m from us?
13. Not long ago Hubble's "constant" was determined to be 26 km/sec/10^6 LY. Using this number, what is an estimate for the age of the universe?
14. Initially the galactic distance measurements were *underestimated* by a factor of 10. How did this effect Hubble's constant and the estimated age of the universe?
15. Show that Eq. (6.19) follows from Eq. (6.18).
16. What will be the value of Hubble's constant in 10^8 years? What was the value of Hubble's constant 10^8 years ago?
17. Show that Eq. (6.24b) follows from Eq. (6.24a).
18. From curves similar to those shown in Fig. 6.10a, astronomers can put an upper limit on the value of q_0 of 2. Based on this observation, what is the maximum present mass density of the universe?
19. Show how $TV^{2/3} = \text{const}$ follows from $PV^{5/3} = \text{const}$. Are the two constants the same?
20. Say that we have two cylinders, both at $T = 400$ °K, one filled with point particles and the other with photons. If each is allowed to expand adiabatically to eight times its original volume, what would be the final temperature of each cylinder?
21. •Estimate the energy density of the 3 °K fireball at a wavelength of 2 cm. Compare your answer to the value obtained from the black body curve in Fig. 6.13.

CHAPTER 7

Cosmology II—Evolution of the Big Bang

But do not think, O patient friend,
Who reads these stanzas to the end,
That I myself would glorify . . .
You're just as wonderful as I,
And all Creation in our view
Is quite as marvellous as you.

Robert Service, *The Wonderer*

7.1. In The Beginning . . .

The competition among cosmological models has left us with only one reasonable picture of how it was, how it is and how it will all end—the big bang cosmology. This appears to be the only cosmology at the present time which survives the constraints of simple observations (Sec. 6.1), the Hubble expansion (Sec. 6.2), and the cosmic fireball (Sec. 6.3). What is more, as we shall see, the model successfully explains the formation of the nuclear elements from the fireball (primordial nucleosynthesis) in a manner consistent with the present mass density of the universe, $\rho_0 \sim 2 \times 10^{-28}$ kg/m^3, present temperature of the fireball, 3 °K, and present known concentrations of light elements $A < 7$ (most of the heavier elements presumably are created in supernovae explosions).

We begin, therefore, in the beginning.

7.1.1. Thermal Equilibrium and Particle Creation and Annihilation

Leaving aside any conjecture as to the origin of the primordial components, we now believe that at time $t = 0$, a "singularity in space-time" exploded into a hot dense gas (primordial fireball) composed of radiation and matter–antimatter in thermal equilibrium with one another. The radiation was made up of massless photons, neutrinos, antineutrinos, and perhaps gravitons (the latter generating the gravitational force), whereas the matter–antimatter consisted of massive elementary particles such as weakly interacting leptons (electrons, muons) and strongly interacting hadrons (protons, neutrons, hyperons, pions, kaons, etc.) and perhaps even more basic constituents of matter called quarks.

Obviously this hot cauldron of radiation and particles is too complicated to understand in detail, but assuming that gravity was still controlling the expansion of the universe even at these early times, we presume an expansion time of the form $t_{\text{exp}} \sim (G\rho)^{-1/2} \propto n^{-1/2}$ (recall the discussion in Sec. 6.1.5). On the other hand, particles were interacting with one another and with radiation with a mean free path $l = (n\sigma)^{-1} \propto n^{-1}$. Then, since $l/t_{\text{exp}} \propto n^{-1/2} \to 0$ for very large initial number densities, $l \ll ct_{\text{exp}}$. This indicates that the mean free path of interactions was much less than the radius of the universe at that time, or equivalently, the reaction rate $t_{\text{reac}}^{-1} = c/l$ was much greater than the expansion rate of the universe t_{exp}^{-1}.

Such inequalities define what is meant by thermal equilibrium, for then there were many reactions, say, of the pair annihilation and creation of the form $A + \bar{A} \to B + \bar{B}$ as well as their inverse, $B + \bar{B} \to A + \bar{A}$, and A and B remained at the same temperature. Thermal equilibrium allows us to describe the kinetic energy of such particles in terms of a single temperature $\text{KE} \sim kT$, regardless of the detailed dynamics underlying such processes. As the universe expands, however, the value of n decreases. Since $l/t_{\text{exp}} \propto n^{-1/2}$, given enough time $l \gg ct_{\text{exp}}$ and the mean free path of the interactions will become greater than the radius of the universe. Under these conditions the interactions no longer take place and the particle types A and B are effectively decoupled. Since the particles are then no longer in thermal equilibrium they will be at different temperatures during the remainder of the evolution of the universe.

Furthermore, as the universe expands initially relativistically ($T \propto 1/R$), the temperature eventually drops to the point where

$$kT \sim \text{KE}_A \leqslant m_B c^2 - m_A c^2, \tag{7.1}$$

in which case the process $A + \bar{A} \to B + \bar{B}$ is forbidden to continue by energy conservation if $m_B > m_A$, while the inverse reaction still proceeds. For example, if A are photons and B nucleons, we eventually reach a temperature (as we shall see shortly) where the photons are no longer energetic enough to produce nucleon–antinucleon pairs; the nucleons are quite literally "frozen" out of the photon sea. Thus, as the universe expands, one by one the heavier particles are frozen out of the hot interacting gas, remaining either fixed in number (possibly zero if there were an equal number of B and $\bar{B}$ initially) or disappearing by subsequent decays.

7.1.2. Hadron Era

The first particles to be frozen out of the fireball are the strongly interacting hadrons. In Table 7.1 we list a few of the primary elementary particles, their masses and lifetimes. For $\mathrm{kT} \gtrsim \mathrm{M}_{\mathrm{had}} c^2$, there are many many hadrons and antihadrons in the hot gas. But as kT drops below

$$\mathrm{kT} \sim m_\pi c^2 \sim 140 \text{ MeV} \qquad \text{or} \qquad T \sim 10^{12} \text{ °K},$$

the pion, which is the lightest hadron, can no longer be produced. Through annihilation and decay, it will soon disappear, so that the only hadrons remaining are the stable proton and the (relatively) stable neutron, in thermal equilibrium with the photon and weakly interacting leptons.

While this era is obviously very complicated, thermal equilibrium has allowed us to skirt the complexities, emerging at $T \sim 10^{12}$ °K with charged particles p, $e^\pm$, $\mu^\pm$ and the neutrals γ, ν, $\bar{\nu}$, and n. One of the interesting complexities not understood is why at that time matter, in the form of baryons p and n, has been favored over antimatter, the antibaryons $\bar{p}$ and $\bar{n}$. Since baryon number is conserved (at least in our laboratories), this asymmetry has been preserved down to the present time. Recent research suggests that interactions at the quark level may be responsible for this asymmetry in the early universe. Our estimate in chapter 6 that there are now approximately 10^{78} nucleons in the universe (a big bang universe of finite extent) means that in the hadron era, say at $kT \sim 10^{14}$ °K, there were many more baryons and also antibaryons than 10^{78}, about the same number as photons, but always 10^{78} more baryons than antibaryons.

TABLE 7.1

Particle properties

		Particles	Spin	Mass(MeV/c^2)	Lifetime(sec)
Hadrons	Baryons	p	1/2	938.28	stable
		n	1/2	939.57	10^3
	Mesons	$\pi^+\pi^-$	0	140	10^{-8}
		π^0	0	135	10^{-16}
	Leptons	μ^+, μ^-	1/2	106	10^{-6}
		e^+, e^-	1/2	0.51	stable
		$\nu, \bar{\nu}$	1/2	~ 0	stable
	Photon	γ	1	0	stable

At the end of the hadron era, the dominant reactions are electromagnetic, $\gamma + \gamma \rightarrow e^+ + e^-$ and bremsstrahlung processes such as $e \rightarrow e + \gamma$ and $\gamma + e \rightarrow e$ in the presence of an external field or force. These processes are in equilibrium so that charge conservation requires that the number of photons is also conserved. Moreover, the maximum available energy of the bremsstrahlung photons is related to the temperature of the fireball. That is, $E_\gamma = \mathrm{KE}_i - \mathrm{KE}_f$, where i and f refer to the initial and final states of the charged particles. The photon energy will be a maximum when the KE_f is zero; i.e., $E_{\gamma,\max} = \mathrm{KE}_i$, and averaging over all the charged particles we have

$$\langle E_{\gamma,\max} \rangle = \langle \mathrm{KE}_i \rangle \sim kT. \tag{7.2}$$

Since only photons are in equilibrium with leptons, this equation expresses what we mean by equilibrium once strong interaction processes are suppressed for $kT < m_\pi c^2 \sim 140$ MeV.

An important quantity is the ratio of photons to nucleons at the end of the hadron era. Since the photon energy density E_γ / V obeyed a black body law $\sim aT^4$, the number density was of order

$$n_\gamma \sim (E/V)_\gamma / \langle E_{\gamma,\max} \rangle \sim aT^4/kT = \frac{a}{k} T^3. \tag{7.3}$$

(This relation implies that the number of photons remains constant. To see this we note that $N_\gamma = n_\gamma V$, which from (7.3) yields $N_\gamma \propto T^3 V$. For a photon-dominated fireball $T \propto 1/R$ and therefore $V \propto T^{-3}$. Hence $N_\gamma \propto T^3 T^{-3} \propto$ constant.) Jumping ahead to the present, we try to apply the above formula for n_γ^0 via the present fireball temperature $T_0 \approx 2.7$ °K,

$$n_\gamma^0 \sim \frac{7.6 \times 10^{-16} (2.7)^3}{1.4 \times 10^{-23}} \sim 10^9 (\mathrm{photon/m^3}),$$

whereas the nucleon number density is now

$$n_N^0 = \rho_0 / m_N \sim \frac{10^{-28}}{10^{-27}} \sim \frac{1}{10} (\mathrm{nucleon/m^3}).$$

Since both number densities n_γ and n_N decrease as $V^{-1} \propto T^3$, their present ratio is conserved for all time; it is

$$n_\gamma^0 / n_N^0 \sim aT_0^3 / k n_N^0 \sim 10^9/(1/10) \sim 10^{10} \text{ (photon/nucleon)}. \tag{7.4}$$

The fact that the ratio is so large is the important result. The bremsstrahlung process keeps the free, nonrelativistic nucleons at the end of the hadron era and thereafter ($kT < 100$ MeV) in thermal equilibrium with the photon fireball. We know an isolated adiabatically expanding nucleon gas cools down like $T_m \propto 1/R^2$ (where T_m stands for matter temperature), whereas a photon gas behaves as $T_\gamma \propto 1/R$ (recall the discussion in

Sec. 6.3.3). But *together* in thermal equilibrium $T_m = T \propto 1/R$ because the gas is very "hot," with the 10^{88} photons holding up the 10^{78} nucleons at the higher temperature. The fireball continues to cool as a single black body radiator specified by the temperature $T \propto 1/R$ which links the energetics to the spatial properties of the system in a simple manner. If instead it turned out that there were as many nucleons as photons in the fireball at the end of the hadron era, the equilibrium temperature would fall off somewhere between R^{-1} and R^{-2}, but it could not be determined in any theoretical way. The fireball temperature would not correspond to a simple black body radiator and there would be very little we could learn about the quantitative evolution of the universe for $T < 10^{12}$ °K.

Very recently astro- and elementary particle physicists have attempted to explain both the observed baryon–antibaryon asymmetry and photon to baryon ratio (7.4) on the basis of "proton and antiproton" decays $p \to \pi^0 + e^+$, $\bar{p} \to \pi^0 + e^-$. Such reactions violate both baryon and lepton (but not fermion) number but it is thought that these decays have an extremely long lifetime (10^{31} yr). Thus on the time scale of the universe of 10^{10} years, the proton certainly appears to be stable.

Then as the scenario goes, at the very early time of 10^{-36} sec, i.e., fireball temperature of 10^{28} °K (see (7.6)), the antiproton decay is favored over the proton decay. The result is that a net excess of $\sim 10^{78}$ protons is generated, with a corresponding excess of electrons from $\bar{p} \to \pi^0 + e^-$ decay. These electrons annihilate against the positrons in the fireball—leaving in turn a net excess of $\sim 10^{10}$ photons relative to the net proton number. Although these ideas may seem farfetched, such a theory may in fact be partially tested in the near future when the experimental upper bound for the "stable" proton lifetime is "lowered" to the 10^{31} yr level.

But whatever its origin, the first immediate consequence of our (observed) hot universe is that we can relate the current mass and number densities (ρ_0 and n_0) to the same variables in any earlier era (ρ and n). Because nucleon number is then conserved, we know by definition that $\rho, n \propto 1/R^3$. Since R is always proportional to $1/T_\gamma$, even up to the present time (R is not now proportional to $1/T_m$), we can express ρ and n as follows:

$$\rho = \rho_0 (T/T_0)^3, \qquad n = n_0 (T/T_0)^3. \tag{7.5}$$

A second consequence is that the present fireball remains a single black body radiator, allowing detection as in the manner of Penzias and Wilson.

7.1.3. The Time-Temperature Relation

At the end of the hadron era, the equilibrium reaction $\gamma + \gamma \to \text{lepton} + \text{antilepton}$ requires the hot fireball to consist of roughly equal numbers of interacting particles

$$N_\gamma \approx N_{e^-} \approx N_{e^+} \approx N_{\mu^+} \approx N_{\mu^-} \approx N_{\nu_e} \approx N_{\bar{\nu}_e} \approx N_{\nu_\mu} \approx N_{\bar{\nu}_\mu},$$

plus a "few" nucleons of little consequence to the energy density of the gas. In the above relationships between the various numbers of particles we have noted the experimental fact that the neutrinos associated with the electrons are different from those associated with the muon. In reactions involving the above particles the charged lepton and its associated neutrino always appear together (e with ν_e; μ with ν_μ) so that the lepton number of each lepton–neutrino pair is conserved. In this regard note that the e^-, ν_e, μ^- and ν_μ have a positive lepton number while the e^+, $\bar{\nu}_e$, μ^+, and $\bar{\nu}_\mu$ have a negative lepton number.

The energy density of the photons has the black body form, which at $T \approx 10^{12}$ °K corresponds roughly to the energy density of a neutron star,

$$(E/V)_\gamma = aT^4 \sim 10^{33}\ \mathrm{J/m^3},$$

while the nucleon energy density is a little less than that of a white dwarf star,

$$(E/V)_N = \rho_0 c^2 (T/T_0)^3 \sim 10^{-10} (E/V)_\gamma \sim 10^{23}\ \mathrm{J/m^3}.$$

On the other hand, the leptons all have spin $\frac{1}{2}$, while the photon has spin 1. The Pauli exclusion principle then restricts the number of available states allowed for leptons in the fireball, with the result that the relativistic leptons at $kT \sim 100$ MeV all have a black body type energy density reduced by a factor of $\frac{7}{8}$ (due to a simple integral which need not concern us here),

$$(E/V)_{e^-} = (E/V)_{e^+} = (E/V)_{\mu^-} = (E/V)_{\mu^+} = \left[(E/V)_{\nu_e} + (E/V)_{\bar{\nu}_e}\right]$$
$$= \left[(E/V)_{\nu_\mu} + (E/V)_{\bar{\nu}_\mu}\right] = \tfrac{7}{8} aT^4,$$

where because of the special nature of the neutrino spin the sum of the neutrino and antineutrino energy densities has the same value as the energy density of an individual charged lepton. Henceforth for simplicity we shall combine lepton number according to $N_{\nu_e} + N_{\nu_\mu} \to N_\nu$. The total energy density of the fireball is then

$$(E/V)_{\mathrm{tot}} = (1 + \tfrac{7}{8} \cdot 6) aT^4 = \tfrac{25}{4} aT^4 \approx 6aT^4.$$

Recently a heavy τ lepton and perhaps an associated neutrino have been detected, but since they increase the above $(E/V)_{\mathrm{tot}}$ only slightly, we shall neglect them (see problem 6).

We are now in a position to estimate the expansion time of the fireball (i.e., the universe) at these early times. Recall from Sec. 6.1.5 that in the early universe the expansion time was $t_{\mathrm{exp}} \sim (G\rho)^{-1/2}$. For the relativistic photon and lepton fireball at the end of the hadron era the effective mass density must be replaced by $(E/V)/c^2$, giving an expansion time

$$t_{\rm exp} \sim \sqrt{\frac{c^2}{GE/V}} \sim \sqrt{\frac{c^2}{6GaT^4}} \sim \sqrt{\frac{9\times10^{16}}{6\cdot7\times10^{-11}\cdot8\times10^{-16}T^4}}$$

or

$$t_{\rm exp} \sim \left(\frac{10^{10}\ ^\circ{\rm K}}{T}\right)^2. \tag{7.6}$$

Aside from slight numerical corrections, this time-temperature relation is the same as obtained using general relativity. While we have derived it at $T \sim 10^{12}$ °K, it is approximately valid for T in the range 10^{12} °K $\gtrsim T \gtrsim 10^3$ °K, corresponding to the radiation-dominated period of the universe.

To summarize then using (7.6), 10^{-6} sec into the big bang the fireball temperature has dropped to $T \sim 10^{13}$ °K with $kT \sim 10^3$ MeV high enough to create hadrons spontaneously, which in turn are almost immediately annihilated. In our laboratories all hadrons (except the proton and neutron) decay in 10^{-23}–10^{-8} sec, so by $kT \sim 10^3$ MeV only the light mesons are around to interact with the nucleons. At the end of the hadron era, $kT \sim 10^2$ MeV, $T \sim 10^{12}$ °K and $t \sim 10^{-4}$ sec, the light mesons also disappear.

7.1.4. Lepton Era

At the beginning of the lepton era, there are about six times as many leptons (e^-, μ^-, ν) and antileptons (e^+, μ^+, $\bar{\nu}$) as there are photons in the universe. Almost immediately, however, the muons $\mu^\pm$ disappear via the reaction $\mu^+ + \mu^- \rightarrow 2\gamma$ for $kT \approx m_\mu c^2 \approx 106$ MeV, or $T \sim 10^{12}$ °K. As there are then only four leptons and antileptons left obeying $E/V = \frac{7}{8}aT^4$, while the photons obey $E/V = aT^4$, the total energy density of the fireball at that time is $(E/V)_{\rm tot} = (1 + \frac{7}{8} \times 4)aT^4 = \frac{9}{2}aT^4$, which only slightly alters the early time-temperature relation, $t_{\rm exp} \sim (10^{10}/T)^2$. Thus the time elapsed into the big bang is still $t \sim 10^{-4}$ sec.

The next leptons to be frozen out of the fireball are the neutrinos and antineutrinos. Energetically one might suppose that $e^+ + e^- \rightarrow \nu + \bar{\nu}$ would not shut off until kT $\sim \frac{1}{2}$MeV (or $T \sim 5 \times 10^9$ °K). We show below, however, that this process shuts off at a temperature that is a factor of 10 higher because at the higher temperature (and earlier times) the mean free path of the interaction l becomes greater than the size of the universe R_u.

The size of the universe is related to the expansion time by $R_u \sim ct_{\rm exp}$ and since $t \sim 10^{20}T^{-2}$ we have in MKS units

$$R_u \sim 10^{28}T^{-2} \text{ meters.}$$

On the other hand, the mean free path for these weak interactions is

$l = 1/n\sigma$ where for either electrons or neutrinos, the number density of target particles at temperature T is

$$n \sim aT^4/kT \sim (10^{-15}/10^{-23})T^3 = 10^8 T^3 \text{ (meters)}^{-3}.$$

Also the weak lepton interaction cross section is thought to behave as

$$\sigma \sim 10^{-47}(kT/m_e c^2)^2 \sim 10^{-68} T^2 \text{ meters}^2$$

in MKS units (note that at $kT \sim m_e c^2 \sim \frac{1}{2}$ MeV, $\sigma \sim 10^{-47}$ m^2, similar in size to the weak cross section encountered for solar-neutrino detection). Consequently $l \sim 10^{60} T^{-5}$, and when $l \gtrsim R_u$ both processes $e^+ + e^- \leftrightarrow \nu + \bar{\nu}$ are shut off because the particles can transverse the universe without interacting; that is,

$$10^{60} T^{-5} \gtrsim 10^{28} T^{-2},$$

leading to

$$T^3 \lesssim 10^{32} \qquad \text{or} \qquad T \lesssim 5 \times 10^{10}\ {}^\circ\text{K}.$$

At this point the neutrinos and antineutrinos are not annihilated into oblivion; instead they simply stop annihilating and just expand adiabatically as $T \propto 1/R$ along with the other components of the fireball. In Fig. 7.1 the dotted curve represents the T^{-2} falloff of $R_u \sim ct_{\text{exp}}$, the radius of the expanding universe. Curve l_A represents the T^{-5} dependence of the mean free path for $e^+ + e^- \leftrightarrow \nu + \bar{\nu}$. At $T \sim 5 \times 10^{10}$ °K they intersect; thereafter $l_A > R_u$ and this process is then shut off.

The lepton era comes to an end when $kT \sim \frac{1}{2}$ MeV or equivalently $T \sim 5 \times 10^9$ °K, for then $\gamma + \gamma \rightarrow e^+ + e^-$ no longer goes while its inverse $e^+ + e^- \rightarrow \gamma + \gamma$ proceeds to annihilate all remaining positrons. It is again

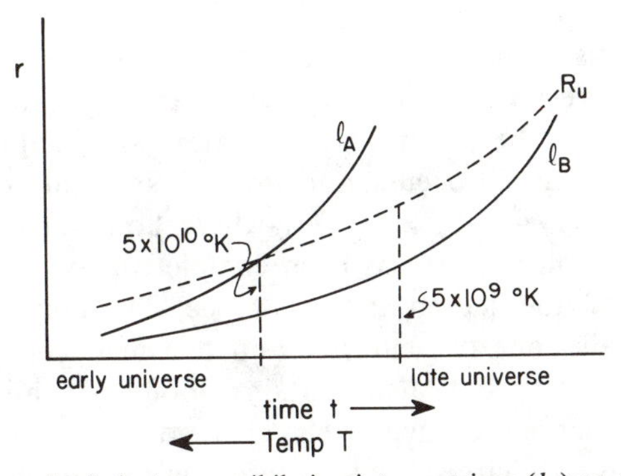

Fig. 7.1. Mean free path of electron annihilation into neutrinos (l_A) vs. temperature (expansion time) of universe. Mean free path of electron annihilation into photons (l_B) vs. temperature (expansion time) of universe.

of interest to ask, as in the neutrino case, if this latter reaction goes without interruption, or is it suppressed until the universe expands sufficiently to encompass its mean free path. For $T \sim 5 \times 10^9$ °K, we have from (7.6),

$$t_{\text{exp}} \sim (10^{10}/5 \times 10^9)^2 \sim 4 \text{ sec}, \qquad R_u \sim ct_{\text{exp}} \sim 10^9 \text{ m},$$

whereas the electromagnetic cross section $\sigma \sim 10^{-28}$ m^2 and number density $n \sim 10^8 (5 \times 10^9)^3 \sim 10^{37}$ electrons/m^3 imply

$$l = \frac{1}{n\sigma} \sim \frac{1}{10^{37} 10^{-28}} \sim 10^{-9} \text{ m}.$$

Clearly $l \ll R_u$, so the reaction $e^+ + e^- \rightarrow 2\gamma$ proceeds rapidly and the positrons soon disappear. This is depicted in Fig. 7.1 by the dotted R_u line and l_B. Note that the cross-over point does not occur in the lepton era. For T between 5×10^{10} °K and 5×10^9 °K, the rapid annihilation $e^+ + e^- \rightarrow 2\gamma$ heats up the photon fireball relative to the neutrino fireball because the latter, no longer in thermal equilibrium with the fireball electrons, cannot be similarly heated. In fact it is not hard to show that $T_\gamma \approx 1.4\, T_\nu$.

The important point in the derivation of $T_\gamma \approx 1.4 T_\nu$, is that the number of electrons plus positrons plus photons before freeze out equals the number of photons after annihilation is completed. The reason is that for each e^+, e^- pair that annihilates, *two* photons are produced. Since the number of particles is proportional to the energy of the particles divided by their average energy we have $N_{\text{before}} = N_{\text{after}}$ or

$$\left(\frac{E}{\langle E \rangle} \right)_{\text{before}} = \left(\frac{E}{\langle E \rangle} \right)_{\text{after}}$$

$$\left(\frac{(1 + 2 \times \frac{7}{8}) a T^4 V}{kT} \right)_{\text{before}} = \left(\frac{a T'^4 V'}{kT'} \right)_{\text{after}}, \tag{7.7}$$

where we have used the appropriate energy density for each object and where T and V are the photon temperature and volume before freeze out and T' and V' are the same quantities after annihilation is over. Since $T_\gamma = T_\nu$ before freeze out occurs, and since T_ν smoothly falls as $1/R$, (7.7) reduces to $T'_\gamma = (11/4)^{1/3} T'_\nu \approx 1.4\, T'_\nu$, where both temperatures are now after the annihilation process has been completed.

At the end of the lepton era, $t \sim 4$ sec, the universe contains two concentric fireballs, one of neutrinos and one made up primarily of 10^{88} photons. The latter fireball also contains about 10^{78} nucleons and more than half that number of negatively charged electrons—just enough to cancel the positively charged protons resulting in a universe with zero net charge.

7.1.5. Neutral Massive Leptons

As stressed in Sec. 6.2.4, perhaps the most interesting problem in cosmology is the "missing mass"—whether somewhere in space there is sufficient undetected mass to close the universe. One such possibility is a neutral massive lepton which we shall denote by L_0 (neutrinos are, we believe, massless and electrons are charged). It turns out, as we shall now show, that cosmology places a limit on the mass of such a beast.

Since L_0 is a neutral lepton, it will only interact weakly with its environment. This means that at a temperature below 5×10^{10} °K it will decouple from neutrinos, just as electrons do, so that $L_0 + \bar{L}_0 \leftrightarrow \nu + \bar{\nu}$ will no longer "go" in either direction. If the mass of L_0 is greater than $k \cdot 5 \times 10^{10}$, that is

$$m_{L_0}c^2 > kT = k(5 \times 10^{10}\ °\mathrm{K}) = 0.5\ \mathrm{MeV},$$

then the production of L_0 via $\nu + \bar{\nu} \rightarrow L_0 + \bar{L}_0$ will have stopped *before* the interaction turns off, leaving only the annihilation reaction $L_0 + \bar{L}_0 \rightarrow \nu + \bar{\nu}$ which will continue until T falls to 5×10^{10} °K. As a result, the neutral massive leptons will for all intent and purpose, disappear. If, however, the mass of L_0 is less than 0.5 MeV, then $\nu + \bar{\nu} \rightarrow L_0 + \bar{L}_0$ will stop *after* the temperature falls below 5×10^{10} °K, and therefore the two-way reaction will continue right up to the moment that the mean free path finally becomes greater than R_u. As a result, L_0 and the neutrinos remain in equilibrium at the neutrino density until they and the neutrinos decouple and then since no further interactions take place, the densities of L_0s will remain at the neutrino density down to the present time.

Conceivably then, if $m_{L_0}c^2 < 0.5$ MeV, the neutral massive lepton might not only exist, it could, if it has sufficient mass, close the universe. All that is required according to (6.21) is that $\rho_{L_0} > \rho_c \approx 6 \times 10^{-27}$ kg/m^3. Since $n_{L_0} = n_\nu$, we have $\rho_L = n_\nu m_{L_0}c^2/c^2$, and we can solve explicitly for the minimum value of $m_{L_0}c^2$ needed for closure,

$$\left(m_{L_0}c^2\right)_{\mathrm{closure}} > \rho_c c^2/n_\nu. \tag{7.8}$$

With a present photon density of $\sim 4 \times 10^8/\mathrm{m}^3$ and with $n_\nu \approx 0.6 n_\gamma$ (see problem 10), (7.8) yields a minimum closure mass of ~14 eV, which is indeed small!

Or instead we could turn the tables on (7.8) and argue that the upper bound for the present cosmic density found from (6.23) to be $\rho_0 \sim 2 \times 10^{-26}$ kg/m^3 (determined by the *observed upper limit* on the deceleration parameter $q_0 \lesssim 2$) requires that $m_{L_0} \lesssim 60$ eV. So far no such leptons have been detected in Earth laboratories.

We did mention at the end of Sec. 4.1.3 that neutrinos may have a mass.

If they do, the neutrinos become our neutral massive leptons and the arguments in the preceding paragraph would place an upper limit on the sum of the masses of the various types of neutrinos at less than 60 eV/c^2.

7.2. Primordial Nucleosynthesis

We now shift our attention away from the dominant photon and lepton fireballs and return to the nonrelativistically expanding cool nucleon gas. While this matter can no longer be destroyed once out of the hadron era, it can shift its appearance by virtue of weak interaction processes such as β decay, $n \rightarrow p + e^- + \bar{\nu}$, which is driven spontaneously by the known n-p rest-energy difference

$$Q = m_n c^2 - m_p c^2 = 939.57 \text{ MeV} - 938.28 \text{ MeV}$$
$$= 1.29 \text{ MeV}.$$

At the same time the neutron and proton can form the deuteron, the only stable n-p bound state with binding energy

$$BE_d = m_n c^2 + m_p c^2 - m_d c^2 = 2.22 \text{ MeV}.$$

Because these two energies, 1.29 MeV and 2.22 MeV, are quite close to the rest-mass energy released in electron–positron annihilation $2 \cdot m_e c^2 \approx$ 1 MeV, we expect deuteron production and e^+e^- annihilation to be competing with one another. To complicate matters further, as soon as a deuteron is formed it tries to combine again with still other nucleons and deuterons to form a more stable bound state, the α particle or helium nucleus with binding energy

$$BE_\alpha = 2m_n c^2 + 2m_p c^2 - m_\alpha c^2 = 28.34 \text{ MeV}.$$

This corresponds to $BE/A = 28.34/4 \approx 7.1$ MeV, so that the α is tightly bound. The deuteron, on the other hand, has $BE/A \approx 1$ MeV and is loosely bound (see Fig. 4.1).

To simplify the following discussion, we shall consider each of these three processes, n-p conversion, deuteron production, and helium production in turn in order to probe the implications of primordial nucleosynthesis.

7.2.1. Neutron-Proton Conversion

At the beginning of the lepton era with $T \sim 10^{12}$ °K, the nucleon gas contains an equal number of protons and neutrons. However, by the middle of the lepton era at $T \sim 5 \times 10^{10}$ °K, the ratio of neutrons to protons drops from 1 to about 0.74. This shift is caused by the slight asymmetry in the n and p masses for the three sets of *competing* weak interaction processes converting n to p and p to n:

$$n \to p + e^- + \bar{\nu} \quad \text{and} \quad p \to n + e^+ + \nu,$$
$$\nu + n \to e^- + p \quad \text{and} \quad \bar{\nu} + p \to e^+ + n$$
$$e^+ + n \to \bar{\nu} + p \quad \text{and} \quad e^- + p \to \nu + n.$$

We want to find the ratio of neutrons to protons in the final state, n/p. In order to do so we must compare heavier neutrons and lighter protons at the same *final* kinetic energy. *Initial* protons in $p \to n$ reactions must then be at a higher kinetic energy than initial neutrons in $n \to p$ reactions by an amount $\Delta \mathrm{KE} = Q = 1.29$ MeV. The final state number ratio is then given by the Boltzmann factor for the particles in the *initial* state,

$$\frac{n}{p} = \left(\frac{N_n}{N_p} \right)_f = \frac{e^{-\mathrm{KE}_{p_i}/kT}}{e^{-\mathrm{KE}_{n_i}/kT}} = e^{-Q/kT}.$$

Thus, for $T \sim 10^{12}$ °K or $kT \sim 10^2$ MeV, we have

$$n/p = e^{-0.0129} \approx 1 - 0.013 = 0.987,$$

whereas at $T \sim 10^{10}$ °K or $kT \sim 1$ MeV, the exponential is $\exp(-1.29) \approx 0.28$, a drastic reduction indeed! A more detailed calculation of this rapidly varying exponential at $T = 1 \times 10^{10}$ °K leads to the ratio

$$(n/p)_{T=10^{10}\,°\mathrm{K}} = e^{-1.29/0.866} \approx 0.225.$$

This latter temperature is singled out because it is the cross-over point (recall Fig. 7.1) for the reversible weak interactions $\nu + n \rightleftarrows p + e^-$ and $e^+ + n \rightleftarrows p + \bar{\nu}$. That is, these processes will cease when the mean free path $l = 1/n\sigma$ becomes greater than the size of the universe, $R_u \sim ct_{\mathrm{exp}} \sim 10^{28} T^{-2}$. As we saw in Sec. 7.1.4, the number density of these relativistic leptons is $n_l \sim 10^8 T^3$, while the weak cross section is only slightly different for nucleons than for leptons,

$$\sigma \sim 10^{-47} \left(\frac{kT + Q}{1\ \mathrm{MeV}} \right)^2 \text{meters}^2$$

in MKS units. Then $l = R_u$ requires a value for T of about 10^{10} °K.

Given $T \approx 10^{10}$ °K and $n/p \approx 0.225$ when $l \approx R_u$, we can continue to trace the behavior of n/p for $T < 10^{10}$ °K by first converting this number ratio to a mass fraction

$$(X_n)_{T=10^{10}\,°\mathrm{K}} = \frac{N_n m_n}{N_n m_n + N_p m_p} \approx \frac{n/p}{1 + n/p} \approx 0.18.$$

For $T < 10^{10}$ °K, l increases faster than $R_u \propto T^{-2}$ so $l > R_u$ effectively stops the slippage of n to p. The only remaining process then free to proceed of its own accord is β decay, $n \to p + e^- + \bar{\nu}$, driven again because $m_n > m_p$. The theory of radioactive decay then tells us that at a time later than $t \sim 1$ sec, corresponding to $t \sim (10^{10}/T)^2$ sec for $T \sim 10^{10}$ °K, the mass fraction decays exponentially:

$$X_n(t) = (X_n)_{T \approx 10^{10}\,°\mathrm{K}} e^{-t/\tau} \approx 0.18 e^{-t(\mathrm{sec})/10^3}, \tag{7.9}$$

where the neutron lifetime is measured to be $\tau \approx 10^3$ sec. We shall return to this calculation shortly.

7.2.2. Deuteron Production

About the same time that the neutron mass fraction begins to diminish appreciably, the remaining neutrons interact strongly with protons to form the deuteron and also photons via

$$n + p \to d + \gamma.$$

Roughly speaking, the inverse process $\gamma + d \to n + p$ is frozen out when the fireball cools down *somewhat below* $\langle E_{\gamma,\max} \rangle \sim kT \sim \mathrm{BE}_d \approx 2.22$ MeV or $T \sim 2 \times 10^{10}$ °K. To see why the freeze out is at a lower temperature, we compute the mean free path for deuteron production and breakup by $l = 1/n\sigma$, where measurements indicate that $\sigma \sim 10^{-32}$ m^2 for both processes and the present number density of nucleons, $n_0 \sim 10^{-1}/\mathrm{m}^3$, requires that $n = n_0(T/T_0)^3 \sim 10^{28}/\mathrm{m}^3$ at $T \sim 10^{10}$ °K. Then $l \sim 10^4$ m $\ll R_u \sim ct_{\exp}$ at $t \sim 1$ sec. Because there are so many interactions within the size of the universe at $T \sim 10^{10}$ °K, even when the average kinetic energy (i.e., kT) of the photons drops below $\mathrm{BE}_d \sim 2$ MeV, there is still the photon fraction $\exp(-\mathrm{BE}_d/kT)$ on the high-energy Boltzmann tail of the black body distribution which has enough energy to drive the breakup reaction $\gamma + d \to n + p$.

At what lower temperature T^* does this breakup reaction freeze out, allowing the deuterons to form unmolested? To determine this reduced temperature we investigate the rate densities for the deuteron production $\mathcal{R}_{\mathrm{I}}$ and breakup $\mathcal{R}_{\mathrm{II}}$. We recall from Sec. 2.3.2 the definition of rate as number density times relative velocity (flux) times cross section. In this case,

$$\mathcal{R}_{\mathrm{I}} = n_n n_p v_{np} \sigma_{\mathrm{I}} \tag{7.10a}$$

$$\mathcal{R}_{\mathrm{II}} = n_\gamma n_d c \sigma_{\mathrm{II}} e^{-\mathrm{BE}_d/kT}, \tag{7.10b}$$

where the exponential factor in $\mathcal{R}_{\mathrm{II}}$ corresponds to the high-energy tail of the black body curve allowing the numerous photons ($n_\gamma/n_N \sim 10^{10}$) to drive the process even when the average photon energy drops below kT. In equilibrium these total rate densities are equal: $\mathcal{R}_{\mathrm{I}} = \mathcal{R}_{\mathrm{II}}$.

To proceed further, we note that the product σv is equivalent to a two particle reaction rate, Γ_{12}, times the "available states in momentum space"

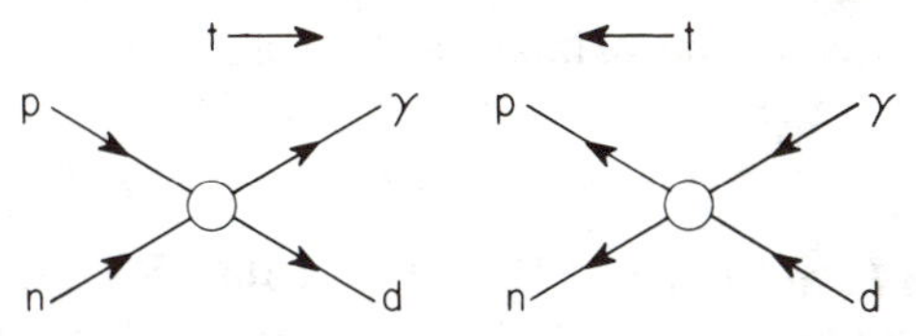

Fig. 7.2. Diagram of reaction production $(p + n \rightarrow \gamma + d)$ and dissociation$(\gamma + d \rightarrow p + n)$.

of the two *outgoing particles*. In Sec. 4.3.3 we learned that these available states are proportional to the "volume" $4\pi p^3/3$ for each particle. Thus we have

$$\sigma_{\mathrm{I}} v_{np} \propto \Gamma_{np}(p_d p_\gamma)^3$$

$$\sigma_{\mathrm{II}} c \propto \Gamma_{\gamma d}(p_n p_p)^3.$$

Then we invoke *time reversal invariance* (detailed balance) to equate the simple two particle reaction rates for the production and breakup processes as shown in Fig. 7.2. That is, we set $\Gamma_{np} = \Gamma_{\gamma d}$ in the above equations to find

$$\frac{\sigma_{\mathrm{I}} v_{np}}{\sigma_{\mathrm{II}} c} \approx \frac{(p_d p_\gamma)^3}{(p_n p_p)^3}.$$

Combining this result with the equilibrium condition $\mathcal{R}_{\mathrm{I}} = \mathcal{R}_{\mathrm{II}}$ and (7.10), we obtain the number fraction

$$\frac{n_n}{n_d} \approx \left(\frac{p_n p_p}{p_d p_\gamma}\right)^3 \frac{n_\gamma}{n_p} e^{-\mathrm{BE}_d/kT}, \tag{7.11}$$

where the cube of the momentum factors then corresponds to $\sigma_{\mathrm{II}} c / \sigma_{\mathrm{I}} v_{n_p}$. In equilibrium, the nonrelativistic neutrons, protons, and deuterons have $KE = p^2/2m \sim kT$ or $p \sim \sqrt{2mkT}$, while for the relativistic photon $E = pc \sim kT$. Making these substitutions into (7.11), recalling that $n_\gamma / n_p \approx 10^{10}$ in the universe and taking $n_n / n_d \sim 1$ so that $\ln(n_n / n_d) \sim 0$ [the latter certainly an accurate approximation scaled to $\ln(n_\gamma / n_p) \approx 23$], the natural logarithm of (7.11) yields,

$$0 \approx \tfrac{3}{2} \ln(m_n c^2 / kT^*) + \ln(n_\gamma / n_p) - \mathrm{BE}_d / kT^*. \tag{7.12}$$

In order to solve this equation for T^*, we shall use a method of successive approximations. As our first approximation we set $kT^* = \mathrm{BE}_d$ (i.e. $T^* \approx 2.6 \times 10^{10}$ °K) and substitute this into the *first* term in (7.12). Evaluating the logarithm terms we obtain from (7.12),

$$0 \approx 9 + 23 - \mathrm{BE}_d / kT^*.$$

Solving, we find that $kT^* = \mathrm{BE}_d/32$ ($T^* \approx 0.80 \times 10^9$ °K). We shall now use this as our second approximation to the solution of (7.12) and plug $kT^* = \mathrm{BE}_d/32$ into the first term. We obtain

$$0 \approx 14 + 23 - \mathrm{BE}_d/kT^*$$

and from this, $kT^* = \mathrm{BE}_d/37$ ($T^* \approx 0.70 \times 10^9$ °K). Notice that the last two approximations are within 13% of one another. Another iteration would yield a value of T^* which differs from the previous one by only $\frac{1}{2}$%. We shall therefore take as our solution to (7.12)

$$kT^* = \mathrm{BE}_d/37 \qquad \text{or} \qquad T^* \approx 0.7 \times 10^9 \text{ °K}. \tag{7.13}$$

Thus the reaction $n + p \rightarrow d + \gamma$ freezes out at a temperature that is well below the one expected from simply equating kT^* to BE_d.

This reduction in the equilibrium nucleosynthesis temperature is altered very little by the inclusion of other competing rates such as the production of helium. Moreover this reduction factor of about 1/37 is not too different for any reaction $A + B \rightarrow C + \gamma$ which takes place in a hot photon bath, with $l \ll R$. In particular the ionization energy for helium is ~25 eV for $\gamma + \mathrm{He} \rightarrow \mathrm{He}^+ + e^-$, so the onset of the He^+ line in the absorption spectra for a star (see Sec. 1.6.2) means that its surface temperature is of the order $kT^* \sim 25$ eV/37 ~ 0.7 eV or $T^* \sim 10^4$ °K.

Returning to deuteron production, when T drops to T^* of (7.13), deuterons are formed out of all the remaining free neutrons almost *immediately* because, as we have seen, $l \ll R_u$ for the production reaction $n + p \rightarrow d + \gamma$. At this time, the relation (7.6) gives $t \sim (10^{10}/0.7 \times 10^9)^2 \sim 200$ sec, so that the n/p ratio is

$$(\mathrm{n/p})_{t\sim 200 \text{ sec}} = \frac{\mathrm{X_n}(200)}{1 - \mathrm{X_n}(200)} = \frac{0.15}{0.85} = 0.18 \tag{7.14}$$

This fraction of neutrons is then immediately locked into the deuteron, to be preserved down to present times. If, per chance, nature had deemed the *n-p* mass difference much larger or the deuteron binding energy much smaller, than the race to form the deuteron would have been lost. Primordial nucleosynthesis consequently could not occur to trap unstable free neutrons into light nuclear elements. While the big bang would then have been a "big bust," ultimately the fusion process within stellar cores combined with supernovae explosions would produce light and heavy nuclear elements but with much smaller concentrations than implied by (7.14).

In the late 1940s, Gamow and collaborators realized the importance of deuteron production in a qualitative way. Their approach in effect turns around the arguments just presented, predicting the present temperature of the fireball T_0. Requiring that the probability for forming one deuteron in $n + p \rightarrow d + \gamma$ is unity (see Sec. 2.3.2),

$$\mathcal{P} = n_N \sigma v \Delta t \sim 1$$

(or greater just to insure that the deuteron is formed), they used the measured $n + p \rightarrow d + \gamma$ cross section $\sigma \sim 10^{-32}$ m^2 and nucleon velocities at 10^9 °K of

$$v = \sqrt{\frac{3kT}{m_N}} \sim 5 \times 10^6 \; m/sec,$$

at $\Delta t \sim 10^2$ sec into the expansion to predict nucleon densities at that time to be

$$n_N = n_N^0 (T/T_0)^3 \sim 10^{24} \text{ nucleons/m}^3.$$

Inputing a present nucleon density of roughly 10^{-1} nucleons/m^3, they predicted that now $T_0 \sim 5$ °K. Their prediction is certainly astounding. What is surprising is that no one looked for the fireball for almost two decades thereafter, and then it was found only by accident (as were pulsars and many other important discoveries in physics)!

7.2.3. Helium Synthesis

As soon as an appreciable number of deuterons accumulate, they are quickly converted into the slightly more stable nuclei He^3 and H^3 and then into the much more stable helium nucleus, the α particle, He^4, through reactions like

$$\begin{aligned} d + d &\rightarrow He^3 + n \qquad & H^3 + d &\rightarrow He^4 + n \\ &\rightarrow H^3 + p & He^3 + n &\searrow \\ d + p &\rightarrow He^3 + \gamma & H^3 + p &\rightarrow He^4 + \gamma. \\ d + n &\rightarrow H^3 + \gamma & d + d &\nearrow \end{aligned}$$

While most of the deuterons and He^3 produced in the early big bang are used up in the formation of helium, a small residual abundance will be left in the fireball, existing today as the measured solar system and interstellar cosmic abundances

$$d/\text{H} \sim 10^{-4}\text{–}10^{-5}, \qquad He^3/\text{H} \sim 10^{-5}.$$

We shall make use of these ratios later. It is interesting to note that interstellar deuterium (by now of course each deuteron has picked up an electron and the neutral element is called deuterium D) is detected by its emission of 92-cm magnetic spin-flip radiation. This is analogous to the spin-flip that produces the 21-cm radiation from neutral hydrogen.

Recall from (7.14) that the n/p number ratio is fixed at about 15% when the deuterons are produced. Since almost all of the neutrons are incorporated into helium, we may compute the He mass fraction ($n_{He} = n/2$, $n_H = p\text{-}n$, where p and n are the proton and neutron number densities before nucleosynthesis),

$$x_{\mathrm{He}} = \frac{N_{\mathrm{He}}m_{\mathrm{He}}}{N_{\mathrm{He}}m_{\mathrm{He}} + N_{\mathrm{H}}m_{\mathrm{H}}} \approx \frac{4n_{\mathrm{He}}}{4n_{\mathrm{He}} + n_{\mathrm{H}}} = \frac{2n/p}{1 + n/p} \approx 0.29 \qquad (7.15)$$

That is, 26% of the primordial matter gas by weight is converted to α particles (the helium nucleus). Since nature provides a bottleneck, a lack of stable nuclei with $A = 5$, 6, and 8 (only Li^7 is stable—see Fig. 4.1), the production of these latter elements is suppressed. For nuclei with $A = 8$, the Coulomb barrier prevents their formation at these temperatures. Consequently, the approximate 29% He abundance existing in the fireball at the end of primordial nucleosynthesis ($t \sim 10^4$ sec) should exist today—and it does, in the surface composition of stars (seen via stellar spectra), in cosmic rays and in globular clusters. The onset of the various processes just discussed can be seen graphically in Fig. 7.3.

The primordial nucleosynthesis processes occur only because of a series of masses, binding energies, "windows," and "bottlenecks" which happen to be just right to allow the production of light nuclear elements. The consequent predictions are borne out today with no apparent contradictions. This is certainly a striking and significant success for the big bang cosmology. Nevertheless it is important to rule out any alternative models for cosmic nucleosynthesis. The only competing nucleosynthesis processes that we know of take place in stars, where it is believed that the deuterium,

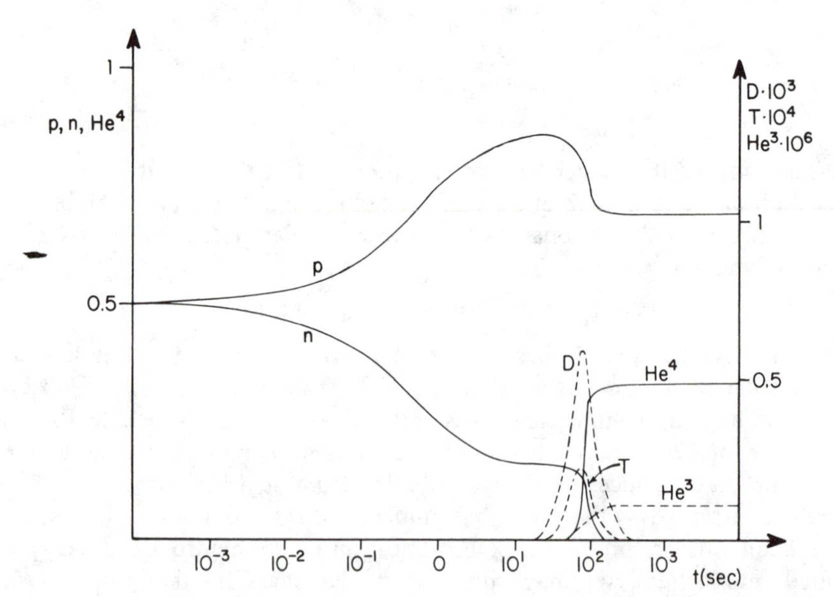

Fig. 7.3. Composition of matter in the early universe.

He^3 and He^4 abundances produced are far less than the measured amounts quoted above. More specifically, we recall from Sec. 4.1.1 that only 5% of the solar mass has been converted to He in 4×10^9 yr, as based upon the ratio $(L/M)_\odot$. Furthermore we know from (6.2) that in the Galaxy, $(L/M)_{\rm gal} \sim \frac{1}{15}(L/M)_\odot$, so that the fraction of He produced in the Galaxy must be

$$x_{\rm He}^{\rm gal} \sim \frac{(L/M)_{\rm gal}}{(L/M)_\odot} X_{\rm He}^\odot \sim \frac{1}{15} \cdot \frac{1}{20} \sim 0.003,$$

far less than the measured 26%. Thus almost all of the D, He^3, Li, and He^4 that has survived today is probably of cosmological origin, presumably produced at $t \sim 10^2 - 10^4$ sec into the big bang.

7.3. The Radiation Era

After taking time out to build up the light elements, we now return to tracking the primary repository of energy in the universe—the photon fireball. From the time $t \sim 4$ sec with $T \sim 5 \times 10^9$ °K when all the positrons are annihilated via $e^+ + e^- \to 2\gamma$, until $t \sim 10^6$ yr with $T \sim 10^3$ °K (after which the energy density of the universe becomes matter-dominated), the energy density is dominated by photons and perhaps leptons. Through bremsstrahlung, Compton scattering, and their inverse processes, the radiation and matter will still remain in thermal equilibrium. The radiation pressure is still too high, however, for matter to contract into galaxies. The hot universe, $n_\gamma / n_N \sim 10^{10}$ requires the fireball to continue to expand adiabatically with $T_m = T_\gamma \propto 1/R$.

7.3.1. Recombination of Hydrogen

When the fireball temperature drops to $\langle E_{\rm max} \rangle \sim kT \sim E_I$, where E_I is the ionization energy of hydrogen, $E_I = 13.6$ eV, the free nonrelativistic electrons and protons finally begin to combine to form neutral hydrogen via the reaction $e^- + p \to {\rm H} + \gamma$. As was the case for deuteron production, the inverse process, the photoelectric effect, has a very large cross section,

$$\sigma_{\gamma + {\rm H} \to e^- + p} \sim 10^{-20} (E_I / kT)^{7/2} \text{ meters}^2$$

in MKS units. At $kT \sim 13.6$ eV or $T \sim 10^5$ °K, we have $\sigma \sim 10^{-20}$ m^2 and along with the nucleon number density of $n \sim n_0 (T/T_0)^3 \sim 10^{13}$ nucleon/m^3 we obtain a mean free path of $l = 1/n\sigma \sim 10^7$ m. On the other hand, $T \sim 10^5$ °K and (7.6) correspond to $t \sim (10^{10}/10^5)^2 \sim 10^{10}$ sec, so

$R_u \sim ct_{\text{exp}} \sim 10^{18}$ m. Clearly $l \ll R_u$ so the many photons from the fireball immediately rip apart all hydrogen atoms at this temperature.

As T falls significantly below $kT \sim E_I$, however, the high-energy tail of the photon black body fireball distribution finally has less energy than is required to drive the breakup photoelectric process. The situation is again similar to deuteron formation, so we may borrow the analysis of Sec. 7.2.2 to obtain the reduced temperature T^* for the recombination of hydrogen:

$$\frac{n_e}{n_{\text{H}}} = \left(\frac{p_e p_p}{p_{\text{H}} p_\gamma} \right)^3 \frac{n_\gamma}{n_p} e^{-E_I/kT} \tag{7.16}$$

so that for n_e/n_{H} of order unity at $T = T^*$, the logarithm of (7.16) is

$$0 \approx \tfrac{3}{2} \ln(2m_e c^2/kT^*) + \ln(n_\gamma/n_p) - E_I/kT^*. \tag{7.17}$$

Solving as in (7.12) by the method of successive approximations we find that

$$kT^* \approx E_I/45 \qquad \text{or} \qquad T^* \approx 3500\ {}^\circ\text{K}. \tag{7.18}$$

The reduction factor for recombination of hydrogen of 45 is not too different from the factor 37 for deuteron production.

At this lower temperature it is important to check if the recombination proceeds at a rapid rate. The cross section is, at $T^* \sim 3500$ °K,

$$\sigma_{e^- + p \to H + \gamma} \sim 10^{-24} (E_I/kT^*)^{5/2} \sim 10^{-20}\ \text{m}^2,$$

while $n_N = n_0(3500/2.7)^3 \sim 2 \times 10^8/\text{m}^3$ at time $t \sim (10^{10}/3500)^2 \sim 10^{13}$ sec. Then $l = 1/n\sigma \sim 10^{12}$ m, which is an extremely long mean free path compared to other reactions so far considered but still much less than the size of the universe at that time, $R_u \sim ct_{\text{exp}} \sim 10^{22}$ m.

Thus at $T^* \sim 3500$ °K, $t \sim 10^{13}$ sec $\sim 10^6$ yr, electrons finally become bound to protons, forming hydrogen atoms. This neutralizing of matter suppresses the bremsstrahlung interaction with the fireball photons and the universe becomes "transparent." Thereafter the thermal equilibrium between radiation and matter is broken. Without this radiation pressure from the fireball, gravity begins to condense local parts of the expanding matter gas into galaxies—but more about that shortly.

7.3.2. Transition to Matter-Dominated Universe

With the nonrelativistic matter gas ($\text{KE}_m \sim E_I/45 \sim 0.3$ eV $\ll m_N c^2 \sim 10^9$ eV) now decoupled from the photon fireball, its temperature falls off like a nonrelativistic adiabatic gas should, $T_m V^{2/3} = \text{const}$, or $T_m \propto 1/R^2$, whereas the radiation temperature of the fireball continues to fall off like $T \propto 1/R$. Likewise the effective photon mass density of the fireball is still

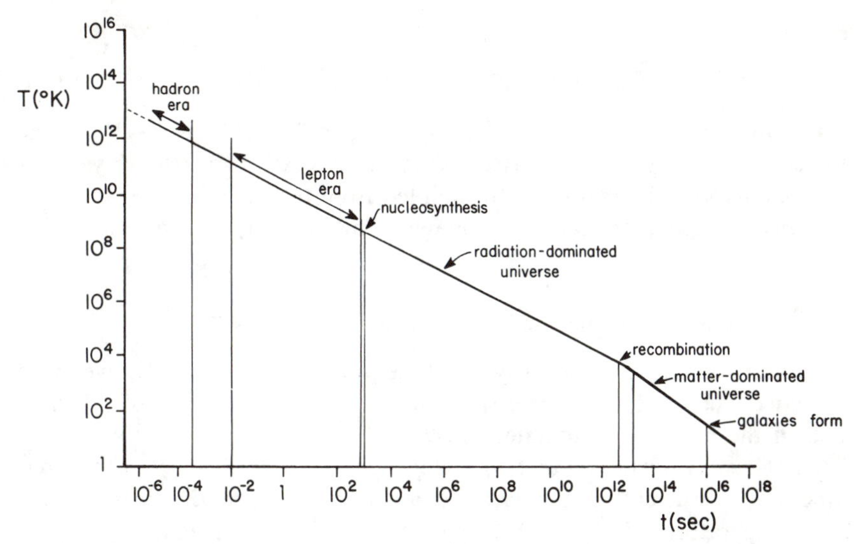

Fig. 7.4. Temperature vs. time and the transition from a radiation-dominated to matter-dominated universe.

$\rho_\gamma = aT^4/c^2$, or for the total radiation gas, photons plus neutrinos ($T_\nu \sim 0.7T_\gamma$),

$$\rho_R = \rho_\gamma + \rho_\nu = aT_\gamma^4/c^2 + \tfrac{7}{4}aT_\nu^4/c^2 \approx 1.5aT_\gamma^4/c^2.$$

The sharp decoupling of radiation and matter is illustrated in Fig. 7.4. Clearly the expansion time of the universe, controlled by matter, will no longer be $t \sim (10^{10}/T)^2$. We shall find the correct relationship between t and T in the next section. Here, however, we would like to reemphasize an important point. Regardless of how T falls off with R we have by definition $\rho \propto 1/V \propto 1/R^3$. Since the t vs. T curve for the photon gas is a straight line on the log-log plot of Fig. 7.4, unlike the t vs. T curve for matter which has a break at $t \sim 10^{13}$ sec, we can use the *photon* temperature T_γ to find the *matter* density ρ_m, that is,

$$\rho_m = \rho_0 (T_\gamma / T_0)^3$$

providing we remember to use the present fireball temperature, $T_0 \approx 2.7$ °K. We made this point earlier but it is worth stressing a second time.

The critical temperature T_γ^c separating the transparent universe from the matter-dominated universe corresponds to $\rho_R = \rho_m$, or

$$T_\gamma^c \sim \frac{\rho_0 c^2}{1.5 a T_0^3} \sim \frac{2 \cdot 10^{-28} \cdot 3^2 \cdot 10^{16}}{1.5 \cdot 7.6 \times 10^{-16} (2.7)^3} \sim 800 \text{ °K} \qquad (7.19a)$$

at time

$$t_{\mathrm{exp}}^{c} \sim (10^{10}/800)^2 \sim 2 \times 10^{14}\ \mathrm{sec} \sim 10^7\ \mathrm{yr}. \tag{7.19b}$$

To summarize, from $T \sim 10^9$ °K, $t \sim 10^2$ sec to $T \sim 3500$ °K, $t \sim 10^6$ yr the universe is radiation dominated; then to $T \sim 800$ °K, $t \sim 10^7$ yr it is transparent with the matter gas decoupled from the fireball. After this time the universe becomes matter dominated. This scenario is depicted in Fig. 7.4.

7.3.3. Radius of the Universe vs. Expansion Time

Up to this point, in the early universe for $T > T_c \sim 10^3$ °K, we have estimated the size of the universe by $R_u \approx ct_{\mathrm{exp}}$ where $t_{\mathrm{exp}} \approx (10^{10}/T)^2$. In fact, in the radiation-dominated era $R \propto T^{-1}$ so that when we substitute $T \propto t^{-1/2}$ we find the correct time dependence to be $R_u \propto t^{1/2}$. To determine the coefficient in the latter expression, we take the present *fireball* temperature of $T_0 \sim 3$ °K and matter-dominated radius of $R_0 \sim 10^{10}$ LY $\sim 10^{26}$ m with

$$R_u = R_0 T_0 / T \sim 3 \times 10^{26} T^{-1}\ \mathrm{m}. \tag{7.20}$$

Then we see from (7.6) that for $T \sim 10^{10} t^{-1/2}$, (7.20) gives

$$R_u \sim 3 \times 10^{16} t_{\mathrm{exp}}^{1/2}\ \mathrm{m}, \qquad T > T_c \tag{7.21}$$

which is still numerically of order $R_u \sim ct$ for $t < t_c \sim 2 \times 10^{14}$ sec. However for $t > t_c$, $T < T_c$, the simple relation (7.6) between time and fireball temperature breaks down. Instead the expansion time of the universe, still approximately

$$t_{\mathrm{exp}} \sim (G\rho)^{-1/2}$$

is controlled by the dominant *matter* density in the universe for $t > t_c$,

$$t_{\mathrm{exp}} \sim \left[G\rho_0 (T/T_0)^3 \right]^{-1/2} \sim \left[10^{-10} \cdot 10^{-28} \cdot 10^{-1} T^3 \right]^{-1/2}$$
$$\sim 10^{19} T^{-3/2}\ \mathrm{sec}. \tag{7.22}$$

Substituting this latter expression into (7.20), we obtain

$$R_u \sim R_0 (G\rho_0)^{1/3} t^{2/3} \sim 10^{14} t^{2/3}\ \mathrm{m}, \qquad T < T_c. \tag{7.23}$$

However, for $t > t_c$, the $t^{2/3}$ dependence of (7.23) will differ numerically from ct, corresponding to the slowing down of the expansion of the universe in the matter-dominated era. This $t^{2/3}$ dependence in the early matter-dominated universe is common to all cosmological models, corresponding to the small t region in Fig. 6.8. (The radiation-dominated era is a mere speck close to $t = 0$.) As time progresses, however, the slope of the curves become different as will the R vs. t dependence for each model. We shall take up this latter point in Sec. 7.5.

Before leaving the radiation era, it is interesting to note that we can look back in time and try to see the universe *then* by observing the distant galaxies *now*. Recall from (6.15) that the Doppler shift measures $z = \Delta\lambda/\lambda_0$ or

$$\lambda = \lambda_0\left(\frac{1 + v/c}{1 - v/c}\right)^{1/2} = \lambda_0(z + 1).$$

Since λ_0 corresponds to a wavelength of an emitted source in past times ($\lambda_0 = \lambda_{\text{then}}$) and λ is the measured wavelength received on Earth ($\lambda = \lambda_{\text{now}}$), we can use the black body fireball with $\lambda_{\text{peak}} \propto 1/T$ to convert the above expression to a temperature relation, viz.,

$$T_{\text{then}} = T_{\text{now}}(z + 1).$$

To look back to $T_{\text{then}} \sim 3000$ °K compared to $T_{\text{now}} \sim 3$ °K means we must detect a $z \sim 10^3$. This is hardly encouraging since the largest z we measure spectroscopically is $z \sim 3$ for a few quasistellar objects. Thus, direct observation today can only reveal the recent matter-dominated history of the universe, well after our galaxy was formed.

7.4. The Matter Era

After the recombination of hydrogen at $T \sim 3500$ °K, the transparent fireball exerts very little radiation pressure on the matter gas and the latter begins to condense into galaxies. The photons become spectators as gravity takes over as the dominant force controlling the evolution of galaxies and stars. In a sense we have come full circle in the story; we are close to the point where we began to explore the interstellar medium in chapter 2.

7.4.1. Galactic Formation

It would be very satisfying if we could report to you that there is a definitive theory of galactic formation within the big bang context, but unfortunately at this time there is not. The conventional wisdom concerning galactic formation is based on the theory of gravitational instability. In the primordial universe one can imagine density fluctuations, as shown in Fig. 7.5, that become the seed for the further gravitational attraction of matter and eventually evolve into galaxies. How large would these density fluctuations have to be in order to lead to the formation of galaxies? At present, the mean density within 3×10^4 LY of the center of our galaxy is

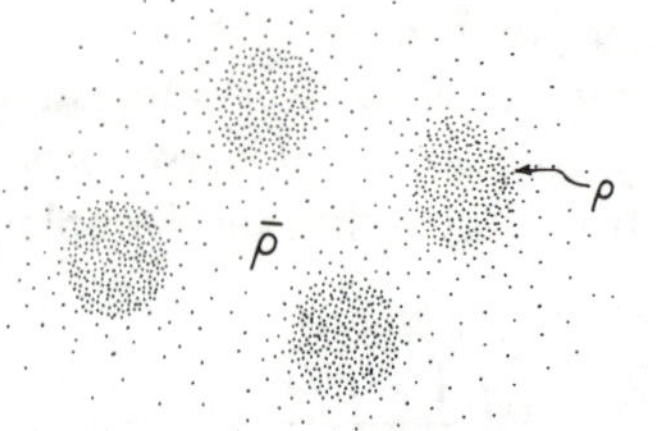

Fig. 7.5. Density fluctuations in expanding fireball.

$\sim 10^{-21}$ kg/m^3 while the mean density of the universe is $\sim 10^{-28}$ kg/m^3. Thus the galaxies represent a density fluctuation over the mean density of the universe $((\rho - \bar{\rho})/\bar{\rho} \equiv \Delta\rho/\bar{\rho})$ of about 10^7. Extrapolating backward in time and taking into account the Hubble expansion of the universe (expansion of course makes contraction more difficult), astrophysicists estimate that a density fluctuation of 1% would be needed at the time of recombination in order to account for the present galactic inhomogeneity. One way to achieve these density fluctuations is through clumping that occurs naturally when particles in thermal equilibrium are in random motion. Unfortunately with 10^{68} particles in a typical galaxy, the average statistical fluctuation of particles in thermal equilibrium is only $\Delta\rho/\bar{\rho} = N^{-1/2} \approx 10^{-34}$! Thus we are in a quandary with any gravitational instability theory: How does one account for the necessarily large primordial fluctuations? Is it possible we are not in thermal equilibrium or are nongravitational forces involved? No one knows!

Nevertheless, if we *postulate* that large initial density fluctuations did exist, then it would be satisfying to explain at least the present observed galactic scales, the average mass $M_{\text{gal}} \sim 10^{11} M_{\odot}$ and the average spacing between galaxies, $R_{\text{IG}} > 10^{22}$–10^{23} m. To start with, we note that if the primordial increase in local density is to be maintained until we reach the recombination era, the particles must be prevented from spreading out. The relevant interaction is the force between the photons, which are moving outward with the expanding universe, and the electrons which are trying to condense toward the density fluctuation, as shown in Fig. 7.6. Indeed, because of the local increase in gravity, the particles in the fluctuation will expand more slowly than the universe as a whole and hence the electrons will experience an outward drag from the photons (see problem 19.) The appropriate cross section for the interaction is called the Thomson cross section and is given by

$$\sigma_T = 8\pi r_{0,e}^2/3 \sim \tfrac{2}{3} \times 10^{-28}\ \text{m}^2,$$

where $r_{0,e} = k_0 e^2/m_e c^2 \approx 3 \times 10^{-15}$ m is the "classical radius" of the elec-

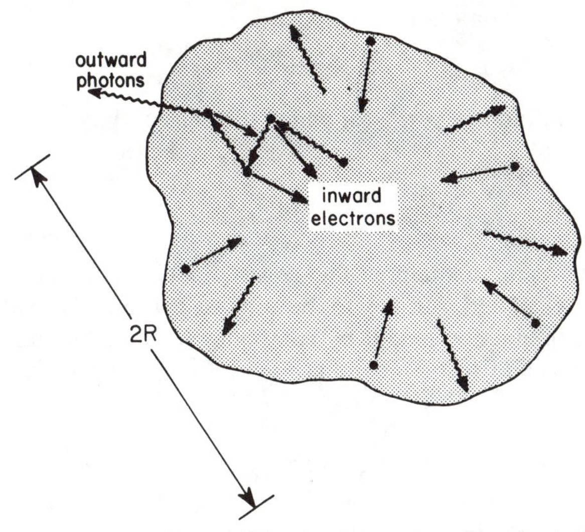

Fig. 7.6. Outward moving photons "Thomson" scattering off condensing electrons.

tron. (Note that with $m_p \sim 2000\, m_e$, the proton Thomson cross section is 10^6 times smaller.) This coupling will disappear at $T \sim 3500$ °K when the photons no longer have sufficient energy to ionize the recombing electrons and protons. At this point any density fluctuations that survive will surely collapse into galaxies. But which will survive to reach the recombination era?

We can get a handle on this problem as follows. As we have just noted, material particles are trying to coalesce around the local peak in density due to their mutual gravitational attraction. Radiation pressure is trying to disperse them. The interacting photons scatter randomly off the gas particles. Now we saw in Sec. 3.3.1 that it takes a long time for a photon to diffuse a distance R because of the random nature of the interaction process. If R is the radius of the region and L the interaction length ($L = 1/\kappa\rho$), then the discussion of Sec. 3.3.1 reduces to the simple formula for the "random walk" diffusion time t_D,

$$t_D = \left(\frac{R}{c}\right)\left(\frac{R}{L}\right).$$

If at the time recombination takes place the diffusion time is less than the age of the universe ($t_D < t_{\text{exp}}$), then the photons will early on wash out the density fluctuations. What we would like is the diffusion time to be greater than the age of the universe ($t_D > t_{\text{exp}}$). Under these conditions the photons will not have had enough time to smooth out the local increase in density and the fluctuation will survive. The expansion time before recombination is given by expression (7.6)

$$t_{\mathrm{exp}} = 10^{20}/T^2,$$

so that the condition $t_D > t_{\mathrm{exp}}$ for which the density fluctuation will survive is

$$\frac{R^2}{cL} > 10^{20}T^{-2}.$$

Thus if the region containing the density fluctuation is large enough the inhomogeneity should survive to become a galaxy. To evaluate R we calculate the interaction length by combining σ_T with the number density of the matter gas at recombination, $n = n_0(T/T_0)^3 \sim (\frac{1}{10})(3500/2.7)^3 \sim 2 \times 10^8\ \mathrm{m}^{-3}$, as

$$L = \frac{1}{n\sigma_T} \sim \frac{1}{2 \times 10^8 \times \frac{2}{3} \times 10^{-28}} \sim 0.7 \times 10^{20}\ \mathrm{m}.$$

The size of a stable density fluctuation then becomes

$$R > \sqrt{cL \times 10^{20} \times T^{-2}} = \sqrt{3 \times 10^8 \times 0.7 \times 10^{20} \times 10^{20} \times (3500)^{-2}}$$
$$\approx 4 \times 10^{20}\ \mathrm{m}. \tag{7.24}$$

This distance should also give us a rough idea of the separation between galaxies, scaled up to the present value

$$R_{\mathrm{now}} \simeq (T_{\mathrm{rec}}/T_0)R \approx \left(\frac{3500}{2.7}\right) \times 4 \times 10^{20} \approx 5 \times 10^{23}\ \mathrm{m}. \tag{7.25}$$

Moreover, given the mass density of matter at recombination $\rho_{\mathrm{rec}} = m_N n \approx 3 \times 10^{-19}\ \mathrm{kg/m^3}$, the total mass within this boundary that should evolve into a galaxy is

$$M \sim 4\pi\rho_{\mathrm{rec}}R^3/3 \sim 4 \times 3 \times 10^{-19}(4 \times 10^{20})^3\ \mathrm{kg} \sim 4 \times 10^{13} M_\odot. \tag{7.26}$$

Clearly the galactic length and mass scale we have derived are too large. Indeed they are more consistent with the evolution of clusters of galaxies rather than an individual galaxy. There is some indication, however, that the determination of galactic masses from rotation curves has vastly underestimated their mass. If this is the case then the galactic scales based on the diffusion of photons may turn out to be quite satisfactory after all. Or, huge clouds may have formed first and then the galaxies fragmented out of these. We shall have to wait to see which situation is correct.

What we have shown above is that for a region of around 4×10^{20} m as we approach the recombination era, density fluctuations will not dissipate. Another question we might ask is: Is this region gravitationally bound so that the inhomogeneity will coalesce into a galaxy? As was the case for stellar formation in chapter 2 and stellar collapse in chapter 5, a gravitationally bound system has

$$E = \mathrm{KE_{tot}} - |\mathrm{PE_{tot}}| < 0.$$

For $\mathrm{PE_{tot}} = \frac{3}{5}GM^2/R = \frac{3}{5}G\rho^2(4\pi/3)^2R^5$ and a matter–gas kinetic energy in equilibrium with the photon fireball *before* recombination, $\mathrm{KE_{tot}} = aT^4 \times 4\pi R^3/3$, then E will be greater than zero for

$$R < R_{\max} = \frac{3T_{\mathrm{rec}}^2}{4\pi\rho_{\mathrm{rec}}}\left(\frac{5a}{3G}\right)^{1/2} \sim 10^{23}\ \mathrm{m}.$$

As was expected, the pressure of the fireball in equilibrium with the matter gas prevented gravitational condensation of the density fluctuation.

For $T < 3500$ °K *after* the decoupling of the fireball, the KE is dominated by the thermal motion of the matter gas itself, $\mathrm{KE_{tot}} = \frac{1}{2}m_{\mathrm{H}}\langle v^2\rangle \times N$, where $\langle v^2\rangle$ is determined by the temperature as $\frac{1}{2}m_{\mathrm{H}}\langle v^2\rangle = \frac{3}{2}kT$ or

$$\mathrm{KE_{tot}} = \tfrac{1}{2}\rho\frac{4\pi}{3}R^3\left(\frac{3kT}{m_{\mathrm{H}}}\right).$$

Setting $E = 0$ we find an $R_{\min}$ for which gravitational collapse will take place at $T \sim 3500$ °K,

$$R > R_{\min} = \left(\frac{15kT_{\mathrm{rec}}}{8\pi G\rho_{\mathrm{rec}}m_{\mathrm{H}}}\right)^{1/2} \approx 10^{18}\ \mathrm{m}.$$

Since R is greater than 10^{18} m by (7.23), all the surviving density fluctuations are of sufficient size to be gravitationally bound and eventually collapse into galaxies.

Finally let us estimate the galactic formation time. Following the procedure used to estimate the collapse time of stellar clouds in Sec. 3.1.9, i.e. the free-fall time t_{ff}, for the cloud to shrink to 90% of its original size

$$t_{\mathrm{ff}} \approx (R^3/5GM)^{1/2} \approx (G\rho)^{-1/2},$$

we find that galactic clouds at $T \sim 3500$ °K with $\rho \sim 3 \times 10^{-19}$ kg/m^3 give

$$t_{\mathrm{ff}} \sim \left(\frac{1}{6.7 \times 10^{-11} \times 3 \times 10^{-19}}\right)^{1/2} \sim 2 \times 10^{14}\ \mathrm{sec} \sim 10^7\ \mathrm{yr}.$$

This is consistent with estimates of galactic formation in 10^8 years.

7.4.2. The Present Fireball

To keep to the cosmological story we next jump to the present and investigate the effects of the cosmic fireball on today's universe. First we

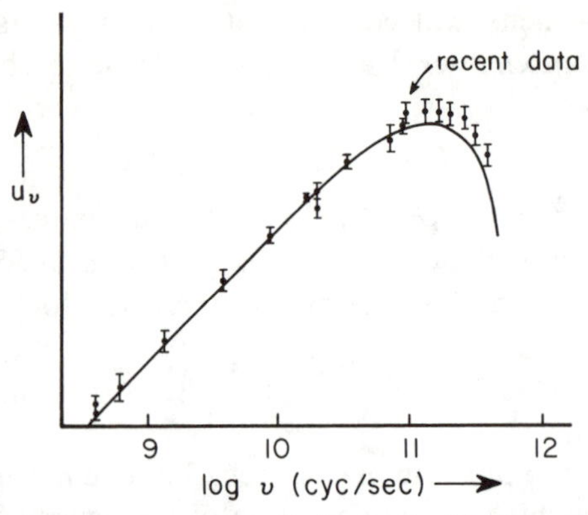

Fig. 7.7. Black body curve for microwave radiation at $T \sim 2.7$ °K emphasizing recent data in the high frequency region.

reaffirm that a single black body spectrum of $T_0 = 2.7$ °K really does exist. The initial measurements depicted in Fig. 6.13 have since been extended to map out most of the long wavelength or small frequency part of the black body curve as shown in Fig. 7.7. Moreover, recent measurements using balloon-borne instruments for the first time show that the black body curve indeed turns over for $\nu > \nu_{\text{peak}} \sim 1.5 \times 10^{11}$ cycle/sec or $\lambda < \lambda_{\text{peak}} = c/\nu_{\text{peak}} \sim 2$ mm.

Since Fig. 7.7 plots energy density per unit frequency range u_ν vs. frequency ν, we can obtain the total energy density of the present fireball by calculating the area under the curve in Fig. 7.7, i.e., $E/V = \sum u_\nu \Delta\nu$. Alternatively, integral calculus does this for us, giving the familiar mass density relation for the fireball,

$$\rho_\gamma^0 = aT_0^4/c^2 \approx \frac{7.6 \times 10^{-16}(2.7)^4}{9 \times 10^{16}} \approx 5 \times 10^{-31} \text{ kg/m}^3. \qquad (7.27)$$

This further confirms the fact that the universe is now matter dominated since $\rho_\gamma^0 \ll \rho_m^0 = \rho_0$. Recall from (6.5) that the energy density of starlight in the neighborhood of the solar system is $\rho_{\text{SR}} \sim 10^{-3} \rho_{\text{lum}} \sim 10^{-31}$ kg/m^3, only slightly less than the present fireball density. Fortunately, we can distinguish between these two types of radiation because their spectra do not overlap; the fireball is a single black body with $\lambda_{\text{peak}} \sim 2 \times 10^{-3}$ m whereas starlight peaks in the visible region, $\lambda_{\text{peak}} \sim 10^{-6}$ m. Thus there is little starlight contamination over the fireball spectrum, 1 m $< \lambda < 10^{-4}$ m. Why the present fireball and starlight densities should be comparable is not really understood. It would appear to be a numerical coincidence for the big bang cosmology but it might be fundamental in other models.

An important connection exists between the primordial nucleosynthesis period $T \sim 10^9$ °K and now, $T_0 \sim 3$ °K. Given the present cosmic deuterium, and helium abundances, presumably the remnant of the cosmic nucleosynthesis process, it is possible to pinpoint the number density of nucleons at $T \sim 10^9$ °K by simultaneously satisfying the rate equations and cross section relations for each nucleosynthesis reaction. It turns out that the number density is insensitive to the helium concentration, but very sensitive to the deuterium concentration. As we can see from Wagoner's 1973 analysis plotted in Fig. 7.8, the abundance ratio (present abundance to big bang abundance) for deuterium ϕ_D changes by a factor of $\sim 10^4$ when the deuterium concentration changes by only a factor of $\sim 10^2$, where the number density n_0 has been transformed to the present mass density of the universe via $\rho_0 = m_N n_0 (T_0/T)^3$. Since the deuterium produced in the big bang is not expected to be reduced by more than a few percent in stellar processes ($d + d \rightarrow He^3 + \gamma$), the value of ϕ_D should be approximately one. Since the ϕ_D vs. ρ curve is so steep, setting $\phi_D = 1$ ($\log \phi_D = 0$) places an upper limit on ρ_0 of

$$\rho_0 \approx 6 \times 10^{-28} \text{ kg/m}^3, \tag{7.28}$$

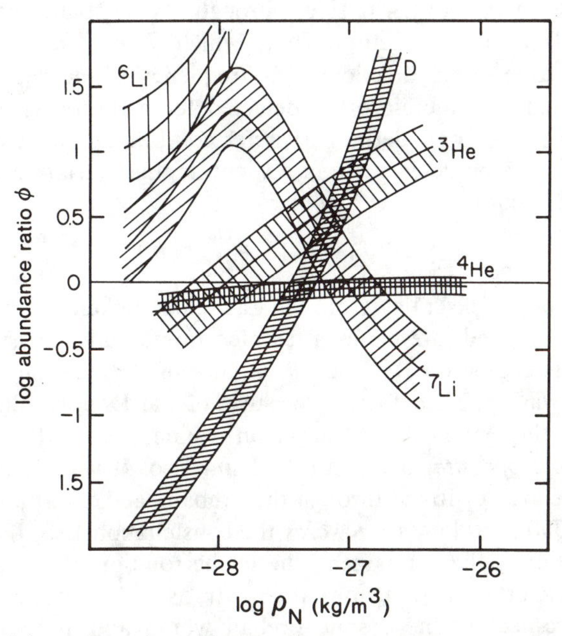

Fig. 7.8. Plotted is ϕ, the ratio of the present cosmic abundance of an element to the primordial abundance generated in the big bang, vs. ρ_N, possible values of the current baryon density.

very close indeed to ρ_0 as found from luminous galactic measurements, $\rho_0 \approx 2 \times 10^{-28}$ kg/m^3! We cannot expect exact agreement between these two numbers, however, because the luminous ρ_0 implicitly depends upon the square of the measured Hubble constant H_0, while ρ_0 found from Fig. 7.8, i.e. (7.28) is linked to the dynamical details of the big bang cosmology. Certainly this close agreement argues convincingly for the validity of the big bang model. If no other source of energy density is ultimately found, then this agreement also argues strongly for an open universe. However, the existence of neutral massive leptons, for example, with $m_{L_0} \sim 14$ eV could close the universe; yet a closed universe would not destroy the link between the observed luminous matter density and the nucleon mass density (7.28) required to produce the present deuterium concentration.

Finally, given this ρ_0 determined by deuterium on Fig. 7.8, astronomers are then able to use the curves of He^3, Li^6, Li^7 also plotted in Fig. 7.8 to indicate the extent that processes other than cosmic nucleosynthesis are involved in the production of these light elements.

7.4.3. Present Implications of the Fireball

The present fireball provides us with other interesting physical consequences. Because the Earth rides through the fireball, the background radiation black body spectrum from which T is determined, will not appear the same in all directions. The frequency of the detected radiation will be shifted and therefore the detected temperature will be shifted as well. Thus, the fireball defines a preferred (but not absolute) reference frame which can be detected from a slight angular variation of the fireball temperature according to the Doppler shift

$$T(\theta) = T_0\left(1 + \frac{v}{c}\cos\theta\right).$$

In 1977 a group of scientists from the Lawrence Berkeley Laboratory, using a specially equipped U-2 plane, attempted to measure the asymmetry term in the above equation, $(v/c)\cos\theta$. A nonzero value on the order of 10^{-3}–10^{-4} was expected to exist because of the Doppler shift due to the rotation of the Milky Way galaxy in which the Earth is moving at $\sim$300 km/sec *opposite* to the constellation Leo. It was determined, however, that we are "drifting" through the fireball "aether" at an approximate speed of $\sim 390 \pm 60$ km/sec *towards* the constellation Leo. If one subtracts from this velocity the component due to the rotation of our galaxy ($\sim$300 km/sec in the opposite direction), the net translational motion of the Milky Way with respect to the cosmic fireball would appear to be about 600 km/sec. It also turns out that apart from this anisotropy, the cosmic radiation background is remarkably homogeneous (i.e., no $\cos^2\theta$ term or higher in the above equation). While the latter is strong support for the fundamental cosmological principle, the surprisingly large velocity of our galaxy through the fireball may create problems for the big bang theory.

Lastly, very high-energy cosmic-ray protons can slam into the "slowly moving" cosmic fireball and photoproduce pions via the reaction

$$p_{\text{fast}} + \gamma_{2.7} \to n_{\text{slow}} + \pi^+.$$

Since the rest mass of the proton is less than the sum of the rest masses of the neutron plus pion, that is $m_p < m_n + m_\pi$, the reaction cannot "go" unless the proton is sufficiently energetic. The energy at which $p + \gamma \to n + \pi^+$ begins to go is called the threshold energy. In the "lab" frame the neutron and pion come out as a unit at threshold, as shown in Fig. 7.9, so that the initial momentum, primarily due to the fast proton $p_i \approx p_p$, must also be the momentum of the recoiling $n\pi^+$ ball, p_f. On the other hand, energy conservation $E_p + E_\gamma = E_f$ implies, since $E^2 = p^2c^2 + (mc^2)^2$ for a massive particle, that $(E_i + E_\gamma)^2 = E_f^2$ or

$$p_i^2c^2 + \left(m_pc^2\right)^2 + 2E_{p_i}E_\gamma + E_\gamma^2 \approx p_f^2c^2 + (m_n + m_\pi)^2c^4. \tag{7.29}$$

Using $p_f \approx p_i$, $(m_n + m_\pi)^2 \approx m_n^2 + 2m_nm_\pi$ since $m_\pi \sim 140\ \text{MeV}/c^2 \sim \frac{1}{7}m_n$ and $E_\gamma \sim 3kT \sim 10^{-3}\ \text{eV} \ll E_p$, the threshold energy in (7.29) becomes

$$E_{p_i} \approx \left(\frac{m_nc^2}{E_{\gamma_{2.7}}}\right)m_\pi c^2 \sim \frac{10^9}{10^{-3}}10^8 \sim 10^{20}\ \text{eV!} \tag{7.30}$$

It turns out that the upper end of the measured cosmic ray spectrum for protons is quite near 10^{20} eV and one should therefore expect to see a "kink" at this energy as shown in Fig. 7.10, since fast protons are then absorbed out of the cosmic ray beam and converted to slow neutrons and pions via this photoproduction reaction.

Again it is important to show that the above reaction has a high probability of occurrence within the confines of the present universe. The measured threshold photoproduction cross section is $\sigma(\gamma p \to n\pi^+) \sim 10^{-32}\ \text{m}^2$ and the "target" photons have a density of 10^{10} times the present

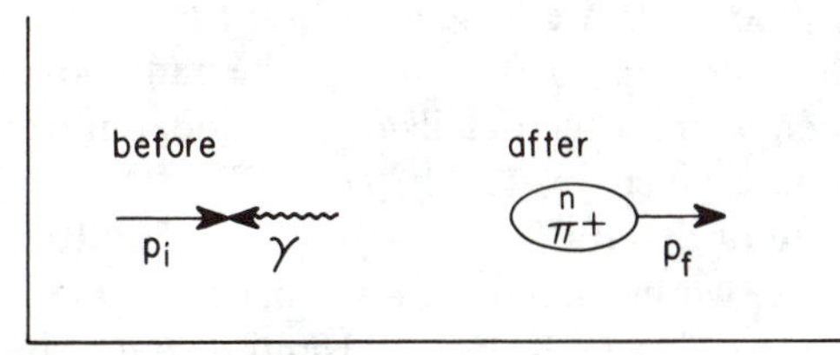

Fig. 7.9. Collision between fireball photon and cosmic ray proton at threshold for producing $n + \pi^+$.

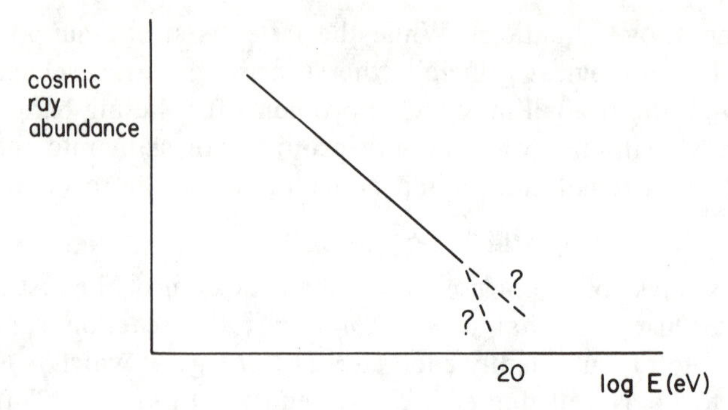

Fig. 7.10. Cosmic ray abundance vs. energy (measured on a log scale).

density of nucleons, $\frac{1}{10}m^{-3}$, or equivalently $n_\gamma \sim aT^3/k \sim 10^9$ photon/m^3. Then the mean free path for this process is

$$l = \frac{1}{n_\gamma \sigma_{\gamma p \to n\pi^+}} \sim \frac{1}{10^9 10^{-32}} \sim 10^{23}\ \text{m}.$$

Since $l \ll R_u \sim 10^{26}$ m, this reaction $p + \gamma_{2.7} \to n + \pi^+$ should occur "frequently," perhaps once every other galaxy for any given cosmic ray proton of energy greater than 10^{20} eV. Consequently, future cosmic ray measurements should detect such a kink in the cosmic ray abundance curve.

7.5. In The End . . .

Having traced the history of the universe up to the present time in the big bang model, we now jump ahead to see what is in store for the big bang universe in the distant future. The effects of general relativity no longer can be ignored; accordingly we give a brief discussion of it when necessary.

7.5.1. Time Evolution of the Future Universe

Returning to the question of the time development of the universe since the initial big bang at time $t = 0$, we have seen that t goes like $t_{\text{exp}} \sim (10^{10}/T)^2$ sec from the hadron era up to the end of the radiation era, $t \sim 10^6$ yr. Thereafter t_{exp} depends upon the details of the model universe in question, with a rough estimate given by (7.22), $t \sim 10^{19} T^{-3/2}$ sec. This gives the present time estimate of $t_0 \sim 10^{19}(3)^{-3/2} \sim 10^{18}$ sec$\sim 3 \times 10^{10}$ yr. In order to follow t in the future, we must zero in on a better estimate of the present time t_0. This we do by referring to Fig. 7.11 and distinguishing between the three possible evolutions: A, B, and C. It is clear that in each case an upper

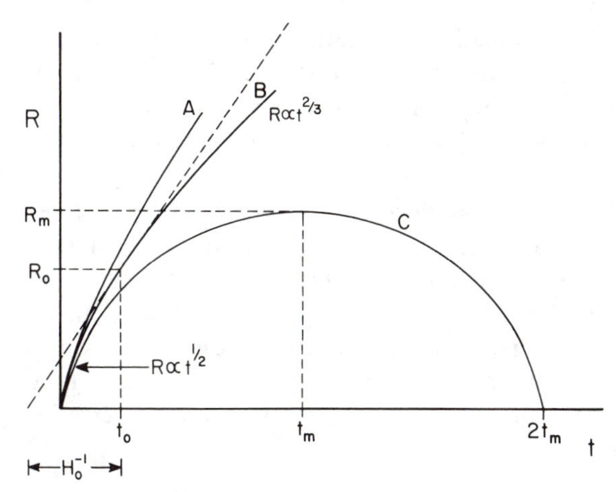

Fig. 7.11. Radius of the universe vs. expansion time, stressing the time dependence in the early and late big bang universe.

bound for the present age is less than the straight line Hubble slope ($R = vt$)

$$t_0 < H_0^{-1} \approx 18 \times 10^9 \text{ yr},$$

consistent with the present radioactive dating of the Earth, 4.5×10^9 yr, and the Galaxy, $\sim 10 \times 10^9$ yr. The energy conservation statement of Sec. 6.2.4, $v^2 - 2GM/R = v_0^2 - 2GM/R_0$ can be recast in the form

$$v^2 = H_0^2 R_0^2 \left(\frac{\rho_0 R_0}{\rho_c R} + 1 - \frac{\rho_0}{\rho_c} \right), \tag{7.31}$$

where $v_0 = H_0 R_0$, $M = 4\pi R^3 \rho/3$, $\rho R^3 = \rho_0 R_0^3$, and ρ_c is the critical mass density $\rho_c = 3H_0^2/8\pi G$. General relativity leads to the same dynamical equation, the solution of which follows from integral calculus. We find approximate solutions of (7.31) by considering special examples for the three big bang evolutions.

7.5.1.1. Case A: Open Universe ($\rho_0 < \rho_c$ or $q_0 < \frac{1}{2}$)

Present data such as the luminous density of galaxies and the (primordial) nucleosynthesis abundances both lead to $\rho_0/\rho_c \sim 1/30$. This implies that the universe is open and will expand forever. Setting $\rho_0/\rho_c \ll 1$ in the above dynamical equation (7.31), we get $v \approx H_0 R_0$ even for $R \gtrsim R_0$. Thus we see from Fig. 7.11 that R approaches and is now quite near the Hubble line $R = vt$, with the present time

$$t_0 \approx H_0^{-1} \approx 18 \times 10^9 \text{ yr}. \tag{7.32}$$

In the future, the fireball will remain decoupled from matter, the universe density will become more matter-dominated but less dense with $\rho_m \sim R^{-3}$, $\rho_R \sim R^{-4}$. At some point the possibility of new galaxies forming will become remote and the universe will simply cool off and die . . . .

7.5.1.2. Case B: Einstein–deSitter Universe ($\rho_0 = \rho_c$ or $q_0 = \frac{1}{2}$)

Suppose that the cosmic energy density is an order of magnitude larger, perhaps due to as yet undetected interstellar matter, black holes, or neutral massive leptons. Or possibly the estimate of the ρ_{lum}, the luminous mass density of the universe, and the nucleosynthesis analysis must be modified for some reason. In any case, it is not inconceivable that $\rho_0 = \rho_c$. This condition corresponds to the total energy in the universe being zero, with (7.31) becoming

$$v = \frac{\Delta R}{\Delta t} = H_0 R_0^{3/2} R^{-1/2},$$

a result which also follows from general relativity. This equation can be solved in a manner similar to the adiabatic laws of Sec. 6.3.3 ($PV^{5/3} = \text{const}$) as

$$R(t) = R_0 \left(\frac{3H_0}{2} t \right)^{2/3},$$

corresponding to the $R \propto t^{2/3}$ relation obtained in Sec. 7.3.3, but now stated in terms of Hubble's constant. Thus at the present time $R(t_0) = R_0$ and

$$t_0 = \tfrac{2}{3} H_0^{-1} \approx 12 \times 10^9 \text{ yr}. \tag{7.33}$$

The future of this universe would appear to resemble the open universe, case A.

7.5.1.3. Case C: Closed Universe ($\rho_0 > \rho_c$ or $q_0 > \frac{1}{2}$)

Philosophically it seems a pity that nature came so close to making the universe bound but in the end failed. Suppose that more than the missing mass (or energy) is discovered or that the deceleration parameter and galactic evolution becomes better understood with a resulting $q_0 > \frac{1}{2}$. Such a closed universe will ultimately extend out to a maximum distance R_m at time t_m and then contract into a final catastrophic "big crunch" at time $2t_m$. We approximate this behavior by a simple harmonic sine curve in Fig. 7.11, with

$$R(t) = R_m \sin \omega t,$$

where $\omega = 2\pi/\tau$ and τ is the "period" of the universe, $\tau = 4t_m$. Recalling that the velocity of a harmonic oscillator is $v(t) = \omega R_m \cos \omega t$, we square

$R(t)$ and $v(t)$ and appropriately add to find (since $\sin^2\omega t + \cos^2\omega t = 1$)

$$v^2 + R^2\omega^2 = \omega^2 R_m^2,$$

which says from (7.31) at $R = R_0$ that

$$\omega^2 = \frac{H_0^2}{(R_m/R_0)^2 - 1}. \tag{7.34}$$

At $R = R_m$, $v = 0$ and the dynamical equation (7.34) can be solved for the ratio R_m/R_0,

$$\frac{R_m}{R_0} = \frac{1}{1 - \rho_c/\rho_0}. \tag{7.35}$$

To be specific, let us choose $\rho_0 = 2\rho_c$ or $q_0 = 1$. Then (7.35) and (7.34) imply that $R_m = 2R_0$ and $\omega = H_0/\sqrt{3}$. The period of a harmonic oscillator is $\tau = 2\pi/\omega$ and t_m corresponds to one-quarter of this value. Thus for the case of $q_0 = 1$ the universe expands to its maximum size at time

$$t_m = \frac{\pi}{2\omega} = \frac{\sqrt{3}}{2}\pi H_0^{-1} \approx 50 \times 10^9 \text{ yr}, \tag{7.36a}$$

and the "end" will come with the big crunch in

$$t_{\text{end}} = 2t_m = \sqrt{3}\,\pi H_0^{-1} \approx 100 \times 10^9 \text{ yr}. \tag{7.36b}$$

Further, since $R_0 = \frac{1}{2}R_m$, $\sin\omega t_0 = \frac{1}{2}$ implies that ωt_0 is 30° or $\pi/6$, giving $t_0 = \frac{1}{3}t_m \sim 16 \times 10^9$ yr. It is this latter estimate which is somewhat large. A more accurate evaluation of the general relativistic dynamical equation for $q_0 = 1$ gives for the present age of the universe in this model

$$t_0 \approx 10 \times 10^9 \text{ yr}. \tag{7.36c}$$

With hindsight we note from (7.32), (7.33) and (7.36c) that all three big bang model universes predict a present age which differs from one another by less than a factor of two. Thus all other dynamical predictions of the big bang will not depend critically upon the present age of the universe.

7.5.2. Role of General Relativity in Cosmology

We have posed the problems and solutions of cosmology in a manner independent of general relativity. It is well to point out, however, that general relativity is necessary to explain why the radius of the universe in the matter era measures both the size of the photon and neutrino fireballs as well as the matter gas. This might seem confusing because the massless fireball particles move at the speed of light, as does $R_u \sim ct$ in the radiation-dominated era. On the other hand, in the matter-dominated era we saw that R_u is slowing down, for example to $R_u \propto t^{2/3}$ in an Einstein–deSitter universe. Even though the fireballs are expanding as $T_\gamma \propto 1/R$, $T_\nu \propto 1/R$ and the matter gas as $T_m \propto 1/R^2$, the fireballs are not racing ahead of the matter gas; the Rs are the same but $T_m < T_\gamma, T_\nu$. How can this be? General relativity tells us that the massless-fireball photons and neutrinos have an effective mass $E_{\gamma,\nu}/c^2$ which is influenced by the gravitational attraction from the matter gas and other particles in the fireballs. The result is that out near the "edge" of the universe, space is curved as the "straight line" trajectories of photons and neutrinos begin to bend back on themselves. Alternatively we may describe the photons by their frequencies, which are redshifted far away from a strong gravitational source like the universe itself (recall the discussion of black holes in Sec. 5.5.1). We repeat this discussion but in a slightly different context.

Quantitatively the amount of curvature is measured by the gravitational potential felt by one photon or neutrino, which in dimensionless units is

$$\phi = \frac{GM}{c^2 R}\ .$$

For a photon at the edge of the Earth (i.e. on the surface),

$$\phi_E = \frac{GM_E}{c^2 R_E} \sim \frac{6.7 \times 10^{-11} \cdot 6 \times 10^{24}}{9 \times 10^{16} \cdot 6.4 \times 10^6} \sim 10^{-9},$$

while for the Sun and our galaxy, $\phi \sim 10^{-6}$. For white dwarfs $\phi_{\mathrm{WD}} \sim 10^{-4}$, for neutron stars $\phi_{\mathrm{NS}} \sim 10^{-1}$, while for black holes the Schwarzschild radius implies

$$\phi_{\mathrm{BH}} = \frac{GM_{\mathrm{BH}}}{c^2 R_{\mathrm{BH}}} \geqslant \tfrac{1}{2}\ .$$

Only when $\phi \gtrsim \frac{1}{2}$ does general relativity play a dominant role.

Now we believe that our present universe has about $M_u \sim 10^{51}$ kg with $R_u \sim 10^{10}$ LY $\sim 10^{26}$ m, so that the gravitational potential of the entire universe is

$$\phi_u = \frac{GM_u}{c^2 R_u} \sim \frac{10^{-10} \cdot 10^{51}}{10^{17} \cdot 10^{26}} \sim 10^{-2}$$

and general relativity begins to play a role for photons on the "edge" of the universe. For an Einstein–deSitter universe, the total energy vanishes, implying

$$\phi_{\mathrm{ED}} = \frac{GM_u}{c^2 R_u} = 1 \rightarrow \tfrac{1}{2}\ ,$$

where the factor of $\frac{1}{2}$ follows from general relativity considerations. If the universe is closed, we may treat it as a huge black hole, trapping photons on the "edge" with (see Figs. 5.21 and 5.22)

$$\frac{GM_u}{R_u} m_\gamma > m_\gamma c^2$$

or

$$\phi_{cu} = \frac{GM_u}{c^2 R_u} > 1 \rightarrow \tfrac{1}{2},$$

where again the factor of $\frac{1}{2}$ is due to general relativity. We can also state this inequality in terms of Hubble's constant by replacing c/R_u in ϕ_{cu} by H (i.e., $v = HR$) and using $M_u = 4\pi\rho_u R_u^3/3$, we get

$$\rho > \frac{3H^2}{8\pi G} = \rho_c,$$

which is indeed the condition for a closed universe (see Sec. 6.2.4).

7.5.3. General Relativity and Quantum Mechanics

If the universe is open, i.e., $\rho < \rho_c$ as it now appears, then the universe will continue to cool off and expand forever. Eventually it will not be possible for inhomogeneous galactic clumps to form nor stars to be created. The universe will then die, a cold death at that—the "big freeze." Perhaps quantum mechanics will ultimately play a role in the final low-temperature limit.

On the other hand, if the universe is closed, i.e., $\rho > \rho_c$, we must look forward to a hot big crunch. In this case, the events of the late universe will be a time-reversed history of the early universe. Recall that when $T > 10^{13}$ °K, i.e., $kT \sim 10^3$ MeV, all possible elementary particles will be created and almost instantly annihilated in the intense heat. The black hole nature of this small but extremely dense universe will continue to trap radiation and all other particles as the temperature and density increase even more. How does it all end? Classically, the closed universe and black hole stars end in oblivion.

It is now speculated that quantum mechanics may conspire with general relativity to alter drastically the final outcome of a closed universe and black hole stars. Hawking has shown that small black holes about the size of a proton but with a density of 10^{57} kg/m^3 (recall black hole star densities are $>10^{18}$ kg/m^3), presumably produced in the big bang, have temperatures of order 10^{11} °K or $kT \sim 10$ MeV. Such objects, while having a large gravitational potential

$$\phi = \frac{Gm}{c^2 r} \sim \frac{G4\rho r^2}{c^2} \sim \frac{3\times 10^{-10}\times 10^{57}\times 10^{-30}}{9\times 10^{16}} \sim 3,$$

capable of trapping photons in the classical black hole sense, are also small enough to obey the laws of quantum mechanics. In this case $kT \sim 10$ MeV is high enough to produce electron–positron pairs ($2m_e c^2 \sim 1$ MeV) at the edge of the black hole. For such strong gravitational fields, general relativity requires one of these particles to be sucked towards the center of the black hole, while quantum mechanics allows the other to tunnel out of the black hole. In about 10^{10} years these primordial black holes radiate themselves out of existence, ending in a final explosion which could, in principle, be detected from Earth if they are no further away than the closest stars.

As for the closed universe, perhaps the resulting black hole collapse will be limited by quark matter, or possibly the big crunch will pack the matter energy so close that at some point the uncertainty principle $\Delta p \Delta x \sim \hbar = h/2\pi$ should lead to an outward pressure against gravity in roughly the same way that the uncertainty principle pressure supports an electrostatic atom. The quantum analog *of the proton* "electromagnetic Compton wavelength" $\sqrt{\alpha}\,\hbar/m_p c = \sqrt{k_0 e^2 \hbar/m_p c^3} \sim 10^{-17}$m is the gravitational "Planck length" formed from $G, \hbar$, and c:

$$l^* = \sqrt{\frac{G\hbar}{c^3}} \sim 3\times 10^{-35}\text{ m.} \tag{7.37a}$$

One may also form a fundamental quantum gravitational mass and mass density from G, $\hbar$, and c:

$$m^* = \sqrt{\frac{\hbar c}{G}} \sim 5\times 10^{-8}\text{ kg} \tag{7.37b}$$

$$\rho^* = \frac{c^5}{\hbar G^2} = \frac{m^*}{l^{*3}} \sim 10^{96}\text{ kg/m}^3. \tag{7.37c}$$

Thus when $\rho^* c^2 \sim aT^4$ or $T \sim 10^{32}$ °K, quantum mechanics should alter the classical big crunch in some significant manner. While little is understood about this density region to date, perhaps quantum gravity could cause a reversal, turning the big crunch into another big bang. If so, then the (closed) universe might be oscillating (bouncing) endlessly. Such "cosmic bounces" form perhaps a more acceptable philosophical premise underlying the theory of cosmology than a single big bang occurring at a single time and for God only knows what reason . . . .

"THE BIG BANG? BELIEVE ME, IT WAS VERY, VERY, VERY, VERY, VERY, VERY BIG."

Problems

1. At what temperature will the neutron and proton freeze out of the fireball?
2. Write down the reaction equation for the process that keeps the photons and nucleons in thermal equilibrium after the nucleons freeze out of the photon fireball.
3. Find the approximate number density of photons at the time the π-mesons freeze out.
   Extrapolating from the current nucleon number density, find the number density of nucleons at the π-meson freeze-out temperature.
4. At what time will the nucleons be frozen out of the fireball?
5. In each reaction, using lepton number conservation, identify each "ν" as an electron neutrino or antineutrino or as a muon neutrino or antineutrino.

   a. $\pi^+ \rightarrow \mu^+ + \text{"}\nu\text{"}$

      $\pi^- \rightarrow \mu^- + \text{"}\nu\text{"}$

   b. $\mu^+ \rightarrow e^+ + \text{"}\nu\text{"} + \text{"}\nu\text{"}$

   c. $p \rightarrow n + e^+ + \text{"}\nu\text{"}$

d.
$$K^0 \rightarrow \pi^+ + e^- + \text{“}\nu\text{”}$$
$$K^0 \rightarrow \pi^- + \mu^+ + \text{“}\nu\text{”}$$

6. Recent experiments indicate that another charged lepton (the $\tau^{\pm}$) and its associated neutrino (ν_τ) may exist.
   a. If so, find X in $(E/V)_{\text{tot}} = XaT^4$ at the beginning of the lepton era.
   b. Find the new expression for t vs. T which replaces (7.6).
7. What would be the cross-over point for l_A and R_u in Fig. 7.1 if the weak interaction cross section were 1000 times larger?
   What consequences would this have for the photon and neutrino fireball temperatures?
8. Referring to Fig. 7.1, approximately at what fireball temperature will l_B be greater than R_u?
9. Show that (7.7) reduces to $T_\gamma = (\frac{11}{4})^{1/3} T_\nu$.
10. In the text we indicated that at present the ratio n_ν / n_γ is approximately 0.6. Show where the factor of 0.6 comes from. When did the ratio first take on this particular value?
11. What would the mass of the neutral massive lepton have to be in order to close the universe if there were *three* species of neutrinos $(\nu_e, \nu_\mu, \nu_\tau)$?
12. Show that for $q_0 \leqslant 2$, it follows that the mass of the neutral massive lepton must be less than 60 eV.
13. Show that (7.12) follows from (7.11).
14. If (7.13) had turned out to be $kT^* = \text{BE}$ instead of $kT^* \approx \text{BE}/37$, what would be the percentage of He in the universe?
15. What percentage of a star's mass would have to burn in order to produce enough He to compare to the amount produced in primordial nucleosynthesis?
16. Iterate (7.12) a third time to find kT^*/BE.
17. Find the *present* value of the matter temperature, T_m. How does it compare to the current value of the photon temperature, 3 °K?
18. In discussing (7.15) we noted that $n_{\text{H}} = p - n$. Why? In (7.15) where do the factors of 4 come from in the middle ratio? Why is there a factor of 2 in the numerator and a factor of 1 not 2 in the denominator of the last ratio in (7.15)?
19. If there were a total of six neutrinos and antineutrinos instead of four, what would be the cross-over temperature where $\rho_m = \rho_\gamma$?
20. Let us see if we can find the expression for the force retarding the condensation of a density fluctuation (sometimes called the Thomson drag). We assume the photons from the fireball are moving outward with velocity v relative to the infalling electrons.
   a. Show that the rate density for the interactions between photons

and electrons is given by

$$\mathcal{R} \approx v n_{e1} \sigma_T \frac{(E/V)_{\text{rad}}}{\langle E_\gamma \rangle}.$$

b. The energy of a photon is related to its momentum by $E = pc$. Show that the outward force per unit volume on the inward-falling electrons is given by

$$\frac{\delta F}{\delta V} \approx \frac{v}{c} n_{e1} \sigma_T (E/V)_{\text{rad}}.$$

21. If the Thomson cross section for the scattering of photons off free electrons were 100 times larger, how much mass would there be in a galactic cluster?
    What would be t_{ff} for such a cluster?
22. Clearly show how the relation $T(\theta) = T_0(1 + v\cos\theta/c)$ follows from a Doppler-shifted black body spectrum.
23. Show that (7.30) follows from (7.29).
24. How would the value of the threshold energy of Eq. (7.30) change if the reaction between cosmic ray protons and fireball photons were

$$p_{\text{fast}} + \gamma_{2.7} \rightarrow n_{\text{slow}} + \pi^+ + \pi^- + \pi^+?$$

25. Show how (7.34) follows from (7.31) for $R = R_0$.
26. If the universe is barely closed with $\rho \gtrsim \rho_c$, calculate the minimal fireball temperature $T_{\min}$ at the time the universe will begin to contract.
27. Evaluate the gravitational potential felt by a photon at the surface of the Sun, a galaxy, a white dwarf, and a neutron star.
28. Check the dimensions of (7.37a)–(7.37c).

Appendix A

Basic Constants

Physical Constants

Speed of light	$c = 3.00 \times 10^8$ m/sec
Gravitation constant	$G = 6.67 \times 10^{-11}$ N − m²/kg²
Planck's constant	$h = 6.625 \times 10^{-34}$ J-sec
	$= 4.135 \times 10^{-15}$ eV-sec
	$\hbar = 6.582 \times 10^{-16}$ eV-sec
	$hc = 12.40 \times 10^{-7}$ eV-m
Boltzmann's constant	$k = 1.38 \times 10^{-23}$ J/°K
	$= 0.862 \times 10^{-4}$ eV/°K
Stefan–Boltzmann constant	$\sigma = 5.67 \times 10^{-8}$ J/m²-°K⁴-sec
Black body constant	$a = 4\sigma/c = 7.56 \times 10^{-16}$ J/m³-°K⁴
Charge of electron	$e = -1.602 \times 10^{-19}$ Coul
Electrostatic constant	$k_0 = 9 \times 10^9$ N-m²/Coul²
Electron volt	1 eV $= 1.602 \times 10^{-19}$ J
Avogadro's number	$N_0 = 6.023 \times 10^{23}$ molecules/mole
Mass of electron	$m_e = 9.11 \times 10^{-31}$ kg
	$m_e c^2 = 0.511$ MeV
Mass of proton	$m_p = 1.673 \times 10^{-27}$ kg
	$m_p c^2 = 938.3$ MeV
Mass of neutron	$m_n = 1.675 \times 10^{-27}$ kg
	$m_n c^2 = 939.6$ MeV

Astronomical Constants

Astronomical unit	AU $= 1.496 \times 10^{11}$ m
Parsec	pc $= 206{,}265$ AU
	$= 3.26$ LY
	$= 3.086 \times 10^{16}$ m
Light year	LY $= 6.324 \times 10^4$ AU
	$= 0.307$ pc
	$= 0.946 \times 10^{16}$ m

Sidereal year	$1 \text{ yr} = 365.26$ days
	$= 3.16 \times 10^7$ sec ($\approx \pi \times 10^7$ sec)
Mass of Sun	$M_\odot = 1.99 \times 10^{30}$ kg
Radius of Sun	$R_\odot = 6.96 \times 10^8$ m
Luminosity of Sun	$L_\odot = 3.827 \times 10^{26}$ J/sec
Distance Earth to Sun (radial)	$R_{ES} = 1$ Au
	$\cong 200\ R_\odot$
Mass of Earth	$M_E = 5.98 \times 10^{24}$ kg
Equatorial radius of Earth	$R_E = 6.378 \times 10^6$ m

Appendix B

Derivation of the Fundamental Kinetic Theory Equation

The derivation based on Fig. B.1 begins with the pressure generated by each particle, $P_i = F_i/A$, where A is the area of the wall and F_i the force the ith particle exerts on the wall. To find F_i, we note that a particle hitting the "x" wall transfers momentum only in the x direction, $\Delta p_x = m_i(v_{xi,\text{in}} - v_{xi,\text{out}}) = 2m_i v_{xi}$, since in an elastic collision with the "x" wall we have $v_{xi,\text{out}} = v_{xi,\text{in}}$ and $\Delta p_y = 0$. Because KE and momentum are conserved the magnitude of the x component of velocity remains constant as the particle ricochets around the container. Therefore, the time needed to hit the x wall, rebound off the far wall, and return, is $\Delta t = 2L_x/v_{xi}$. The average force then exerted by this one particle *over* Δt *is*

$$\langle F_{xi} \rangle = \frac{p_{xi}}{\Delta t} = \frac{2m_i v_{xi}}{(2L_x/v_{xi})} = \frac{m_i v_{xi}^2}{L_x}\,.$$

For N particles, we have

$$\langle F \rangle = \sum \langle F_{xi} \rangle = \frac{m}{L_x}(v_{x1}^2 + v_{x2}^2 + v_{x3}^2 + \ldots) = \frac{m}{L_x}\left(\sum_{c=1}^{N} v_{xi}^2\right),$$

where we have assumed each particle has the same mass. The pressure on the x wall, F_x/A, is then

$$P_x = \frac{m\sum v_{xi}^2}{AL_x} = \frac{m\sum v_{xi}^2}{V}\,,$$

or

$$P_x V = m\sum_{i=1}^{N} v_{xi}^2. \tag{B.1}$$

Clearly all particles participate in the pressure, but it is the sum of the squares of the x components that determines P [and not the square of the sum, $(\sum_i v_{xi})^2$]. To get the total number of particles into (B.1), we multiply

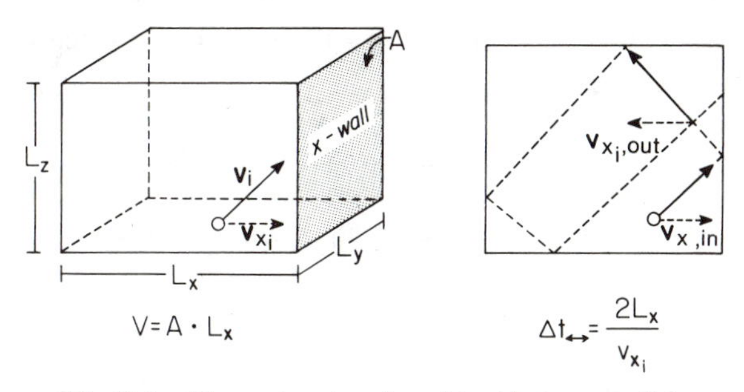

Fig. B-1. Momentum transferred by ideal gas particles.

the numerator and denominator by N,

$$P_x V = mN \left[\frac{\sum_{i=1}^{N} v_{xi}^2}{N} \right].$$

As discussed in Sec. 1.3, the term in the bracket is called the mean square average speed and is denoted by $v_{\text{rms},x}^2$, so that

$$P_x V = mNv_{\text{rms},x}^2. \tag{B.2}$$

But the measured pressure on any face, regardless of its orientation, is the same. The reason should now be obvious. If the particles are moving about randomly, which is one of our conditions, then $v_{\text{rms},x}^2 = v_{\text{rms},y}^2 = v_{\text{rms},z}^2$, since there can be no difference in any aspect of random motion between arbitrary directions like x, y, and z.

So far we have concentrated on the components of each particle's velocity. These are related to the total velocity via $v_i^2 = v_{xi}^2 + v_{yi}^2 + v_{zi}^2$. It then follows that

$$\sum v_i^2 = \sum v_{xi}^2 + \sum v_{yi}^2 + \sum v_{zi}^2,$$

so that

$$\frac{\sum v_i^2}{N} = \frac{\sum v_{xi}^2}{N} + \frac{\sum v_{yi}^2}{N} + \frac{\sum v_{zi}^2}{N}$$

and therefore,

$$v_{\text{rms}}^2 = v_{\text{rms},x}^2 + v_{\text{rms},y}^2 + v_{\text{rms},z}^2.$$

Since all rms components are equal, the last relation above reduces to

$$v_{\text{rms}}^2 = 3v_{\text{rms},x}^2.$$

Plugging this into (B.2) we get (dropping the subscript x since P is the same on all faces),

$$PV = \frac{mNv_{\rm rms}^2}{3} . \tag{B.3}$$

This is the fundamental kinetic theory equation which we discussed in Sec. 1.3.

Appendix C

Derivation of the Doppler Effect

It is easier to see how the expressions quoted in Sec. 1.7.1 are derived if we make a simple model for producing EM radiation. You will recall that an accelerating charge will emit an electric and magnetic vector which propagate through space with $v = c$ such that the cross product relationship between **E** and **B** gives

$$\text{(direction of } \mathbf{E}) \times (\text{direction of } \mathbf{B}) = \text{direction of } \mathbf{c},$$

and where, in MKS units, $E = cB$. Let us then imagine a charge e attached to a massive body M by a spring with the charge oscillating vertically, as shown in Fig. C.1. If the charge oscillates up and down with frequency ν_0, then the conditions producing the radiation will repeat with period $\tau_0 = 1/\nu_0$. Since the radiation propagates with velocity c, two points that are spatially $\lambda_0 = c\tau_0$ apart will see the same **E** vector (magnitude and direction) at the same time. If the source moves towards us, however, as shown in Fig. 1.22, the situation changes a bit. At $t = 0$ we will assume **E** is at a maximum magnitude pointing up. The radiation emitted at $t = 0$ propagates outward at $v = c$, independent of the motion of the source. At $t = \tau_0$, although **E** is again maximum and up, the source has moved a distance $L = u\tau_0$, where u is the speed of the source. Therefore, the spatial separation between identical **E** vectors (which to our observer will be the wavelength of the radiation) is

$$\begin{aligned} \lambda' &= c\tau_0 - L = c\tau_0 - u\tau_0 \\ &= c\tau_0(1 - u/c) \qquad \begin{pmatrix} \text{source moving} \\ \text{toward observer} \end{pmatrix} \end{aligned}$$

Since $\lambda_0 = c\tau_0$, we finally have $\lambda' = \lambda_0(1 - u/c)$, where since $\lambda' < \lambda_0$, the wavelength has been "blue" shifted. If the source is moving away from us, then clearly

$$\lambda' = \lambda_0(1 + u/c), \qquad \begin{pmatrix} \text{source moving} \\ \text{away from observer} \end{pmatrix}$$

and the wavelength will appear "red" shifted.

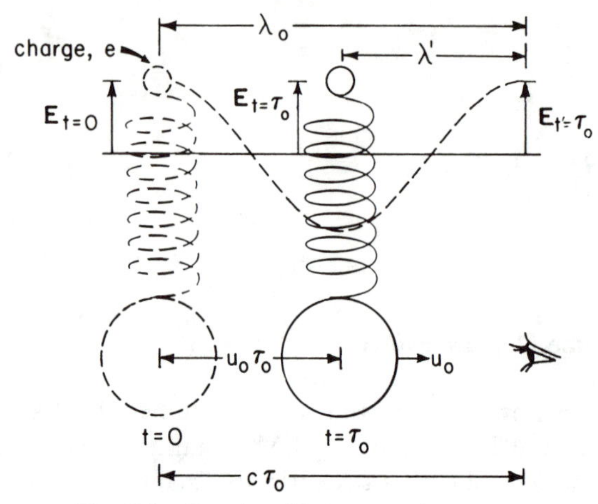

Fig. C-1. Doppler shift for a moving source.

We leave it as an exercise for the reader to show that (see Problem 16 in chapter 1)

$$\lambda' = \lambda_0 \frac{1}{(1 \pm u/c)}$$

when the source is stationary and the observer is moving.

GLOSSARY OF PRIMARY SYMBOLS

a	acceleration, black body constant, orbital radius, semimajor axis
a_0	Bohr radius for hydrogen
A	area; nucleon number
$\mathring{A}$	angstrom = 10^{-10} m
α	fine structure constant; alpha particle
$\mathbf{B}$	magnetic field
BE	binding energy
β	parallax angle
c	speed of light, core (as subscript)
γ	photon
d	distance; deuteron
δ, Δ	increment
e	electron; proton charge; 2.71828 . . .
eV	electron volt
E	electric field, energy

E_I	ionization energy of H
$\mathcal{E}$	energy
ϵ	rate of nuclear energy generated per mass
F	force; Fermi level (as subscript)
f	final as index
G	gravitation constant, gravity (as subscript)
gal	galaxy (as subscript)
$h, \hbar = h/2\pi$	Planck's constant
H	hydrogen
H	general Hubble constant
H_0	present Hubble constant
He, He^4	helium
θ	angle
i	index label; initial as index
I	current; moment of inertia, ionization (as subscript)
°K	degrees Kelvin
k	Boltzmann constant
k_0	electric force constant
κ	absorption coefficient (m^2/kg)
KE	kinetic energy
⟨KE⟩	average kinetic energy
L	length; luminosity; orbital angular momentum
l	length; angular momentum quantum number, mean free path = interaction length
$\mathcal{L}$	luminosity density (area) = energy flux (J/m^2-sec)
$\mathcal{L}_{\lambda,\nu}$	energy flux per unit wavelength (or unit frequency)
$\Delta L/\Delta r$	luminosity gradient
LY	light year
λ	wavelength
M	mass (usually of a star)
m	mass, meter
M_r	mass inside radius r
$\Delta M/\Delta r$	mass gradient
μ	mean molecular weight in units of proton mass, magnetic dipole moment, muon
n	particle number density, neutron, energy quantum number
N	number of particles; neutron number
N	Newton
ν	frequency; neutrino
P	pressure
$\Delta P/\Delta r$	pressure gradient
PE	potential energy

$\langle PE \rangle$	average potential energy
p	momentum, proton
$\mathcal{P}$	probability
pc	parsec
π	pion; 3.14159 . . .
q	deceleration parameter, charge
QHE	quasihydrostatic equilibrium
R	radius (fixed) of large object
r	variable radius, radius of small object
ρ	mass density (ρ_{gal} of galaxy; ρ_{H^+} of ionized hydrogen; ρ_{GC} of galactic cluster; ρ_{cos} of universe; ρ_0 present ρ_{cos}; ρ_{lum} luminous; ρ_{SR} of stellar radiation; ρ_n of neutrons; ρ_c critical to close universe)
$\mathcal{R}$	reaction rate density
$\odot$	solar subscript
S	spin angular momentum, surface (as subscript)
σ	cross section; Stefan–Boltzmann constant
σ_T	Thomson cross section
T	temperature
τ	period
$\Delta T/\Delta r$	temperature gradient
t	time
t_{exp}	expansion time
t_{ff}	free fall time
U	internal energy
u	energy density, velocity
V	volume
v	velocity
Φ	magnetic flux
ϕ	dimensionless gravitational potential
ϕ_ν	neutrino flux
ω	angular velocity $\Delta\theta/\Delta t$ or frequency $2\pi\nu$
X	mass fraction
x	mass fraction of hydrogen
y	mass fraction of helium
Z	proton number
z	mass fraction of heavy elements; Doppler line shift $\Delta\lambda/\lambda_0$

Reading List

For those who wish to read further, the articles and books that contributed to the material in *Physics of Stellar Evolution and Cosmology* are listed below. In order to facilitate the correlation between this text and the reference material we have included the titles of the articles and have noted in parentheses to which chapter in this textbook the material refers.

Bachall, J. N. and Sears, R. L., "Solar Neutrinos," *Annual Review of Astronomy and Astrophysics*, **10**, 25(1972). (Chap. 4)

Bachall, J. N. and Davis, R. Jr., "Solar Neutrinos: A Scientific Puzzle," *Science*, **191**, 264(1976). (Chap. 4)

Berry, M., "Principles of cosmology and gravitation," Cambridge University Press, London, 1976. (Chap. 6, 7)

Blumenthal, G. R. and Tucher, W. H., "Compact X-Ray Sources," *Annual Review of Astronomy and Astrophysics*, **12**, 23(1974). (Chap. 5)

Bok, B. J., "The Birth of Stars," *Scientific American*, **227**, No. 2, 48(1972). (Chaps. 2, 3)

Brandt, J. C. and Williamson, R. A., "The 1054 Supernova and Native American Rock Art," *Archaeoastronomy*, **1**, S1(1979). (Chap 5)

Brecher, K., Leiber, A., and Leiber, E., "A Near Eastern Sighting of the Supernova Explosion of 1054," *Nature*, **273**, 728(1978). (Chap. 5)

Cameron, R. C., *Introduction to Space Science: Stellar Evolution*, Gordon and Breach, New York, 1968. (Chaps. 3, 4)

Cox, J. and Giuli, R., *Principles of Stellar Structure: Vol.* I, II, Gordon and Breach, New York, 1968. (Chaps. 3, 4, 5)

Field, G. B., "Intergalactic Matter," *Annual Review of Astronomy and Astrophysics*, **10**, 227(1972). (Chap. 6)

Gamow, G., "The Evolutionary Universe," *Scientific American*, **195**, No. 3, 136(1956). (Chap. 7)

Gerola, H. and Seiden, P., "Stochastic Star Formation and Spiral Structure of Galaxies," *Astrophysical Journal*, **223**, 129(1978). (Chap. 3)

Giacconi, R., "Progress in X-Ray Astronomy," *American Journal of Physics*, **44**, 121(1976). (Chap. 5)

Gordon, M. A. and Burton, W. B., "Carbon Monoxide in the Galaxy," *Scientific American*, **240**, No. 5, 54(1979). (Chap. 2)

Gorenstein, P. and Tucker, W., "Supernova Remnants," *Scientific American*, **225**, No. 1, 74(1971). (Chap. 5)

Gott, J. R., Gunn, J. E., Schramm, D. N., and Tinsley, B. M., "Will the Universe Expand Forever?," *Scientific American*, **234**, No. 3, 62(1976). (Chaps. 6, 7)

Gott, J. R., "Recent Theories of Galaxy Formation," *Annual Review of Astronomy and Astrophysics*, **15**, 235(1977). (Chap. 7)

Greenberg, J. M., "Interstellar Grains," *Scientific American*, **217**, No. 4, 106(1967). (Chap. 2)

Gursky, H. and van der Heuvel, E., "X-Ray Emitting Double Stars," *Scientific American*, **223**, No. 3, 24(1975). (Chap. 5)

Harrison, E. R., "Standard Model of the Early Universe," *Annual Review of Astronomy and Astrophysics*, **11**, 155(1973). (Chap. 7)

Harrison, E. R., "Why the Sky is Dark at Night," *Physics Today*, **27**, 30(1974). (Chap. 6)

Harwit, M., *Astrophysical Concepts*, John Wiley and Sons, New York, 1973. (Chap. 3)

Hawking, S. W., "The Quantum Mechanics of Black Holes," *Scientific American*, **236**, No. 1, 34(1977). (Chaps. 5, 7)

Heiles, C., "Physical Conditions and Chemical Constitution of Dark Clouds," *Annual Review of Astronomy and Astrophysics*, **9**, 293(1971). (Chap. 2)

Herbig, G., "The Youngest Stars," *Scientific American*, **217**, No. 2, 30(1967). (Chaps. 2, 3)

Herbst, W. and Assousa, G., "Supernovas and Star Formation," *Scientific American*, **241**, No. 2, 138(1979). (Chap. 3)

Hewish, A., "Pulsars," *Scientific American*, **219**, No. 4, 25(1968). (Chap. 5)

Hewish, A., "Pulsars," *Annual Review of Astronomy and Astrophysics*, **8**, 265(1970). (Chap. 5)

Jastrow, R. and Thomson, M. H. *Astronomy: Fundamentals and Frontiers,* John Wiley and Sons, New York, 1974. (Chaps. 1, 2, 3, 4.)

Jones, B. J. T., "The Origin of Galaxies," *Review of Modern Physics*, **48**, 107(1976). (Chap. 7)

Larson, R. B., "Processes in Collapsing Interstellar Clouds," *Annual Reviews of Astronomy and Astrophysics*, **11**, 219(1973). (Chaps. 2, 3)

Meadows, A. J., *Stellar Evolution*, Pergamon Press, London, 1968. (Chaps. 3, 4, 5)

Miller, J. S., "The Structure of Emission Nebulas," *Scientific American*, **231**, No. 4, 34 (1974). (Chap. 2)

Motz, L. and Duveen, A., *Essentials of Astronomy*, Columbia University Press, New York, 1971. (Chaps. 2, 3, 4)

Novotny, E., *Introduction to Stellar Atmospheres and Interiors*, Oxford University Press, London, 1973. (Chaps. 3, 4)

Orear, J. and Salpeter, E. E., "Black Holes and Pulsars in an Introductory Physics Course," *American Journal of Physics*, **41**, 1131(1973). (Chap. 5)

Ostriker, J. P. and Gunn, J. E., "On the Nature of Pulsars," *Astrophysics Journal*, **157**, 1395(1969). (Chap. 5)

Ostriker, J. P., "The Nature of Pulsars," *Scientific American*, **224**, No. 1, 48(1971). (Chap. 5)

Pacini, F. and Rees, M. J., "Rotation in High-Energy Astrophysics," *Scientific American*, **228**, No. 2, 98 (1973). (Chap. 5)

Peebles, P. J. E., *Physical Cosmology*, Princeton University Press, Princeton, 1971. (Chaps. 6, 7)

Percy, J. R., "Pulsating Stars," *Scientific American*, **232**, No. 6, 67 (1975). (Chap. 4)

Reddish, J. C., *Physics of Stellar Interiors*, Crane-Russak Co., New York, 1974. (Chap. 3)

Rees, M. J. and Silk, J., "The Origin of Galaxies," *Scientific American*, **222**, No. 6, 26(1970). (Chap. 7)

Reeves, H., "On the Origin of the Light Elements," *Annual Review of Astronomy and Astrophysics*, **12**, 437(1974). (Chap. 7)

Ruderman, M. A., "Solid Stars," *Scientific American*, **224**, No. 2, 24(1971). (Chap. 5)

Ruderman, M. A., "Pulsars: Structure and Dynamics," *Annual Review of Astronomy and Astrophysics*, **10**, 427(1972). (Chap. 5)

Salpeter, E. E., "Central Stars of Planetary Nebulae," *Annual Review of Astronomy and Astrophysics*, **9**, 127(1971). (Chap. 4)

Sargent, W. L., *Extragalactic Observational Astronomy from High Energy Astrophysics*, MIT Press, Cambridge, 1972. (Chap. 6)

Schwarzchild, M., *Structure and Evolution of the Stars*, Dover Publications, New York, 1958. (Chaps. 3, 4)

Sciama, D. W., *Modern Cosmology*, Cambridge University Press, London, 1971. (Chaps. 6, 7)

Seiden, P., Schulman, L., and Gerola, H., "Stochastic Star Formation and the Evolution of Galaxies," *Astrophysical Journal*, **232**, 702(1979). (Chap. 3)

Shu, F. H., "Spiral Structure, Dust Clouds and Star Formation," *American Scientist*, **61**, 524(1973). (Chaps. 2, 3)

Silk, J., "Cosmic Black Body Radiation," *Astrophysics Journal*, **151**, 459(1968). (Chap. 6, 7)

Smoot, G. F., Gorenstien, M. V., and Muller, R. A., "Detection of Anisotropy in the Cosmic Black Body Radiation," *Physical Review Letters*, **39**, 898(1977). (Chap. 6, 7)

Steigman, G., "Cosmology Confronts Particle Physics," *Annual Review of Nuclear and Particle Science*, **29**, 313(1979). (Chap. 7)

Stodolkiewicz, J., *General Astrophysics*, American Elsevier Publishing Company, New York, 1973. (Chaps. 3, 4, 5)

Taylor, R. J., *The Stars: Their Structure and Evolution*, Wykeham Publications, London, 1974. (Chaps. 3, 4)

Taylor, J. H., Fowler, L. A. and McCulloch, P. M., "Measurements of General Relativistic Effects in the Binary Pulsar PSR1913 + 16," *Nature*, **277**, 437(1979). (Chap. 5)

Turner, B. E., "Interstellar Molecules," *Scientific American*, **228**, No. 3, 51(1973) (Chap. 2)

Turner, M. S. and Schramm, D. N., "Cosmology and Elementary-Particle Physics," *Physics Today*, **32**, No. 9, 42(1979). (Chap. 7)

Wagoner, R., "Test for the Existence of Gravitational Radiation," *Astrophysical Journal*, **196**, 163(1975). (Chap. 5)

Wagoner, R., "Big Bang Nucleosynthesis Revisited," *Astrophysical Journal*, **179**, 343(1973). (Chap. 7)

Watson, W. D., "Interstellar Molecule Reactions," *Reviews of Modern Physics*, **48**, 513(1976). (Chap. 2)

Webster, A., "The Cosmic Background Radiation," *Scientific American*, **231**, No. 2, 26(1974). (Chaps. 6, 7)

Weeks, T., *High Energy Astrophysics*, Chapman and Hall Limited, London, 1969. (Chap. 5)

Weinberg, S., *Gravitation and Cosmology*, John Wiley and Sons, New York, 1972. (Chaps. 6, 7)

Weinberg, S., *The First Three Minutes*, Basic Book, New York, 1977. (Chap. 7)

Weisskopf, V. F., "Of Atoms, Molecules and Stars," *Science*, **187**, 605(1975). (Chap. 5)

Wheeler, C. J., "After the Supernova, What?" *American Scientist*, **61**, 42(1973). (Chaps. 4, 5)

Zeilik, M., *Astronomy: The Evolving Universe*, Harper and Row, New York, 1976. (Chap. 2)

Zel'dovitch, Y. B. and Novikov, I. D., *Relativistic Astrophysics, Vol. 1: Stars and Relativity*, University of Chicago Press, Chicago, 1971. (Chaps. 5, 6, 7)

Zuckerman, B. and Palmer, P., "Radio Radiation From Interstellar Molecules," *Annual Review of Astronomy and Astrophysics*, **12**, 279(1974). (Chap. 2)

Subject Index